INTERNATIONAL RIVER WATER QUALITY

INTERNATIONAL RIVER WATER QUALITY

Pollution and restoration

Edited by

G.A. Best, T. Bogacka and E. Niemirycz

E & FN SPON
An Imprint of Chapman & Hall

London · Weinheim · New York · Tokyo · Melbourne · Madras

Published by E & FN Spon, an imprint of
Chapman & Hall, 2–6 Boundary Row, London SE1 8HN, UK

Chapman & Hall, 2–6 Boundary Row, London SE1 8HN, UK

Chapman & Hall GmbH, Pappelallee 3, 69469 Weinheim, Germany

Chapman & Hall USA, 115 Fifth Avenue, New York, NY 10003, USA

Chapman & Hall Japan, ITP-Japan, Kyowa Building, 3F, 2-2-1 Hirakawacho, Chiyoda-ku, Tokyo 102, Japan

Chapman & Hall Australia, 102 Dodds Street, South Melbourne, Victoria 3205, Australia

Chapman & Hall India, R. Seshadri, 32 Second Main Road, CIT East, Madras 600 035, India

Distributed in the USA and Canada by Van Nostrand Reinhold, 115 Fifth Avenue, New York, NY 10003, USA

First edition 1997

© 1997 E & FN SPON

Typeset by Saxon Graphics Ltd, Derby

Printed in Great Britain by St Edmundsbury Press, Bury St Edmunds, Suffolk.

ISBN   0 419 21540 9

Apart from any fair dealing for the purposes of research or private study, or criticism or review, as permitted under the UK Copyright Designs and Patents Act, 1988, this publication may not be reproduced, stored, or transmitted, in any form or by any means, without the prior permission in writing of the publishers, or in the case of reprographic reproduction only in accordance with the terms of the licences issued by the Copyright Licensing Agency in the UK, or in accordance with the terms of licences issued by the appropriate Reproduction Rights Organization outside the UK. Enquiries concerning reproduction outside the terms stated here should be sent to the publishers at the London address printed on this page.

The publisher makes no representation, express or implied, with regard to the accuracy of the information contained in this book and cannot accept any legal responsibility or liability for any errors or omissions that may be made.

A catalogue record for this book is available from the British Library

Library of Congress Catalog Card Number: 97–66023

∞ Printed on permanent acid-free text paper, manufactured in accordance with ANSI/NISO Z39.48-1992 and ANSI/NISO Z39.48-1984 (Permanence of Paper).

*Dedicated to
David Dunnette*

CONTENTS

List of contributors		ix
Acknowledgements		xiii
Introduction		xv
1	The Willamette River in Oregon: a river restored?	1
2	The Vistula River in Poland: environmental characteristics and historical perspective	11
3	Water quality in the Vistula basin	21
4	The Tualatin River: a water quality challenge	33
5	Changes in the water chemistry of the Vltava River from 1959 to 1993	39
6	An integrated approach: catchment management planning for the River Tyne	45
7	Water quality variability in the Agüera Stream watershed at different spatial and temporal scales	55
8	Sources of pollutants and current status of water quality in the lower reaches of the Neman and Pregel rivers	69
9	Acid mine-water pollution from an abandoned mine	73
10	The effect of pollutants from the Vistula River on the water quality of the Bay of Gdansk	81
11	Metal loadings to the Baltic Sea from Poland's rivers	89
12	Surface water pollutants and problems with their analysis	95
13	Monitoring the water quality of the Radunia River	105
14	The influence of industrial and municipal waste on the quality of the environment in Poland, taking Gdansk region as an example	111
15	Action taken by the Gdansk Municipality to reduce the loads of pollutants discharged into Gdansk Bay	121
16	Monitoring of volatile organic compounds (VOCs) in the surface and sea waters of Gdansk district	127
17	Cause analysis of microbiological pollution around beaches along the Gdansk coast	133
18	Leaching of nutrients, heavy metals and pesticides from agricultural land in Sweden	139

19	Nutrient non-point pollution in experimental watersheds in Poland	145
20	The analysis of nutrient concentration–water flow relationships in a lakeland river	157
21	A process for developing water quality criteria under the US federal Clean Water Act	165
22	Differences between Polish standards and EC directives on water quality	173
23	Water quality study design for the Willamette Basin, Oregon, using a geographic information system	181
24	The role of benefit–cost analysis in water conservation planning	191
25	Exploring possibilities for an international water quality index applied to river systems	205
26	River water quality modelling in Poland	211
27	Mathematical modelling of soil nitrate and phosphate leaching from small agricultural catchments in northern Poland	221
28	Water quality modelling of the Rega River and surrounding coastal waters	233
29	Environmental protection in the Vistula River valley	243
30	Biological barriers in wastewater treatment	251
31	Hydrobotanical systems: characteristics and examples in Poland	257
32	A multifunctional water management information system (MIS)	267
33	The state, management, use and protection of natural water resources in the Province of Gdansk	273
34	National environmental monitoring programme	281
35	The legal framework of water quality management in England and Wales	291
Index		303

CONTRIBUTORS

C. Avent
Oregon State University, Portland, Oregon, USA

B. Barrett
Middlesex University, London, UK

A. Basaguren
University of the Basque Country, Bilbao, Spain

M. Biziuk
Technical University of Gdansk, Poland

J. Błażejowski
University of Gdansk, Poland

P. Blažka
Hydrobiological Institute, Czech Science Academy, České Budějowice, Czech Republic

T. Bogacka
Institute of Meteorology and Water Management, Gdansk, Poland

R.J. Bogdanowicz
University of Gdansk, Poland

N. Brink
Swedish University of Agricultural Sciences, Uppsala, Sweden

R. Ceglarski
Institute of Meteorology and Water Management, Gdansk, Poland

D. B. Chandler
Oregon Health Sciences University, Portland, Oregon, USA

C. Cude
Portland State University, Portland, Oregon, USA

K. Czerwionka
Technical University of Gdansk, Poland

J. Dojlido
Institute of Meteorology and Water Management, Warsaw, Poland

D. Dunnette
Portland State University, Portland, Oregon, USA

A. Elósegui
University of the Basque Country, Bilbao, Spain

A. Franklin
Portland State University, Portland, Oregon, USA

E. González
University of the Basque Country, Bilbao, Spain

D. Gorlo
Technical University of Gdansk, Poland

D. Grodzicka-Kozak
Environmental Protection Department, Gdansk, Poland

M. J. Gromiec
Institute of Meteorology and Water Management, Warsaw, Poland

G. Gross
Portland State University, Portland, Oregon, USA

D. Hammerton OBE
University of Paisley, Scotland, UK

J. Hartmann
Portland State University, Portland, Oregon, USA

D. Hayteas
Portland State University, Portland, Oregon, USA

E. Heybowicz
Institute of Meteorology and Water Management, Gdansk, Poland

J.E. Jackson
United Sewerage Agency, Hillsboro, Oregon, USA

W. Janicki
Technical University of Gdansk, Poland

W. Jarosiński
Institute of Meteorology and Water Management, Katowice, Poland

E. Jaśniewicz
Institute of Meteorology and Water Management, Wroclaw, Poland

T. Jenkins
Portland State University, Portland, Oregon, USA

C. Johnston
Northwestern School of Law, Portland, Oregon, USA

Z. Kajak
Institute of Ecology, Polish Academy of Sciences, Lomianki, Poland

M. Kędzia
Institute of Meteorology and Water Management, Wroclaw, Poland

J. Kopáček
Hydrobiological Institute, Czech Science Academy, České Budějowice, Czech Republic

J. Kopeć
Municipal Government of Gdansk, Poland

R. Korol
Institute of Meteorology and Water Management, Wroclaw, Poland

P.J. Kowalik
Technical University of Gdansk, Poland

D. Kozak
Environmental Protection Department, Voivodeship of Gdansk, Poland

E.V. Krasnov
Kaliningrad State University, Kaliningrad, Russia

M. Kulbik
Technical University of Gdansk, Poland

I. Kulik-Kuziemska
Technical University of Gdansk, Poland

K. Leben
Portland State University, Portland, Oregon, USA

A. Lewandowski
Geoscience and Marine Research & Consulting Co. Ltd, Sopot, Poland

J. Lyngdal
Portland State University, Portland, Oregon, USA

D. Marks
Portland State University, Portland, Oregon, USA

G. McMurray
Department of Environmental Quality, Portland, Oregon, USA

A. Mierzwiński
National Inspectorate of Environmental Protection, Warsaw, Poland

C. Morganti
Portland State University, Portland, Oregon, USA

N.J. Mullane
Department of Environmental Quality, Portland, Oregon, USA

J. Namieśnik
Technical University of Gdansk, Poland

E. Niemirycz
Institute of Meteorology and Water Management, Gdansk, Poland

H. Obarska-Pempkowiak
Technical University of Gdansk, Poland

K. Olańczuk-Neyman
Technical University of Gdansk, Poland

B. Ołdakowski
Geoscience and Marine Research & Consulting Co. Ltd, Sopot, Poland

M.S. Ostojski
Regional Board of Water Management, Gdansk, Poland

Ż Polkowska
Technical University of Gdansk, Poland

J. Pozo
University of the Basque Country, Bilbao, Spain

L. Procházková
Hydrobiological Institute, Czech Science Academy, České Budějovice, Czech Republic

T. Quin
Portland State University, Portland, Oregon, USA

M. Robakiewicz
Institute of Hydroengineering, Polish Academy of Sciences, Gdansk, Poland

R. G. Sakrison
Department of Ecology, Olympia, Washington, USA

W. Szczepański
Institute of Meteorology and Water Management, Katowice, Poland

R. Taylor
Institute of Meteorology and Water Management, Gdansk, Poland

M.A. Uhrich
US Geological Survey, Water Resources Division, Portland, Oregon, USA

M.P. van der Vat
Delft Hydraulics, c/o Institute of Hydroengineering, Polish Academy of Sciences, Gdansk, Poland

A. Walkowiak
Inspectorate for Environmental Protection, Gdansk, Poland

T. Warburton
Environment Agency, North-East Region, Newcastle, UK

W. Wardencki
Technical University of Gdansk, Poland

D.A. Wentz
US Geological Survey, Water Resources Division, Portland, Oregon, USA

D.M. Zaporozhsky
Kaliningrad State University, Kaliningrad, Russia

B. Zygmunt
Technical University of Gdansk, Poland

ACKNOWLEDGEMENTS

The editors wish to thank everyone who contributed towards the preparation of the manuscript and, in particular, the following.

Z. Makowski (Chapters 2, 11 and 34); the granting Agency of the Czech Republic, staff of the Hydrobiological Institute (Chapter 5); University of the Basque Country, the Basque Government (Chapter 7); National Committee for Scientific Research (Poland), J. Rybinski (Chapter 11); Technical University of Gdansk (Chapters 12 and 16); University of Gdansk, management of the National Inspectorate of Environmental Protection in Gdansk (Chapter 20); members of the Technical Advisory Committee, the Policy Advisory Committee and the technical subcommittee on temperature of the Department of Environmental Quality (Portland) (Chapter 21); Maria Sklodawska- Curie Joint Fund II, Ministry of Environmental Protection, Natural Resources and Forestry (Poland), US Environmental Protection Agency (Chapter 26); staff of the Institute of Meteorology and Water Management (Gdansk), R. Taylor (Chapter 27); G. Howells (Chapter 35).

INTRODUCTION

In 1992, Professor David Dunnette of Portland State University, Oregon was a visiting professor to the Technical University of Gdansk, Poland. As a result of this visit, an informal agreement was reached on co-operation between the two institutions. Professor Piotr Kowalik of the Technical University introduced David Dunnette to Professor Teresa Bogacka and Dr Elżbieta Niemirycz of the Institute of Meteorology and Water Management, which has one of its offices in the city of Gdansk, and they also became involved in the agreement.

One of the main outcomes of this co-operation was the establishment of two International River Quality Symposia, which were held in Portland in March 1994 and Gdansk in June 1994. At these symposia, many speakers from the three principal sponsoring organizations were supported by others from different parts of the world. The aims of the meetings were:

1. to compare the quality of the rivers in Oregon with those in Poland;
2. to identify solutions to pollution problems; and
3. to share experiences and expertise.

It was also intended that opportunities for international co-operation in research, education and information transfer between individuals and institutions would be identified and exploited. The focus of the meetings was to make a comparison between the Willamette River in Oregon, which has improved substantially in recent years, and the Vistula River in Poland, which has suffered from a steady deterioration over the last 30 years.

At the Gdansk symposium, the problems causing the pollution of waters in the host country was the main theme. This book contains the majority of those proceedings. At the Portland symposium, the emphasis was on the many aspects affecting the quality of the Oregon rivers – from legislation, through monitoring strategies and modelling, to nature conservation. The proceedings of the Portland symposium are published separately [1].

There has been a remarkable change in the attitude to the environment in Poland and other eastern European countries in the last few years. An OECD report in 1995 [2] revealed that the proportion of investments in the environment to the GDP was two and a half times greater in 1992 than in 1985 and amounted to the equivalent of US $1 billion. This investment has resulted in a decline in the emission of air pollutants and a halt in the steady deterioration of the quality of water. A key factor was the setting up of an Ecofund financed by the USA, Switzerland, Finland and France, who wrote off parts of their debts from Poland in return for Poland investing an equivalent amount in Ecofund projects. These projects have the aim of enabling Poland to meet its international obligations to the quality of the environment and to meet the European Union's standards for emissions of pollutants.

Since the OECD report, the impetus has continued. For example, 1995 was designated the Year of Clean Technology in Poland and, by 1995, more than a quarter of the 1200 sewage treatment plants planned under the Water Act of 1991 had been built [3]. Surprisingly, only a small proportion, about 5%, of the finance for environmental improvements comes from other countries. Most comes from charges for the use of the water supply and from fines for non-compliance with discharge regulations. The main contribution that the West is making to Poland is in management expertise and a recent example of this is the adoption of the idea of catchment management for Poland's rivers.

The Polish Government adopted a new environment programme in 1991. The priorities include improving the supply of clean drinking water, halving the pollutant discharges and reducing the pollutant load to the Baltic Sea by 50%. Underlying the whole programme is the need to ensure sustainable development to comply with the aims of the 1992 Rio Conference on the state of the global environment.

In this book the proceedings of the Gdansk symposium have been grouped into different topics. Chapters 1–8 are case studies of rivers from different parts of the world; the key chapters describe the Willamette and Vistula rivers but there are also descriptions of rivers in Russia, Spain, the Czech Republic and the UK. Chapters 9–17 deal with some specific pollutants. These range from acid mine-water pollution to volatile organic constituents and include an assessment of the loading of toxic metals entering the Baltic Sea from Poland.

Poland has a large agricultural industry and the run-off of nutrients and pesticides from the land has a marked effect on the quality of the receiving waters. There are a number of contributions on this subject (Chapters 18–20) including an interesting comparison with the situation in Sweden. Many water scientists are trying to find ways of assessing water quality in an objective way and some of these are described (Chapters 21–25). However, as well as devising water quality indices, there is also much work being done on mathematical modelling of water quality and a number of different approaches have been grouped together in Chapters 26–28.

There are a variety of presentations (Chapters 29–31) on biological aspects of water quality and these range from the use of flora to purify wastewater through the bacteriological quality of the Baltic coast, to an overall assessment of the ecological value of the Vistula River corridor. Chapters 32–35 deal with the issues of legislating against water pollution and how to optimize the planned improvements in water quality using a limited budget.

The two International River Quality Symposia brought together scientists separated by language, by thousands of miles and living in countries with very different environments, both in terms of their quality and their economies. The result was a valuable exchange of information and contacts which will ultimately be of benefit to the aquatic environment, particularly in Poland. The tragedy is that David Dunnette was not able to see the results of what he started in the winter of 1992 in Gdansk because he died of viral pneumonia early in 1995. The books of the symposia proceedings are dedicated to his memory and we hope that the spirit of friendliness and co-operation that he extended to all participants will continue in the future.

Gerry Best, Teresa Bogacka and Elżbieta Niemirycz
February 1996

REFERENCES

1. Dunnette, D. and Laenen (eds) (1996) *River Quality: Dynamics and Restoration*, Lewis Publishers
2. OECD (1995) Environmental Performance Review: Poland OECD Publications Service, Paris.
3. Newmark, E. (1995) Changing states for Polish water Water and Waste Treatment, **38**(8) 36–7.

THE WILLAMETTE RIVER IN OREGON: A RIVER RESTORED?

by N.J. Mullane

1.1 INTRODUCTION

The Gdansk–Portland International River Quality Symposia brought together many different and complex scientific and administrative aspects of river quality protection. The symposia examined, among other things, river ecology, water chemistry, water quality monitoring and modelling, and the impact of nutrients and organic chemicals on river systems. They also examined the methods used to regulate pollution sources to achieve river restoration and maintain high water quality.

The restoration efforts on the Willamette River in the State of Oregon, USA involved many of these scientific elements. This case study illustrates the interplay between the public's desire to protect and eventually enjoy a river resource, and the use of the river as a receiving stream for treated and untreated wastewater discharges. An examination of this river clean-up will provide some insights into the necessary blend of scientific information and the formation and implementation of public policy. The regulatory approach established an essential foundation for public, industry, municipality and state interaction throughout the restoration. Carefully considered regulatory actions were supported by scientific information describing river conditions and evaluating whether the controls implemented were having the desired effect. Follow-up monitoring and evaluation led to increasingly higher levels of treatment until clean-up goals were met. Basin water quantity management, to control massive flooding, played a key role in the restoration, but leaves lingering questions about how it affects anadromous fish migration.

The words, in the title of this chapter, 'a river restored', suggest that the clean-up work has been completed. However, restoration is dynamic, not static. Time brings increasing population, expanding industries, agricultural growth, better understanding of pollution impacts and advances in sampling techniques, all of which continually affect water quality. Therefore to restore a river must mean constant vigilance to maintain the goals established and achieved as time brings new challenges and information on potential threats.

This examination begins with a brief historical review of how a river was destroyed through neglect and indifference. It will describe how citizens rallied behind an initiative petition effort to force the state, cities and industries to take the steps necessary to restore the river.

1.2 DESCRIPTION OF THE RIVER BASIN

The State of Oregon is located in the north-west corner of the USA. It is the tenth largest state, with a total area of approximately 97 000 sq miles (251 000 km^2) and a population of almost three million people. The whole of the Willamette River lies within the state and drains approximately 12 000 sq miles (31 000 km^2) or 12% of its land area

2 The Willamette River in Oregon: a river restored?

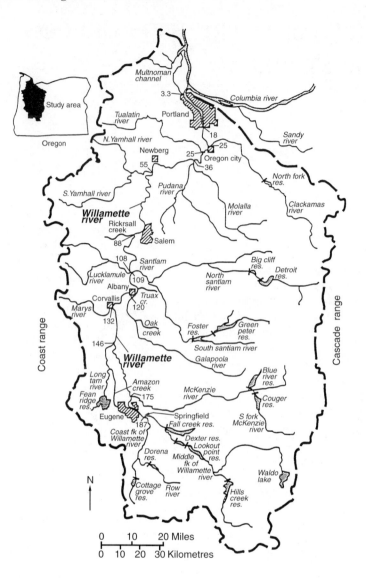

Figure 1.1 Willamette River basin study area

(Figure 1.1). The river is the 13th largest in the USA. The river valley is some 160 miles (257 km) long and 75 miles (121 km) wide. The basin contains the majority of Oregon's urban and industrial complexes and is home to over 2 million people. The basin has a diverse natural resource base supporting healthy recreational, agricultural, forestry and fishing activities. A vast array of agricultural products including numerous vegetable and fruit crops, tree, shrub and flower nurseries, vineyards, livestock and grass seed are grown in the valley. Recreational opportunities abound throughout the basin. The valley flanks are heavily forested and the timber industry provides building materials for a growing economy. The Willamette basin also contains a rapidly growing electronics industry.

The river itself supports a long list of beneficial uses:

- Public domestic water supply
- Private domestic water supply
- Industrial water supply
- Irrigation
- Livestock watering
- Anadromous fish passage
- Salmonid fish rearing
- Salmonid fish spawning
- Resident fish and aquatic life
- Wildlife and hunting
- Fishing
- Boating
- Water contact recreation
- Aesthetic quality
- Hydro-power
- Commercial navigation and transportation.

Figure 1.2 Willamette River riverboats.

It teems with people during the summertime and supports a strong fisheries industry throughout the year, particularly during various salmon migrations. On a typical summer's day one can watch waterskiers, rowers, sailors and windsurfers moving up and down the river. Drinking water is also taken from the river by several communities. All these activities and expectations place a heavy burden on the river.

But to fully appreciate what the river provides today one has to travel back in time several decades and examine the condition of the river in the early 1900s.

1.3 HISTORY OF RIVER POLLUTION

At the turn of the century, the river was a commercial highway through the heart of the natural resource-rich Willamette valley. The river provided an easy and accessible mode of transportation to move people and products up and down the valley. For example, huge log rafts floated down the river to the many mills which extended along its length, and riverboats transported people (Figure 1.2).

The river was not looked upon as a natural resource, but as a commercial resource, something to be used, but not necessarily respected. It was seen as a place to discard the unwanted wastes of a rapidly growing society. It received untreated wastes from a large number of different sources. Cities poured untreated sewage into the river throughout its length. Commercial operations such as slaughterhouses and food processors discharged untreated process-water into the river. Critics of the time suggested the municipalities and industries treated the river as an open sewer. The historical accounts would seem to bear this out. These discharges rapidly exhausted the river's capacity to accept and assimilate pollutants (Figure 1.3).

The river was described as ugly, filthy and intolerable. Waste was thrown down the riverbank without regard to its impact. Workmen refused to work on riverside construction projects because of the stench and fear of illness if they were to fall

Figure 1.3 Pollution problems in the lower Willamette River

into the water. People could not swim or indulge in recreational activities in or on the river. One historical account described how people shunned the river. This was manifested in how they built their homes facing away from the river.

Communities along the river were forced to look elsewhere for drinking water supplies. Drinking water could no longer be abstracted from the river without extensive treatment. Salmonids could no longer survive in the river. In fact, in tests, most fish suffocated in a matter of minutes after being exposed to the river water.

The Willamette River had once before provided readily available drinking water, a vast fisheries industry, numerous recreational resources and other beneficial uses (section 1.2). But the river could not withstand the onslaught of human sewage and industrial waste products and continue to support these uses. It became a river choked with the refuse of a growing society.

The river was destroyed through neglect and indifference. Even with this obvious pollution and terrible loss of resource, no action was taken by the Government, cities or industries to stop the contamination. After a long period of inaction by the state legislature, the citizens of the state rallied behind an initiative petition (Figure 1.4) leading to the passage into law of the Water Purification and Prevention of Pollution Act in 1938. The Act established the Oregon State Sanitary Authority (which in 1969, became the Department of Environmental Quality).

Figure 1.4 Citizen rally

1.4 APPROACH TO POLLUTION CONTROL

The Sanitary Authority was given the responsibility for bringing water pollution under control and restoring the river. It began by documenting water pollution conditions in the river and establishing specific scientific criteria upon which to evaluate river quality. Sampling was conducted for dissolved oxygen (DO), turbidity, bacteria, pH, suspended solids, biochemical oxygen demand, temperature and other parameters. Dissolved oxygen, because it is essential for the support of aquatic life, became the most important criterion. It was chosen as the criterion upon which restoration success would be evaluated [2]. Field teams where sent up and down the river to gather information on DO. In these early studies, DO levels in the Willamette, where it discharged into the Columbia River, were less than 0.5 mg/l. In the years that followed, the conditions continued to degrade to the point where DO was not detected at all during some test periods [1].

1.4.1 WASTEWATER TREATMENT

The causes and results of the river quality decline are not difficult to document. The discharge of untreated sewage and wastewater from numerous cities and industries was exhausting the river's assimilative capacity. The Sanitary Authority wanted the cities and industries to install treatment facilities. The first step was to hold discussions with them to explain the results of the river testing and issue administrative orders to the municipal authorities along the river to install primary treatment facilities. While these facilities were being installed, the Sanitary Authority continued to sample the river to monitor river conditions and document any changes. The sampling results showed that municipal primary treatment would not achieve the goal of restoring the river. The next step was to require industries to install primary treatment as well. As industries installed these treatment facilities, the river was again monitored to document the changes; again the results showed little change.

Next, the Sanitary Authority ordered the municipal authorities to install secondary treatment and the industries to remove 85% of their pollutant load [3, 4]. These controls were installed by the late 1960s, some 30 years after the initiative petition, but the effort was beginning to show results. The river's assimilative capacity was being restored. Dissolved oxygen targets were achieved throughout the river during low flow conditions in 1969, 1970 and 1971 [5]. In 1972, the Willamette River clean-up became a national success story in the USA. The goals of bringing fish back into the river and providing for recreation in and on the river had been achieved. The most important visible sign of success was the return of the salmon and steelhead runs up the river [5]. People could again indulge in recreational activities in and on the water.

It is important to note that in 1972, when the Willamette River was responding to three decades of concerted efforts by municipalities, industries, citizens and the Oregon State Government, the Clean Water Act was passed by Congress establishing the federal water quality protection programme. The Clean Water Act established a federal goal that waters should be fishable and swimmable by 1985. Municipal dischargers were required to install secondary treatment, and industrial dischargers were required to meet specific pollutant removal targets and mandatory industrial effluent limits. A permit programme was established whereby dischargers were required to implement specific levels of treatment. Permit violations were enforceable by the federal and state water pollution-control agencies, as well as private citizens. The potential for private prosecution, through Section 505 of the Clean Water Act, brought considerable pressure to bear upon the regulated community as well as the regulatory agencies to fulfil their responsibilities.

1.4.2 FLOW AUGMENTATION

One very important element of river management which played a key role in the restoration was flow augmentation. The normal flow of the Willamette River varies widely from tens of thousands of cubic feet per second (several hundred cubic metres per second) in the winter and spring to less than 3,000 cubic feet per second (85 cubic metres

Figure 1.5 General flooding in the Willamette Valley

per second) in the summer. The summer low-flow conditions presented a tremendous challenge to restoring river quality. The amount of water in a river at any given time has a very big effect on the ability of a river to assimilate waste products. To achieve clean-up goals without flow augmentation would have been extremely difficult and expensive. The lower a river's flow the greater the waste treatment requirements.

At the same time cities and industries were improving wastewater treatment, the US Army Corps of Engineers received Congressional authorization to build a flood-control system in the Willamette basin [6]. The system was intended to protect people and property from flood damage, and provide irrigation water and flows for river navigation (Figures 1.5 and 1.6). Thirteen dams were constructed in the basin including eleven

Figure 1.6 'High and dry' in the Willamette Valley

primary dams and reservoirs, and two re-regulating dams:

- Hills Creek
- Lookout Point (including the Dexter re-regulating dam)
- Fall Creek
- Cottage Grove
- Dorena
- Cougar
- Blue River
- Fern Ridge
- Green Peter
- Foster
- Detroit (Figure 1.7) (including the Big Cliff re-regulating dam)

The reservoir system worked by storing spring rain and snow-melt flood waters for release during the late spring and summer months. The operational strategy therefore augmented the normal summer low flows. The released stored water provided for irrigation needs during the growing season, and recreational opportunities within the reservoirs and rivers in the form of summertime boating and fishing.

The summertime releases from the reservoirs provided much needed flows to assimilate the newly treated wastewater discharges. Without these flows, the tremendous efforts by cities and industries to install the required treatment facilities would not have been fully successful during the critical summer low-flow period [7].

Figure 1.7 US Army Corps of Engineers' Detroit Dam

1.4.3 RESTORATION: A CONTINUING PROCESS

It has been over 20 years since the Willamette River was 'restored'. During this time Oregon has continued to grow. More people and industries have come, more houses have been built, more land has been disturbed and eroded and more impervious surfaces have been created [8]. Agricultural production continues to grow. Chemicals continue to be used. River quality monitoring has become more sophisticated. Analytical techniques give much more detailed inventories of chemicals present in water as well as fish tissues. As more advanced analysis is undertaken, not surprisingly, more contaminants are found [9]. These contaminants may have been present decades ago, but only now are they being detected and questions being asked as to their effect on the health of the river.

1.5 CO-OPERATIVE WATER QUALITY STUDY

Citizens, municipalities, and industries have again become concerned over water quality in the river. This concern for the river manifested itself in 1991, when the state legislature requested the Department of Environmental Quality to initiate a comprehensive water-quality study of the river [10]. This study was much too expensive for the state to implement on its own so it solicited the support of the municipalities and major industries along the river. A co-operative study was initiated in 1991. The aim of this study was to provide information on how well the river was doing. The study was divided into two phases.

Phase I reviewed and summarized existing data, developed and evaluated tools to measure water quality, and developed methods to predict water-quality conditions. This latter task was the central part of the co-operative effort. All parties involved wanted to know whether the river was being affected. Priorities were established on data collection in Phase I and these data were integrated into models to see if present conditions could be described and future conditions predicted. Below is a summary of the results of the major study components.

1.5.1 TOXIC MODELLING COMPONENT

The toxic modelling component relied on existing data collected from several different studies with various levels of basin coverage. Consequently, the results were complete with numerous data gaps being identified. There was very little the researchers could state conclusively about toxic throughout the basin. High priority was given to the collection of toxic data in Phase II of the study.

1.5.2 DISSOLVED OXYGEN MODELLING COMPONENT

This proved to be one of the most fruitful parts of Phase I. Dissolved oxygen data, collected during a synoptic survey, were used to calibrate a DO – nutrient – phytoplankton growth model.

After adjustments were made for sediment oxygen demand (SOD), the final model fitted the steady-state conditions well. The model predicted that during low-flow periods:

- minimum DO conditions occur in Newberg Pool;
- the most significant DO uptake is from sediments and the degradation of organic compounds from point sources and tributaries; and
- the amount of river flow is a significant factor controlling DO concentrations and phytoplankton biomass.

Researchers want to monitor SOD specifically to confirm the SOD rates used in the Phase I modelling.

1.5.3 BACTERIOLOGICAL MODELLING

This was one of the more disappointing study components. The data collected during Phase I were not sufficient to calibrate the model. Many questions remain unanswered as a result of the bacteriological modelling study. Many more bacteria were in the river than was predicted by the model. Considerable work remains to identify sources, both point and non-point, and re-run the model.

1.5.4 NON-POINT SOURCE LOADING

Considerable time and effort went into the evaluation and selection of non-point source models. Several different models were examined, each with its own advantages and disadvantages. The most important element of these models was the amount of data needed to predict non-point source pollutant loadings accurately. The modelling of non-point sources was achieved by intensely developing non-point source information for one small watershed. This model was then extended to a sub-basin and then to the entire Willamette River basin. The reduction in resolution for each subsequent extrapolation, however, became a key sticking point for general acceptance of this modelling approach. It was apparent that more specific water-quality monitoring data would be needed from many different sites throughout the basin to develop better modelling capabilities and prediction of non-point source impacts.

1.5.5 ECOLOGICAL SYSTEM INVESTIGATION

Of the many Phase I study components perhaps the most significant was the ecological system investigation which was added to the study at the insistence of the environmental community representative. The ecological investigations were a series of different tests and mini-studies designed to gather information on the health of the aquatic systems. The original objection to these tests centred on the ability of the tests actually to determine the causes of different impacts observed. But in the final analysis, the tests were very useful in indicating where potential ecological problems existed. Researchers were still left with the task of finding the causes of some disturbing results, but, without the investigation they would have had precious little to show for their efforts and little information on the overall health of the systems. The river was divided into four regions for the ecological studies (Figure 1.8) and the studies are described in more detail below.

(a) Benthic community assessment

The first ecological system study centred on the benthic communities. Traditional kick-net and sediment-grab samples were collected. Both of these techniques used the US Environmental

8 *The Willamette River in Oregon: a river restored?*

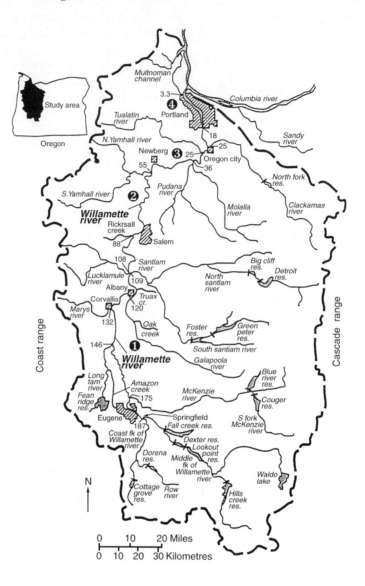

Figure 1.8 Ecological study regions for the Willamette River (Region I – Corvallis headwaters; Region II – Corvallis to Newberg Pool; Region III – Newberg Pool to Willamette Falls; Region IV – Willamette Falls to mouth (Portland Harbor).

Protection Agency (US EPA) rapid bioassessment protocols (RBP) to evaluate benthic community responses to habitat and water quality. The tests indicated that water quality degradation, and not physical habitat degradation, impairs the communities.

(b) Fish community health assessment

Several different tests were conducted to indicate whether the fish communities were impaired. The first two looked at community diversity and fish health. Communities were sampled to determine type and quantity of fish species and these were recorded. Selected fish in each study region were also autopsied. The results showed some regions were affected by pollution but the results were inconclusive. The control site was affected as well as downstream sites. Questions about fish migration and specific point-source effects need to be answered. More sampling stations and additional species need to be studied in the future to provide more confidence in the results.

The third fish community test seemed to yield the most interesting and maybe the most challenging data. The test involved the examination of skeletal abnormalities. These tests showed skeletal deformities in 1% to 2% of those fish tested in Regions I, II and IV. In Region III, however, deformities were found in over 13% of the fish tested. These results are interesting for a number of reasons, one of which is that Region IV is downstream in Portland Harbor and Region III is the first significant sink on the river. Why were test results in the harbour lower than in the upstream site? Could Newberg Pool, Region III, be the first point on the river where pollutants collect? These and other questions will need to be addressed in Phase II.

1.6 RIVER BASIN MANAGEMENT

The results of the work conducted in the original restoration and during the later co-operative water-quality study point to a simple fact. Protection of the Willamette River and maintenance of its beneficial uses needs a comprehensive management approach. This requires the co-operation of dischargers, consumers and recreational users: those discharging pollutants directly into the river (from point sources); the indirect dischargers from non-point sources (agriculture, forestry and the urban areas); those taking water from the river for use as drinking water, industrial water and irrigation water; and those using the river for recreational purposes both in and on the water. All these users need to join together to identify their needs and to balance these against what the river is capable of providing. Then it is up to all of them to ensure the goals that are set are maintained.

1.7 CONCLUSIONS

The clean-up of the Willamette River illustrated how citizens, industries, municipalities and government can work together to restore a highly polluted river. Salmon returned to a river which once suffocated them in a matter of minutes. Two decades later, attention has again focused on the Willamette River. The river basin's population has increased, its industries and urban areas have expanded, and its rural land has experienced continued use and development. A co-operative water-quality study was undertaken to determine the health of the river. Early results indicate that, again, an examination of people's impacts on the river and the development of a new management strategy are needed if the river is to be truly restored and provide for its many beneficial uses.

REFERENCES

1. Oregon State Sanitary Authority (1964) Report on water quality and waste treatment needs for the Willamette River.
2. Oregon State Water Resources Board (1965) Lower Willamette River Basin.
3. Oregon State Water Resources Board (1969) Oregon's long range requirements for water.
4. Gleeson, G.W. (1972) *The Return of a River – The Willamette River, Oregon.*
5. Oregon State Sanitary Authority (1967) Water quality standards, Willamette River.
6. US Army Corps of Engineers (1989) Willamette River basin reservoir system operation.
7. US Army Corps of Engineers (1991) Willamette River basin review reconnaissance study.
8. Water Resources Commission (1992) Willamette basin plan.
9. Oregon Department of Environmental Quality (1992) Oregon water quality status assessment report.
10. Oregon Department of Environmental Quality (1993) Willamette River basin water quality study – summary report.

THE VISTULA RIVER IN POLAND: ENVIRONMENTAL CHARACTERISTICS AND HISTORICAL PERSPECTIVE

by E. Niemirycz

2.1 DESCRIPTION OF THE RIVER BASIN

The Vistula River is the largest Polish river and also one of the 30 largest rivers in the world [1]. After the Neva River in Russia, the Vistula is the second largest river in the Baltic Sea basin, in terms of flow. Its source is in the mountains of southern Poland (Figure 2.1) and it flows northeast, passing by Kraków, Warsaw, and then northwest into the Baltic Sea. Near Warsaw the Vistula is a typical lowland river. The mouth of the Vistula comprises the Vistula Delta, which consists of two branches: the Nogat flowing into the Vistula Lagoon and the Leniwka, the main stream of the Vistula, flowing directly into the Baltic. These two branches are navigable at their mouths. The Vistula Delta (Żuławy), with a total area of 1740 km², is used for agriculture.

The Vistula is 1047 km long and 300–1000 m wide in its middle and lower parts, with a total drainage area of approximately 194 700 km² (over 50% of the total area of Poland). The channel slope is gentle, from 0.1% to 0.31%. The river flow is variable from year to year as can be seen from Figure 2.2. It is also evident that the dry weather in central Europe in the 1980s and 1990s had a marked effect on the Vistula water flow. Variations in the annual cycle of the river flow can also be observed [2]. The maximum spring flows (March and April) can

Figure 2.1 Physical map of Poland.

cause floods along the whole length of the Vistula, while the maximum autumn flows are severe only in the upper part (the Carpathian foothill region). The minimum flows occur in August, September and November – the 'Polish golden fall', when rainfall is low. In general during the winter season (from the second half of December to the end of

International River Water Quality. Edited by Gerry Best, Teresa Bogacka and Elżbieta Niemirycz. Published in 1997 by E & FN Spon, London, ISBN 0419215409

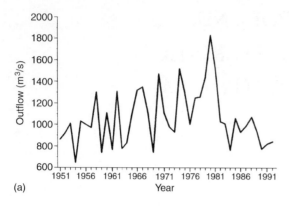

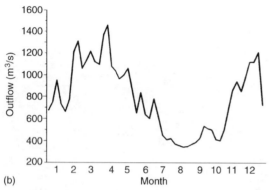

Figure 2.2 Hydrological conditions of the Vistula River at the mouth cross-section (a) average annual flow during 1951–92; (b) flow during 1992.

February) much of the Vistula is frozen over. The average river flow at the lowest reach is 1000 m³/s. By comparison, the average flow near Warsaw, i.e. in its middle part, is 520 m³/s.

The changes in the natural environment of the Vistula basin in the last 50 years are closely connected with historical changes, which have occurred in three post-World War II periods. These periods affected the economic development of the Vistula basin. During the first period, from the mid-1940s to the mid-1950s, the reconstruction of the country, destroyed by five years of war, took place. In the next period, between the 1960s and early 1980s, the Government's investment policy was unsound and inconsistent for many reasons; there was also a disregard for principles of environmental protection: this resulted in the degrada-

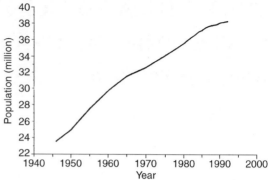

Figure 2.3 Population of Poland 1946–92.

tion of the natural environment. During these periods there were serious consequences for human health. For example, between 1946 and 1980 the population of Poland increased by about 50% (Figure 2.3). This growth resulted in a great increase in the consumption of drinking water, much of which originated from the Vistula River. The third period is from the early 1980s to the present day and is typified by increasing awareness of the poor quality of the environment in Poland and the need for investment in treatment technologies.

At present the land use in the Vistula basin is characterized by: 48.8% arable land; 26.4% forest; 15.7% grassland; 1.4% orchards and 7.7% other categories [3]. Despite a significant growth in afforestation after World War II, the present area of forest in Poland is 2.3 km² per 1000 inhabitants. (This can be compared with Canada, where there is 170.3 km²; the USA – 11.7 km²; Germany – 1.3 km²; and France – 2.7 km².) Deforestation has resulted in greater variability in the river flow [1] and a high erosion rate (16 t/km² per year) [4]. The erosion rate is also affected by the soils in the Vistula basin (Figure 2.4).

The Vistula basin is populated by 22 million people, about 60% of whom are concentrated in urbanized agglomerations. The remainder are farmers. The nutrient inputs from agriculture include surface and underground run-off from fields, meadows and pastures, as well as from settlements without sewerage systems [5, 6]. The abatement of discharges of nutrients from non-caused by phosphorus from phosphates, which are contained in detergents used for cleaning and washing [8]. A change of detergent composition could result in a reduction of phosphorus load in wastewaters of up to 30%.

Manure is also a great threat to the water environment [9]. The manure from large intensive farms is applied to soils without adequate control and causes pollution of surface waters and groundwaters, as well as soil degradation.

In Poland, of the 4414 major industrial plants, only 52% (2297) are equipped with purification facilities [10]. Some 353 industries discharge their wastewaters directly into surface waters, while 1764 discharge their wastes through the municipal sewerage systems or directly into the soil. Wastewaters, especially from chemical, metallurgical, and pulp and paper industries, are characterized by high concentrations of pollutants, some of which are persistent and toxic substances.

Another major problem is the salinity of the Vistula waters, mainly caused by coal mines situated in the Upper Silesian region (Figure 2.2). As a result, water from some parts of the Vistula cannot be used for industrial purposes because it causes corrosion of metallic parts.

2.2 HISTORICAL PERSPECTIVE

The name 'Wisła' (Polish for Vistula) appears in legends, chronicles and various descriptions dating as far back as Roman times. The Vistula is very deeply associated with the history of Poland, with the mentality of its people, its literature, folk-lore and national emotions. The river – symbolized as 'the queen of Polish rivers' – has always inspired Polish writers and composers [11].

The changes in foreign trade in the 15th century encouraged the export of such goods as corn and timber, and from this arose the need to find a suitable means of conveyance. The shipping of the goods by river turned out to be more suitable than transporting them by land. The landowners used the Vistula and its tributaries for trade purposes. This resulted in modifications to the river to ensure navigability of the river route, the building of larger craft and the organization of floating of goods,

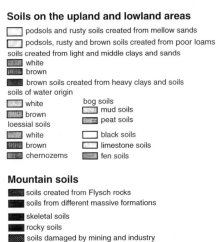

Figure 2.4 Soils in the Vistula River basin.

point sources is difficult to achieve [7]. However, a number of measures can be taken, e.g. improvement of village sanitation (by providing villages with sewerage systems), creation of plant buffer zones and construction of small treatment plants. It is also possible to reduce the nutrient inputs from agriculture by adjusting the quantity and timing of fertilization application to suit the needs of the plants and soil [4]. Surface water pollution is also particularly timber, to the sea ports, all on a scale previously unknown. In their economic calculations the landowners concluded that under the prevailing circumstances the river route was the cheapest conveyance, the more so as there was free servile labour at hand.

The interest in the navigability of Polish rivers by the Seym (parliament) of the Nobility's Republic up to the 17th century resulted in the passing of resolutions ordering regulation of the Vistula and its tributaries. The demand for shipbuilding timber and timber by-products such as wood ash and potash used as dyes for fabrics, was fully met in the 18th century by the great estates of the south-eastern territories of Poland. The goods were first carried by vehicles (horse and cart) as far as the rivers Bug and San. There, at a river port, they were loaded into boats. With the advantage of cheap labour provided by the serfs, the goods arriving at the river ports were organized from quite a remote distance away (over 300 km). The largely feudal organization of export involved a major part of Poland as the market and this required the building of river ports and yards for shipbuilding.

At the same time, there came into being the specialized profession of a raftsman, a servile sailor, and this gave rise to the Vistula environmental songs, legendary tales and raftsmen's ceremonies.

The continental blockade raised during the Napoleonic wars and the general economic crisis after them, as well as the commercial policy of some of the European states in the first half of the 19th century, led to a marked decline in the importance of the Vistula for transporting goods. After Poland's last partition and since the 19th century, the foreign occupiers of Polish territories quite often tendentiously aimed at creating favourable conditions for including the occupied area in their own economic system. The railway had by now become a competitor to the Vistula transport route since it enabled agricultural products to be delivered to the whole of the occupied territory, e.g. Prussia, at greater speed. In the 17th century the port of Gdansk had exported about 80% of Polish corn brought along the river, and this figure still applied until the middle of the 18th century; by 1893 it was barely 37% and in 1912 only 12%.

During the inter-war period (1918–39) and especially in the 1930s, scientists, hydrologists and engineers frequently presented plans for the regulation of the Vistula in order to reactivate its former economic importance. There was also a proposal to build a Silesia–Vistula canal for the export of Polish coal. However, another cheaper scheme was preferred, namely to build a railway trunk line across Bydgoszcz and Kościerzyna to Gdynia. As a result of financial conditions in pre-World War II Poland, it became obvious that after World War I, the Polish Government would have trouble obtaining capital investment for this large undertaking or for further regulation of the river. Foreign countries were also unwilling to invest in its enterprises. However, the establishment of the modern port of Gdynia was a significant achievement and, subsequently, the Polish shipbuilding industry developed greatly.

At present, scientists and engineers in Poland are divided over the restoration (regulation) of the Vistula as a way of transportation. The opponents are afraid of the devastation of the river's natural environment with its unique flora and fauna as well as the deterioration of the self-purification process.

2.3 RIVER POLLUTION FROM 1960 TO 1992

Water quality of the Vistula River has deteriorated during the last 40 years. Much of Poland's heavy industry is located in the drainage area of the upper Vistula, so the river becomes polluted almost from its source.

The concentrations of pollutants are high, even at the mouth of the river, after they have been reduced by a long process of self-purification (Table 2.1). Monitoring of the quality of the Vistula at its mouth has been carried out by the Institute of Meteorology and Water Management, Department of Water Protection in Gdansk since 1972 whilst the data for the 1960s was taken from the archival collection of the Provincial Inspectorate of Environmental Protection in Gdansk [12].

2.3.1 MINERAL SUBSTANCES

The main mineral substances occurring in the Vistula water are chlorides and sulphates (Figures

Table 2.1 Mean annual concentrations and variability coefficients at the mouth of the Vistula in 1992.

No	Parameter	Mean value	Variability coefficient
1	Air temperature (°C)	7.9	98.8
2	Water temperature (°C)	7.6	85.8
3	Colour (mg/l)	19.7	19.1
4	Turbidity (mg/l)	6.8	24.8
5	pH (pH)	7.7	5.9
6	Dissolved oxygen (mg/l)	10.2	18.4
7	BOD_5 (mg/l)	5.4	32.9
8	COD–Mn (mg/l)	7.4	28.9
9	COD–Cr (mg/l)	25.6	36.2
10	TOC (mg/l)	41.9	46.2
11	Ammonium nitrogen (mg/l)	0.62	64.9
12	Nitrite nitrogen (mg/l)	0.019	73.7
13	Nitrate nitrogen (mg/l)	1.91	59.3
14	Kjeldahl nitrogen (mg/l)	1.42	30
15	Total nitrogen (mg/l)	3.35	28
16	Phosphate phosphorus (mg/l)	0.15	46.6
17	Total phosphorus (mg/l)	0.21	43.3
18	Chlorophyll-a (µg/l)	21.7	104.7
19	Dissolved matter (mg/l)	594	21.8
20	Suspended matter (mg/l)	15.7	102.8
21	Chlorides (mg/l)	155.9	38.1
22	Sulphates (mg/l)	74.3	13.6
23	Total hardness (mg/l)	292.8	15
24	Calcium (mg/l)	86.1	13
25	Magnesium (mg/l)	13	13.3
26	Sodium (mg/l)	70.6	40.1
27	Potassium (mg/l)	5.3	21.8
28	Iron (mg/l)	0.35	75.7
29	Manganese (mg/l)	0.11	75.5
30	Zinc (mg/l)	0.031	81.2
31	Cadmium (mg/l)	0.0008	94
32	Copper (mg/l)	0.007	71.2
33	Lead (mg/l)	0.005	66.4
34	Mercury (µg/l)	0.118	143.9
35	Chromium (µg/l)	1.5	62.3
36	Nickel (µg/l)	3.03	52.9
37	Phenols (mg/l)	0.006	156.1
38	Anion detergents (mg/l)	0.082	25.2
39	PCB (ng/l)	7.88	63.1
40	DDT (µg/l)	0.009	101.8
41	DDD (µg/l)	0.003	156
42	DDE (µg/l)	0.005	160
43	Total DDT (µg/l)	0.017	80.8
44	DMDT (µg/l)	0.001	0
45	γ-HCH (µg/l)	0.006	40.1
46	Coli index	0.0188	136.4
47	Waterflow (m³/s)	842.5	38.5

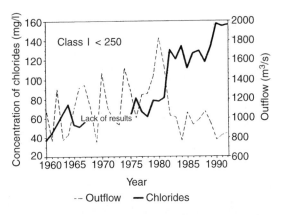

Figure 2.5 Concentration of chlorides v. river outflow 1960–92.

human activities. The concentrations of mineral components have steadily increased over many years [12–14]. However the reported concentrations are not toxic to aquatic organisms and do not affect the self-purification process. The mean concentration of these components is inversely proportional to water flow, so that dilution by precipitation is the main effect on the concentration of mineral substances in the Vistula water.

2.3.2 PH

The average pH values range from 6.9 to 8.5 in the Vistula water and do not exceed the upper limits

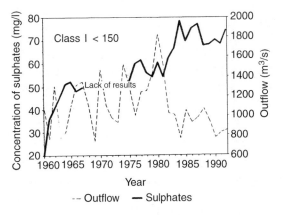

2.5 and 2.6). The sources of these substances are the geological strata in the Vistula basin as well as

Figure 2.6 Concentration of sulphates v. river outflow 1960–92.

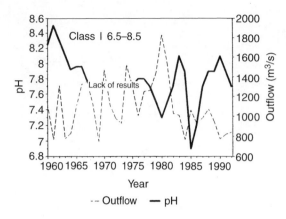

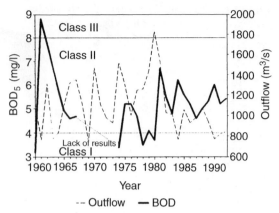

Figure 2.7 River pH v. outflow 1960–92.

Figure 2.8 Biochemical oxygen demand v. river outflow 1960–92.

for class I, according to Poland's water classification scheme [15]. These values are typical for natural surface waters (Figure 2.7). It was found that because of the large buffer capacity of the Vistula waters, the wastewater discharges do not have a significant effect on the pH in the river.

2.3.3 ORGANIC SUBSTANCES

The concentrations of organic compounds, expressed as BOD_5, COD–Cr and COD–Mn, at the mouth of the Vistula River (Figures 2.8–2.10) put the water into the range of class II. The BOD_5 values reported for the 1980s and 1990s are higher than those for the 1970s. However, the concentration of organic matter does not adversely affect the dissolved oxygen levels at the mouth of the river. The dissolved oxygen concentration of over 9 mg/l (Figure 2.11) indicates that the lowest reaches of the Vistula water are well aerated. The results in recent years are lower than those for the mid-1980s, which is causing some concern.

2.3.4 NUTRIENTS

The increased inputs of nitrogen and phosphorus from the anthropogenic sources (agriculture, industry, urbanization) into surface waters accelerate the eutrophication process. As a result the biological equilibrium in the water ecosystem is disturbed. During the period 1960–92 a significant increase in concentration of nitrogen compounds in the Vistula River was observed (Figures 2.12–2.15). In the 1960s the mean concentration of ammonium nitrogen for the Vistula at Gdansk ranged from 0.2 to 0.5 mg/l. This value increased to the level 0.4–0.8 mg/l in the 1980s. Within the same period, the mean concentration of nitrates increased from trace–1.2 mg/l to 0.1–2.5 mg/l. Neither of these forms of nitrogen exceeded the upper limits for class I. Only the concentration values of nitrites were high, within the range of class II. Although the contribution of nitrites to the total nitrogen amount is small, on average below 1%, it cannot be ignored: this form of nitrogen is regarded as harmful to human health.

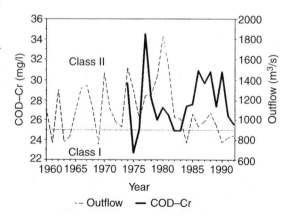

Figure 2.9 Chemical oxygen demand COD–Cr v. river outflow 1960–92.

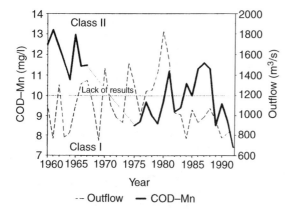

Figure 2.10 Chemical oxygen demand COD–Mn v. river outflow 1960–92.

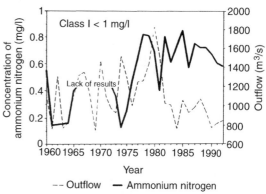

Figure 2.12 Concentration of ammonium nitrogen v. river outflow 1960–92.

Phosphorus in its various forms has a greater impact on pollution of the Vistula River than nitrogen. The concentration of phosphate phosphorus has increased significantly from the values of the early 1980s (Figures 2.16 and 2.17). The mean concentrations of total phosphorus are in the range of class II and class III. The hydrological conditions have a strong effect on the nutrient concentrations in the Vistula water as shown by a decrease in nutrient concentration during the summer with low precipitation and rapid plant growth. Despite this, the correlation between nitrogen concentration and river flow is difficult to express mathematically.

2.3.5 METALS

The concentrations of metals such as iron, manganese, lead, zinc, copper and cadmium at the Vistula mouth do not in general exceed the upper limits for class II (Figures 2.18 and 2.19).

2.4 CONCLUSIONS

1. The Vistula River, one of the biggest European rivers, is the source of water for 22 million Poles, who live in its catchment area. In the past it has had a significant influence on Poland's transport and economy and, even today, still has a strong influence.

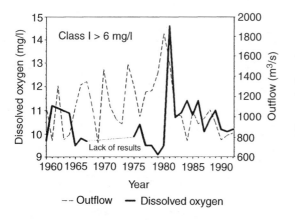

Figure 2.11 Concentration of dissolved oxygen v. river outflow 1960–92.

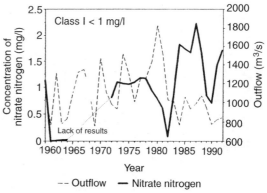

Figure 2.13 Concentration of nitrate nitrogen v. river outflow 1960–92.

18 *The Vistula River in Poland*

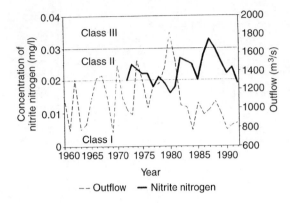

Figure 2.14 Concentration of nitrite nitrogen v. river outflow 1960–92.

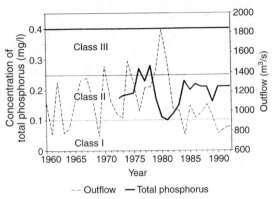

Figure 2.17 Concentration of total phosphorus v. river outflow 1960–92.

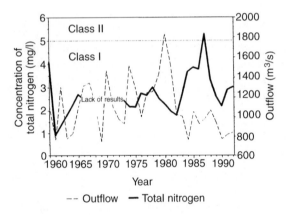

Figure 2.15 Concentration of total nitrogen v. river outflow 1960–92.

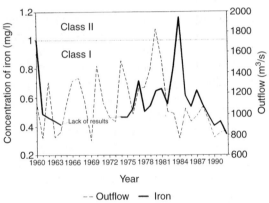

Figure 2.18 Concentration of iron v. river outflow 1960–92.

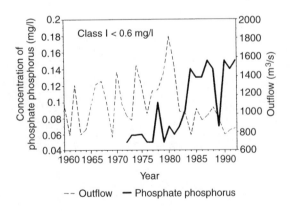

Figure 2.16 Concentration of phosphate phosphorus v. river outflow 1960–92.

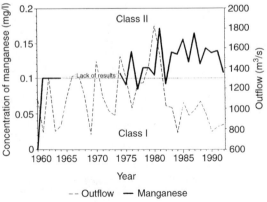

Figure 2.19 Concentration of manganese v. river outflow 1960–92.

2. The water quality of the Vistula River has deteriorated during the last 40 years. Much of the heavy industry is located in the drainage area of the upper Vistula, but concentrations of pollutants are also high at the mouth of the river.
3. The high density of industry in the river basin and insufficient number of treatment plants make the Vistula River one of the most significant sources of pollutants transported into the Baltic Sea.

ACKNOWLEDGEMENTS

I am obliged to Dr Zbigniew Makowski for his help in computer data preparing.

REFERENCES

1. Polish Central Statistical Office (GUS) (1992, 1993) Statistical yearbooks, Warsaw.
2. Niemirycz, E. and Taylor, R. (1992) Land-use and nutrients to surface waters. Institute of Meteorology and Water Management, Maritime Branch, Gdynia.
3. Environmental protection (1993) Polish Central Statistical Office (GUS).
4. Kajak, Z. (1992) The River Vistula and its Floodplain Valley (Poland): Its Ecology and Importance for Conservation in *River Conservation and Management*, (eds P.J. Boon, P. Calow and G.E. Petts), John Wiley & Sons Ltd, Chichester, pp. 35–49.
5. Taylor, R., Bogacka, T., Rybiński, J. *et al.* (1992) Influence of diffusion agricultural pollution on surface water quality. Ministry of Environmental Protection, Natural Resources and Forestry, Warsaw.
6. Niemirycz, E. and Taylor, R. (1993) Source division of the load of eutrophying substances. IAWQ, Kraków.
7. Niemirycz, E., Taylor, R. and Makowski, Z. (1993) Threat of nutrients to surface waters. National Inspectorate of Environmental Protection (PIOS).
8. Roman, M. (1993) Reduction of phosphorus load in surface waters, Mat. Symp. Intern. Assoc. on Water Quality, Kraków.
9. Obarska-Pempkowiak, H. (1994) State of environmental protection in Poland. Sem. Sensors in Environmental Protection 94–114, Technical University of Gdansk.
10. Korol, R., Niemirycz, E. and Szczepański, W. (1992) Water Quality of main Polish rivers, Vistula and Oder [Odra] in 1990, ICWS report the Netherlands.
11. Gierszewski, S. (1982) *The Vistula River in the History of Poland*, Maritime Publishers, Gdansk.
12. Voivodeship Department of Environmental Protection in Gdansk (from 1960) Archival data.
13. Niemirycz, E. and Borkowski, T. (1993) Riverine Input of Pollutants in 1992, Environmental conditions in the Polish zone of the Southern Baltic Sea during 1992, Maritime Branch Materials, Institute of Meteorology and Water Management, 205–207.
14. Januszkiewicz, T., Kowalewska, K., Szarejko, N. and Żygowski, B. (1974) Effect of fertilization on the Vistula water quality, IMGW.
15. Ministry of Environmental Protection, Natural Resources and Forestry (1991) Legislation Gazette No. 116. Regulation on water classification, 5 Nov.

FURTHER READING

1. Rybiński, J., Niemeirycz, E. and Makowski, Z. (1992) Pollution load, marine pollution (2). National Scientific Committee on Oceanic Research, Polish Academy of Sciences, No. 61, 25–52, Gdansk.

WATER QUALITY IN THE VISTULA BASIN

by J. Dojlido

3.1 INTRODUCTION

The Vistula River is located in Poland and neighbouring countries; its source is in the Czech Republic (Table 3.1). By definition, however, the main Vistula starts from the outflow of the Przemsza river and all downstream sampling sites are described in terms of the distances from this point. Upstream of the Przemsza, the river is known as the Small Vistula and the sampling sites are defined in terms of the distance upstream from the start of the main river.

The Vistula is an unregulated river for the major part of its length of 1047 km, with many meanders and islands. As a result of the political situation in Poland in the 19th century, at a time when most of the rivers in Europe were regulated, the Vistula was partly regulated but only in its upper reaches. There are two impoundments on the Small Vistula: Wisła Czarne (at 96.8 km) and Goczałkowice (at 42.8 km). Three major weirs exist on the upper Vistula: Laczany (at 38.5 km), Dabe (at 81 km) and Przewoz (at 92.2 km). The middle part of the Vistula is in a natural condition and there is very little regulation in the lower part. Włocławek Reservoir is situated 674 km from the start of the main river.

3.2 RIVER WATER ABSTRACTION

Poland is now using a large amount of water for various purposes. The changes in water abstraction for the period 1950–91 are shown in Figure 3.1. These abstractions are mostly surface water intakes. The Vistula River water, in spite of its polluted state, is used for many purposes.

Table 3.1 Vistula River catchment data

Extent of Vistula catchment area	Area (km^2)	Population (million)
Poland	169 100	20.8
Belarus	11 100	1.1
Ukraine	12 600	1.8
Czech Republic and Slovakia	1 900	0.142
Total	194 700	23.8

Goczałkowice Reservoir in the upper reaches is used for potable supply. The next large water intake (about 500 000 m^3/day) is located in Warsaw and the one after that in Płock. There are many other water supply intakes situated on tributaries of the Vistula. The Vistula is used for supplying cooling water to large power plants at Polaniec (1600 MW), Kozienice (2600 MW), Siekierki (2300 MW) and Zeran (1400 MW). Each has a 'flow-through' system of cooling and returns heated water to the river. The use of Vistula water for agricultural irrigation is relatively small scale and navigational uses are minimal, usually for the local transport of goods and passengers. (In medieval times, the river was a heavily navigated river from Kraków to the Baltic Sea.) The Vistula also receives large volumes of wastewater.

3.3 HISTORICAL MEASUREMENTS OF RIVER WATER QUALITY

The first information on the measurements of the quality of the Vistula water are from the 19th century when the famous scientist, Mendeleyew,

---

International River Water Quality. Edited by Gerry Best, Teresa Bogacka and Elżbieta Niemirycz. Published in 1997 by E & FN Spon, London, ISBN 0419215409

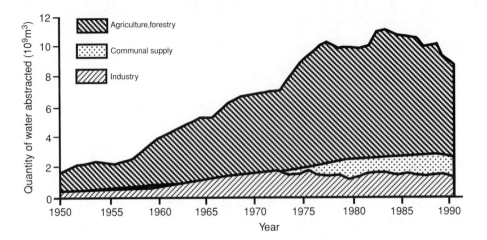

Figure 3.1 Water intake in Poland 1950–91.

analysed the water in 1876 in connection with the treatment of the water for municipal supply. He determined principally the inorganic components in the water such as the cations responsible for the water hardness. More complex monitoring was carried out in the period 1923–4 by Kirkor who measured 14 components every two weeks in samples taken from the river at Warsaw. He found BOD_5 values of 2.2 mg/l, iron 0.2 mg/l and ammonia 0.1 mg/l. His conclusions were that the water was fairly clean but turbid and yellowish in colour. Regular monitoring of the Vistula at Warsaw began in 1926 by the laboratory of Warsaw Water Works. In the period up to 1939, the water was clean but there was evidence that the concentration of iron was increasing. The monitoring carried out in 1945–8 showed that the Vistula was still clean but was becoming highly mineralized. At Warsaw it was good enough for recreation and many beaches were being used for swimming. It was also used for fishing and water sports by the city's inhabitants. Subsequently, the process of Polish industrialization began. Many heavy industries were established, mostly in the upper parts of the river but also along most of its length. The coal mines were expanded and other industries were set up such as petrochemicals, artificial fertilizers, chemicals and power plants with flow-through cooling systems. The construction of treatment plants for wastewater continually lagged behind the construction of industrial premises and new towns, with the result that the pollution of the Vistula increased rapidly.

3.4 THE QUALITY OF THE RIVER WATER FROM 1945 TO 1970

Taking 1945 as the base year when the Vistula water in Warsaw was relatively clean (Table 3.2), it was found that by 1970 nearly every quality parameter had been exceeded (Table 3.3).

The dissolved oxygen (DO) levels were satisfactory despite the increasing organic pollution. This is explained by the natural characteristics of the river with its many bends, falls, beaches and islands. The lowest DO concentration was found in winter when, under the ice cover, it was 2.4 mg/l (Figure 3.2). The ammonia concentration increased between 1945 and 1970 to a maximum value of 4.4 mg/l (Figure 3.3). A seasonal variation was observed with the highest values being measured in December and January when the water was covered by ice. A spectacular increase was noted in the chloride levels (Figure 3.4). The mean concentration increased from 20 mg/l in 1945 to 74 mg/l in

Table 3.2 Water quality of the Vistula River, Warsaw, in 1945

Odour	Colour	COD–Mn (mg/l)	Ammonium (mg/l)	Chloride (mg/l)	Total residue (mg/l)	BOD$_5$ (mg/l)
natural	natural	7	0.01	15	250	2.2

Table 3.3 Changes in water quality of the Vistula River, Warsaw, from 1945 to 1970

| Year | Proportion of results exceeding 1945 levels (%) | | | | | | |
	Odour	Colour	COD–Mn	Ammonium	Chloride	Total residue	BOD$_5$
1950	0	70	2	65	80	80	–
1960	15	90	20	85	98	100	50
1970	100	95	40	100	100	100	95

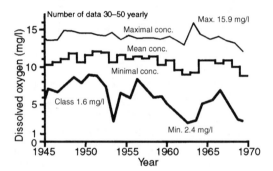

Figure 3.2 Dissolved oxygen concentration in Vistula water in Warsaw 1945–70.

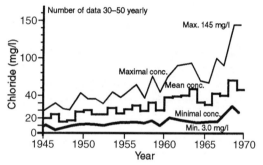

Figure 3.4 Chloride concentration in Vistula water in Warsaw 1945–70.

1969. The reason for this was the growth in the coal-mining industry in the upper part of the Vistula in the Silesia region. It was observed that there was an inverse relationship between the concentration of chloride and river flow.

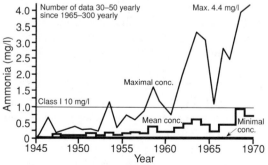

Figure 3.3 Ammonia concentration in Vistula water in Warsaw 1945–70.

3.5 CHANGES IN RIVER WATER QUALITY FROM 1968 TO 1978

The water quality in 1968 and 1978 was compared along the length of the river using the guarantee values (90%-ile), which were calculated from the data collected for each of these two years. The changes in BOD$_5$ were irregular (Figure 3.5). In the middle part of the Vistula, the BOD$_5$ increased but it decreased in the lower reaches. The guarantee BOD$_5$ data show that highest values occurred in the upper reaches of the river with increases in organic pollution downstream of each large town or industrial plant. A large increase in chloride took place in the decade (Figure 3.6), several fold in the upper part, doubling in the middle section and one and a half times in the lower reaches. An increase was also noted for the total dissolved solids. The

24 Water quality in the Vistula basin

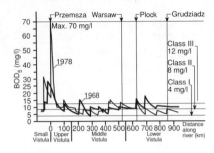

Figure 3.5 Guarantee BOD$_5$ in 1968 and 1978.

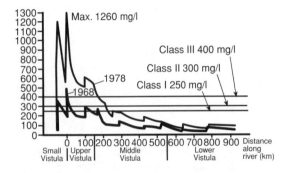

Figure 3.6 Guarantee concentration of chloride in 1968 and 1978.

overall character of the pollution of the Vistula for the years 1968 to 1978 was similar. On this basis, the river can be divided into four sections:

1. the Small Vistula from its source to 40 km – clean;
2. the Small Vistula from 40 km to 0 km and the main river from zero to 20 km – very polluted;
3. the main river from 20 km to about 200 km – decreasing pollution;
4. the Vistula from 200 km to the Baltic Sea – similar levels of pollution with deterioration then recovery below the larger sources of pollution.

3.6 THE QUALITY OF THE RIVER WATER FROM 1989 TO 1992

This evaluation is based on intensive water quality monitoring performed at three sampling stations on the Vistula River and one each on the Bug and Narew rivers.

1. Tyniec, at 63.7 km on the upper part of the Vistula, upstream of Kraków.
2. Warsaw, at 510 km on the middle part of the Vistula, upstream of the discharge of sewage from Warsaw.
3. Wyszków, at 33 km on the Bug River, upstream of Zegrzynskie Lake.
4. Pultusk, at 63 km on the Narew River, upstream of Zegrzynskie Lake.
5. Kiezmark, at 926 km on the Vistula on the lower Vistula, close to the mouth of the river at the Baltic Sea.

The water quality was measured twice weekly in 1989, 1990 and 1991, and once a week in 1992. Approximately 350 data points were collected during the four years of measurements and over 40 parameters determined. The results can be summarized as follows.

- The pH varied from 6.6 to 9.7.
- The DO was lowest in the Vistula at Tyniec where the minimum was zero. The DO of the Vistula at Warsaw varied from 5.4 to 17.4 mg/l (supersaturation). The highest value was recorded for the Bug at Wyszków at 20.3 mg/l. In general the Vistula was well oxygenated but some problems occurred in the upper reaches in winter.
- The highest BOD$_5$ value of 31.8 mg/l was measured in the Vistula in the upper reaches at Tyniec. The BOD decreased along the length of the Vistula. The Bug River contributed a large load of BOD into the system.
- The chloride concentration reached the very high value of 2140 mg/l at the Tyniec site but decreased to 46 mg/l at Warsaw.
- The concentration of dissolved salts was very high in the upper reaches of the Vistula with a maximum value of 4160 mg/l, but lower downstream and in the Bug and Narew rivers.
- The sulphate was also at its maximum level at Tyniec (314 mg/l) and decreased downstream. Similar trends were found for sodium (Figure 3.7). The CaCO$_3$ hardness of the Vistula ranged from 150 to 1000 mg with the highest values in the upper reaches. Similar trends were found for calcium and magnesium (calcium: 33 to 260 mg/l, magnesium: 6.8 to 116 mg/l).

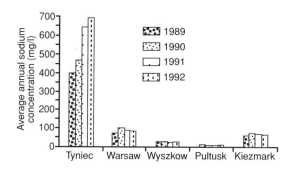

Figure 3.7 Concentration of sodium 1989–92.

- The concentrations of iron and manganese ranged from undetectable to 8 mg/l and 2.4 mg/l respectively. Ammonia levels were very high in the upper reaches of the Vistula, up to 8.3 mg/l but were much lower in other parts of the river, usually below 0.5 mg/l.
- The concentration of nitrate was highest in the middle part of the Vistula (14 mg/l) but the average was between 1.0 and 1.5 mg/l.
- The phosphate concentration was very high at Tyniec with a maximum value of 1.3 mg/l, and a mean value of 0.6 mg/l (these figures are equivalent to 0.42 and 0.2 mg/l phosphate phosphorus respectively). In the middle part of the river, the mean concentration was 0.2 mg/l whilst in the lowest reaches it was less than 0.2 mg/l.
- The total phosphorus concentration ranged from 0.1 mg/l to 16.4 mg/l with the highest value observed for Tyniec.
- The concentration of heavy metals were highest at Tyniec where the maximum concentrations were: chromium: 0.16 mg/l; cadmium: 0.22 mg/l; copper: 0.016 mg/l; and lead: 0.09 mg/l. In other parts of the river the concentrations were much lower, sometimes undetectable. The mean concentrations at Warsaw were: chromium: 0.009 mg/l; cadmium: 0.003 mg/l; copper: 0.01 mg/l; and lead: 0.01 mg/l.
- The concentration of anionic detergents was high at Tyniec (maximum 0.5 mg/l, mean 0.2 mg/l). In other parts, the mean concentration was about 0.05 mg/l.
- The chlorophyll-a concentration was low in the upper reaches of the Vistula, highest at Warsaw (maximum 523 µg/l, mean 98 µg/l) and low at Kiezmark.
- The concentration of γ-HCH was high at Tyniec (max. 0.8 µg/l, mean 0.09 µg/l), and decreased along the length of the Vistula to a mean value at the lowest reaches of 0.006 µg/l.
- Other organochlorine pesticides and PCBs were distributed irregularly but usually the highest concentration was at Tyniec. The maximum values measured were: DDT 0.8 µg/l, DMDT 2.8 µg/l and PCBs 1.1 µg/l. The range of concentrations for pollutants found in the Vistula at the sample points listed above, for the years 1989–92, are shown in Table 3.4. The characteristic values (mean, maxima and minima) for the river at Warsaw, are presented in Table 3.5.

3.7 WATER QUALITY CLASSIFICATION

The quality of water in Polish rivers is differentiated into four classes depending on the water use:

1. class I for drinking water supply;
2. class II for recreation and fish breeding
3. class III for industrial use and for irrigation;
4. class IV no assigned use or unclassified (very polluted).

The classes of water quality of river in the Vistula catchment for 1992 are shown in Figures 3.8 and 3.9. The general classification based on the levels of organic matter and dissolved salts, is shown in Figure 3.8 and this highlights where the most polluted parts are: the Przemsza river, the upper reaches of the Vistula, the Vistula downstream of Bzura, and the San River and the upper part of Bug River. Very saline water is found in the upper and middle reaches of the Vistula. Figure 3.9 presents the classes of the river based on the concentration of nutrients. Particularly polluted parts are: the Przemsza, the upper reaches of the Vistula, and the Wieprz, Narew and Drweca rivers. The classification based on numbers of bacteria (Coli index) show that most rivers are in class IV.

Table 3.6 summarizes the classes of the Vistula catchment based on the general parameters. Of the

Table 3.4 Vistula water quality parameters 1989–92 in Tyniec (T), Warsaw (W) and Kiezmark (K). Results of ca. 350 measurements

Parameter	Min.[a]	Location	Max.	Location
Flow (m^3/s)	31	T	2780	K
Water temperature (°C)	0	T, W, K	28	T
Colour (mg/l Pt)	10	K	150	T
Turbidity (mg/l SiO$_2$)	1	T	220	T
pH	6.3	K	9.7	W
Dissolved oxygen (mg/l)	nd	T	17.4	W
BOD$_5$ (mg/l)	0.5	K	31.8	T
COD–Mn (mg/l)	2.5	W	64.6	T
COD–Cr (mg/l)	7.7	K	433	T
Chloride (mg/l)	46	W	2140	T
Sulphate (mg/l)	29	K	314	T
Dissolved solids (mg/l)	225	K	4160	T
Suspended solids (mg/l)	nd	K	650	T
Hardness (mg/l CaCO$_3$)	165	W	990	T
Calcium (mg/l)	33	W	257	T
Magnesium (mg/l)	6.8	T	116	T
Iron (mg/l)	0.2	T	8.2	T
Manganese (mg/l)	nd	T	2.4	T
Sodium (mg/l)	19	K	1300	T
Potassium (mg/l)	1	T	27.5	T
Ammonia (mg/l N)	nd	W	8.3	T
Nitrate (mg/l N)	nd	W	14	W
Kjeldahl nitrogen (mg/l)	0.1	W	12.8	T
Phosphate (mg/l)	nd	W	1.3	T
Total phosphorus (mg/l)	0.1	W, K	16.4	T
Chromium (mg/l)	nd	T	0.16	T
Cadmium (mg/l)	nd	W, K	0.225	T
Copper (mg/l)	nd	K	0.016	T
Lead (mg/l)	nd	T, K	0.09	T
Phenols (mg/l)	nd	T, W, K	0.07	T
Anionic detergents (mg/l)	nd	T, W, K	0.52	T
Coli (MPN/100 ml)	30	W	25×10^6	T
Chlorophyll-a (µg/l)	nd	T	523	W
γ-HCH (µg/l)	nd	T	0.83	T
DDE (µg/l)	nd	T, W	0.47	W
DDD (µg/l)	nd	T, W, K	5.2	T
DDT (µg/l)	nd	T, W, K	0.79	W
DMDT (µg/l)	nd	T, W	2.8	T
PCB (µg/l)	nd	W, K	1.1	T

[a] nd – not detected

main course of the Vistula 75% is in class IV and 52% of its tributaries are also very polluted. This assessment covers only those rivers monitored in 1992. Some small, clean tributaries and the clean upper reaches of rivers were not monitored or classified. The proportions of the total length of rivers based on different sets of parameters, are shown in Table 3.7.

If the data are compared with those for 1991 some small improvements can be observed. When looking at the situation in 1987, however, there has been a marked improvement in the water quality if

Table 3.5 Characteristic values of water quality parameters for the Vistula River in Warsaw 1989–92

Parameter	Min.[a]	Max.	Mean
Flow (m^3/s)	200	1660	476
Water temperature (°C)	0	25	10.7
Colour (mg/l Pt)	15	60	31
Turbidity (mg/l SiO$_2$)	2	100	25
pH	6.9	9.7	8.4
Dissolved oxygen (mg/l)	5.4	17.4	11.2
BOD$_5$ (mg/l)	0.6	19.2	7.6
COD–Mn (mg/l)	2.5	18.5	9.0
COD–Cr (mg/l)	10	112	45
Chloride (mg/l)	46	460	170
Sulphate (mg/l)	33	110	72
Dissolved solids (mg/l)	230	1160	620
Suspended solids (mg/l)	1	230	32
Hardness (mg/l CaCO$_3$)	165	410	250
Calcium (mg/l)	33	131	70
Magnesium (mg/l)	10	28	19
Sodium (mg/l)	26	160	87
Potassium (mg/l)	3.9	28	7.3
Ammonia (mg/l N)	nd	2.7	0.4
Nitrate (mg/l N)	nd	14	2.6
Kjeldahl nitrogen (mg/l)	0.1	7	1.5
Phosphate (mg/l)	nd	1	0.2
Total phosphorus (mg/l)	0.1	1.9	0.6
Chromium (mg/l)	0.001	0.050	0.009
Cadmium (mg/l)	nd	0.019	0.003
Copper (mg/l)	0.001	0.05	0.01
Lead (mg/l)	0.001	0.063	0.01
Phenols (mg/l)	nd	0.012	0.0017
Anionic detergents (mg/l)	nd	0.29	0.065
Coli (MPN/100 ml)	30	1×10^6	1200
Chlorophyll-a (µg/l)	1.3	523	98
γ-HCH (µg/l)	0.0012	0.56	0.027
DDE (µg/l)	nd	0.47	0.028
DDD (µg/l)	nd	0.15	0.010
DDT (µg/l)	nd	0.79	0.044
DMDT (µg/l)	nd	0.045	0.002
PCB (µg/l)	nd	0.17	0.01

[a] nd – not detected

based on the organic components (BOD and COD) but a deterioration in the concentrations of salinity and nutrient levels.

The changes in water quality classification of all the rivers monitored in Poland in the years 1964–91 are shown in Figure 3.10. This shows a deterioration in water quality since 1964, then a period of stabilization and, recently, a small improvement.

3.8 WATER POLLUTION CONTROL

The main sources of pollution are located in the upper reaches of the Vistula where the river flow is relatively low. The wastewaters from the Silesian industrial region are discharged to the Vistula largely via the Przemsza River which, at its confluence with the Vistula, looks like untreated sewage.

28 *Water quality in the Vistula basin*

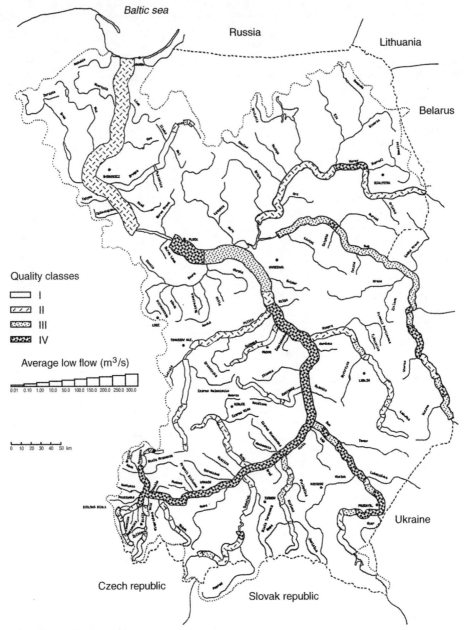

Figure 3.8 Water quality map in 1992 according to parameters: DO, BOD_5, COD–Mn, phenols, chloride, sulphate, dissolved solids and suspended solids.

In Silesia there are many chemical plants, foundries and coal mines.

The other main sources of pollution in the Vistula drainage area are the chemical plants at Oswiecim, Pionki, Bydgoszcz, Sarzyna, Tomaszow and Mazowiecki; the nitrogen plants at Tarnow, Pulawy and Włocławek; pulp and paper mills at Niedomice, Jeziorna, Ostroleka, Kwidzyn, Włocławek and Swiecie, and the petrochemical plant at Płock. The main Vistula receives sewage

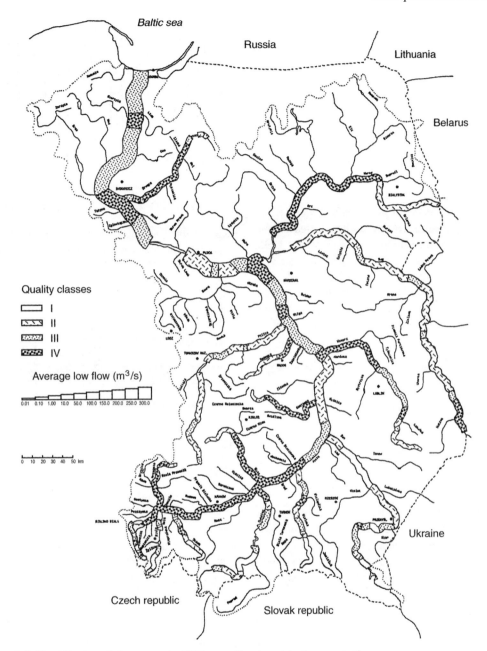

Figure 3.9 Classification of river water in 1992 according to nutrient concentrations.

from the large cities of Kraków, Warsaw, Płock, Włocławek, Toruń and Bydgoszcz, most of which is untreated. The changes in the treatment of wastewater throughout Poland in recent years are summarized in Figure 3.11. This shows that there has been a marked decrease in the volume of wastewater and some improvement in its treatment. The wastewater volumes and the percentage untreated are shown in Table 3.8 and this emphasizes the high percentage of untreated wastewater in the middle

30 Water quality in the Vistula basin

Table 3.6 Classification of river water quality in Vistula drainage area in 1992 (according to all measured chemical parameters)

Class	Length of river (%)	
	Vistula basin	Main course of Vistula
I	3.7	0
II	17.3	0
III	27.3	24.5
IV	51.7	75.5

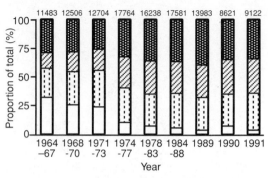

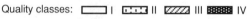

Figure 3.10 Water quality of Polish rivers 1964–91 according to physico-chemical parameters (also shows total lengths of rivers that were monitored).

reaches of the Vistula, from Pilica to Narew. Since 1987, the volume of wastewater has decreased and less of it is discharged untreated. However, in 1992, only 67% of all wastewater in the Vistula basin was treated and, of that, only 33% received biological treatment. The following data show the ratio of the volume of wastewater produced in the Vistula catchment in 1991 compared to the water resources available.

- Outflow of rivers – $25\,708 \times 10^6$ m^3/year
- Water abstracted – 4676×10^6 m^3 year
- Wastewater requiring treatment – 2145×10^6 m^3/year.

The average dilution factor of all wastewater discharged to rivers in the catchment is about 12.

3.9 CONCLUSIONS

The waters in the Vistula basin were very clean in the period 1945–8. Subsequently, pollution started to increase rapidly because of the rapid industrialization in Poland and the lack of treatment facilities. Many industrial plants were set up without any treatment or with grossly inadequate wastewater treatment processes. Cities grew but sewage treatment works were not constructed.

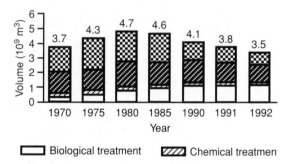

Figure 3.11 Wastewaters discharged to surface waters 1970–92.

Table 3.7 Classification of water quality in the Vistula basin in 1992

Parameter	Length of river in each class (%)			
	Class I	Class II	Class III	Class IV
DO, BOD$_5$, COD–Mn, phenols, Cl, SO$_4$, dissolved solids, suspended solids	19.4	39.1	18.7	22.8
BOD$_5$, COD–Mn, DO	22.5	52.5	14.9	10.0
Cl, SO$_4$, dissolved solids	71.4	14.4	2.9	11.3
PO$_4$, P$_{tot.}$, N–NH$_4$, N–NO$_3$, N–NO$_2$, N$_{tot.}$	6.9	32.2	21.9	39.0
Zn, Cu, Pb, Cd, Hg	74.3	16.1	0	9.6

Table 3.8 Wastewater treatment in the Vistula basin 1987–92

Region	Total volume (10^6 m³/year)			Untreated (%)		
	1987	1991	1992	1987	1991	1992
All rivers in Vistula basin	2533	2145	1927	46	36	33
From source to Dunajec	947	742	663	43	30	27
From Dunajec to Pilica	630	527	420	30	14	18
From Pilica to Narew	381	440	427	92	71	59
From Narew to Baltic Sea	575	436	417	38	34	34

The pollution by organic substances and dissolved salts increased steadily. In the 1970s the problem of excess nutrients and the eutrophication of the waters was noted. However, by the mid-1980s further deterioration of water quality slowed down and then halted.

In the 1990s there is some evidence that some improvement of water quality may be taking place. Even so, in 1992 about 33% of all wastewater in the catchment area of the Vistula River is still not being treated to an adequate degree and the receiving waters are adversely affected.

THE TUALATIN RIVER: A WATER QUALITY CHALLENGE   4

by E. Jackson

4.1 INTRODUCTION

There are many rivers throughout the world that have water quality problems. In the USA one river stands out in the view of water quality managers: the Tualatin River in the north-west of the State of Oregon, USA. It is a river rich with water-quality management issues and where one can ask a number of questions:

- how does one deal with stringent water quality requirements?
- how does one deal with tough new water quality standards?
- how does one deal with a complex compliance schedule that demands precise knowledge, decisions and commitments to accomplish goals?
- and finally how does one do all this in a watershed that is flat, in a river that has low flows and is slow moving, and where 350 000 people want to live?

To answer these questions, the people's commitment to improving water quality and their commitment to maintaining a way of life that includes a clean environment and a robust economy to help pay for the water quality improvements must be considered.

To meet this challenge the Unified Sewerage Agency has devised an innovative and comprehensive approach to water quality management. The accomplishments thus far are not single-purpose solutions nor can they be handled issue-by-issue or by the usual discharge controls. Attention has been focused on the river and its catchment. A long-term strategy has been constructed that is flexible and uses a systems approach. It is an approach that uses a wide array of technical tools and one that is clearly based on the public values for the catchment, the water resource and environmental health.

4.2 THE OUTLINE OF THE PROBLEM

The Tualatin River basin is approximately 1850 km^2 in size (Figure 4.1) The headwaters are on the east slope of Coast Range mountains.

The problem facing the people living in the Tualatin watershed arises from both natural and human causes. The river is slow moving and meanders for approximately 140 river kilometres eastward to its confluence with the Willamette River near Portland. In the lower 40 km, the stream drops approximately 0.3 m. The river flow is provided by surface water run-off during precipitation events and from groundwater discharges. There is no melting snow pack to supplement summer flows. The natural flows during the winter reach peaks of 55–90 m^3/s. The summer natural flows decline to almost nothing. The river's summer flows are now augmented by reservoir releases to maintain an approximate 4 m^3/s flow for pollution abatement.

Added to the problem are 350 000 people distributed on land that is used right up to the edge of the river which prevents natural water quality treatment processes occurring. Forestry is located in the

---

International River Water Quality. Edited by Gerry Best, Teresa Bogacka and Elżbieta Niemirycz. Published in 1997 by E & FN Spon, London, ISBN 0419215409

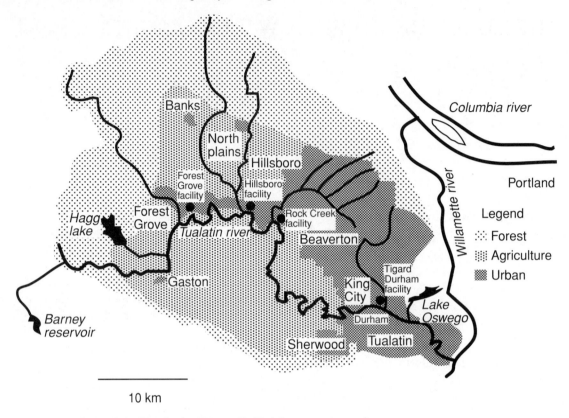

Figure 4.1 The Tualatin River basin. (Source: Unified Sewerage Agency.)

upper reaches of the catchment and also around the agricultural areas. Agriculture of many varieties is located on the flat valley floor. The urban areas are also located on the valley floor, near the lower reaches of the river. The urban influences on the river occur mostly through tributary flow into the main river and permitted waste discharge into the river, including industrial cooling water and treated municipal wastes.

The combination of naturally occurring and imposed conditions produces serious water quality problems. The land uses produce nutrient loads, specifically ammonia–nitrogen and total phosphorus loadings, that are in excess of the assimilative capacity of the river. These cause algal blooms which, in turn, create large diurnal and seasonal dissolved oxygen fluctuations. The river is also subjected to elevated water temperatures, elevated bacteriological contamination from a variety of land uses and turbidity from farming, forestry, and construction activities in the urban areas.

4.3 THE TUALATIN WATER QUALITY STRATEGY

A number of responses have occurred to this water quality challenge. The State of Oregon, under the US Clean Water Act, set total maximum daily load (TMDL) limits for ammonia-nitrogen and total phosphorus. In conjunction with this action, the state established waste load allocations and load allocations for the wastewater treatment plants and land uses respectively.

Many people responsible for water quality management in the catchment, including the United Sewerage Agency for the urban areas, responded to these regulatory requirements. The agency set about to understand the river and the water quality processes operating in it better, modify the treatment processes at the treatment plants, obtain authority to manage stormwater run-off in the urban areas, and use every opportunity to inform the public about its role in the clean-up of the Tualatin River. These efforts form the water quality strategy for the Tualatin.

The Tualatin water quality strategy is based on a holistic catchment approach and not an issue-by-issue approach. Water management decisions are viewed as decisions for all users of water, not just as a matter of waste assimilation. The strategy is also based on water quality management techniques using technical innovation. And finally, it is based on two very important elements: the financial strategy and the public involvement strategy.

4.3.1 PUBLIC INVOLVEMENT STRATEGY

The water quality strategy is dependent upon the public's perception of the value of water. After all, the public is the benefactor of clean water. The value is not in terms of money, but in terms of its necessity to human life. To determine these values, the Unified Sewerage Agency went to the public and asked a series of questions. The results of this effort formed the goals and objectives of the water quality strategy. The public were asked:

- how should the Tualatin River be used?
- should it be ignored?
- should wastes be discharged into it?
- should it become a recreational resource, a wildlife refuge or a channel for navigation?
- what benefits of clean water are important and who benefits?
- should growth be accommodated or curtailed in this basin, since growth is viewed as one of the causes of the problem?
- how much is the public willing to pay for the clean-up and who should [and should not] pay? How much is too much for the average citizen to pay?
- how should the clean-up be financed?
- should high technology be used? Or should natural processes be employed?
- should more source control be used with less pollutant capture and treatment?

To communicate the results of this effort to the public, the Unified Sewerage Agency devised a simple, straightforward description of the goals and objectives. The strategy is discussed in common terms as 'lines of defence', as below.
1. Keep pollutants out of the watershed
2. Keep pollutants out of the wastestream
3. Treat wastes and reuse products
4. Manage river flow.

The idea is to achieve success with the first line. If that is not totally successful, then 'retreat' to the second line and so on. There are many techniques to use at each line of defence. The particular water quality situation will dictate the techniques to be used.

4.3.2 FINANCIAL STRATEGY

The other very important element of the water quality strategy is the financial strategy. Water quality strategies are designed to be implemented. They cannot be implemented without money to do so. Scientists and technicians derive very good strategies, but often tend to forget that it takes money to implement them, thus the need for a financial strategy. The critical elements of the financial strategy are:

- money lender
- convinced public
- political support
- industry support
- growing economy
- large, diverse group of ratepayers.

The most important element of the financial strategy is to convince citizens that **they**, not wastewater treatment plant and industry, are the cause of pollution. The wastes of civilization get into the river and cause the water quality problems (Figure 4.2). Once the public understands this concept, the discussion then shifts to how the public 'invests' in clean water. They invest by funding new treatments plants, new programmes to identify and control wastes, and their own private efforts to control polluted run-off from their homes, businesses, and industries.

This investment then returns to the community via money invested in the community by outsiders

36 The Tualatin River: a water quality challenge

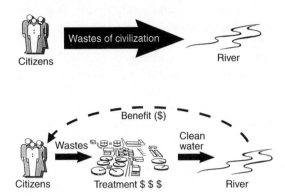

Figure 4.2 Financial strategy. (Source: Unified Sewerage Agency.)

to the community. These outsiders come to the river to take part in recreational activities, bringing their money for food and lodging. New people also move to the community bringing their money to buy homes, to start businesses or build industries (Figure 4.2).

4.4 WATER QUALITY IMPROVEMENTS

The question is asked: What has been accomplished for water quality since this strategy was initiated in 1990?

Advanced wastewater treatment plants have been modified to reduce ammonia discharges to 0.1 mg/l and total phosphorus to 0.07 mg/l (and at times as low as 0.02 mg/l). This has cost approximately US$150 million in improvements; all paid for by the public through fees and service charges, not the government.

A new surface water management programme has been implemented that treats water at the source on industrial and development sites. A phosphorus detergent ban has been instituted to deal with phosphorus coming into the basin. Included in the surface water management programme is a large public education programme. Presentations are made to service groups, to school classes (river rangers programme; approximately 5000 students per year), and community organizations.

A passive education programme is also operating. Examples of this effort include display signs at water quality facilities, information leaflet inserts into newspapers, brochures describing the river recreation opportunities and programmes.

Environmental activism is also encouraged. Activities get neighbours together on projects involving streams and encourages them to become interested in their neighbourhood stream. These include such activities as stencilling on storm drain catchment basins to remind those dumping material down the drains that these basins are connected to the river and not a treatment plant. A number of public meetings are held during the year to encourage environmental activism in the form of information exchange, conducting opinion surveys and setting up various displays at shopping centres and community fairs.

All of these activities are having an effect on the river's ammonia, total phosphorus, and dissolved oxygen levels. There has been a dramatic reduction in ammonia. As one can see in Figure 4.3, the ammonia reduction strategy at the two major treatment plants at Rock Creek and Durham has worked very well. The river's ammonia levels are now well below the total maximum daily load (TMDL) set by the state.

Likewise, a similar reduction in phosphorus has occurred (Figure 4.4). However, it is worth noting that for phosphorus the lower river is still not in compliance with the TMDL. This is associated with non-point source run-off in the urban and agricultural areas. Much work remains to be done in these land use areas.

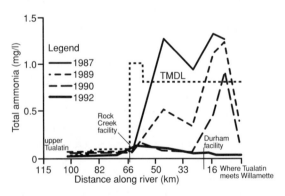

Figure 4.3 Mean ammonia concentration in Tualatin River in month of May, 1987–92. (Source: Unified Sewerage Agency.)

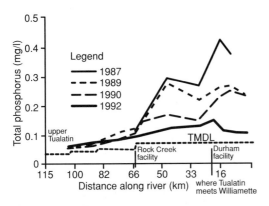

Figure 4.4 Mean phosphorus concentration in Tualatin River in month of May, 1987–92. (Source: Unified Sewerage Agency.)

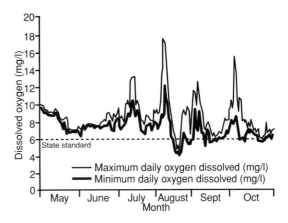

Figure 4.5 Concentration of dissolved oxygen at Lake Oswego Dam, May to October 1993. (Source: Unified Sewerage Agency.)

Ammonia and phosphorus affect the dissolved oxygen in the river by chemical oxidation and encouraging algae growth, respectively. Figure 4.5 displays the 1993 summertime dissolved oxygen, minimum and maximum daily levels relative to state standards at Lake Oswego Dam, a critical place in the river. If the 1987 levels were also displayed, there would be numerous, and very serious, violations of the state standard.

4.5 THE FUTURE

As can be seen, tremendous progress has been made since focusing efforts on water quality improvements in the Tualatin River basin. These efforts have not all been voluntary and have also been in response to regulatory mandates.

The data show that the job is not finished. A trend of improving water quality has been established. The river is understood better. However, one must not become complacent and assume that nothing more will happen to the river.

There are still many challenges on the horizon for the Tualatin River basin. As more people move into the area, they will need to be made aware of their individual responsibilities for water quality control. Associated with this population growth will be increased treated waste loads needing release into the environment.

In addition, as the area grows, there will be an increased demand for potable water sources. How will that population be supported? Can the reuse of treated effluent be used for irrigation to release potable water for drinking and not for watering lawns and flowers?

CHANGES IN THE WATER CHEMISTRY OF THE VLTAVA RIVER FROM 1959 TO 1993

by L. Procházková, P. Blažka and J. Kopáček

Previous publications on the chemistry of the Vltava River and its reservoirs [1,2] have described the fertilization and atmospheric deposition in the catchment. In this contribution, the stabilization of the levels of nutrients and deposition rates and subsequent reductions are reported.

5.1 DESCRIPTION OF THE RIVER

The Vltava rises in the Šumava Mountains near the Czech–German border at an elevation of 1180 m and, after 420 km, it joins the Labe (Elbe) River at the town of Mělnik. There are three main reservoirs on the river, Lipno Reservoir (built in 1959), Orlík Reservoir (1960) and Slapy Reservoir (1954).

The area of Slapy Reservoir is 13.1 km^2, its volume 270×10^6 m^3, the maximum depth 53 m and mean retention time 38 days. The main sampling site is approximately 12 km downstream of the dam close to the Živohošt Bridge at 14°25'0'' east and 49°48'39'' north. The area of the Slapy catchment is 12 900 km^2; about one half is farmland and one third is forested. There are 650 000 inhabitants in the catchment.

The Orlík Reservoir dam is just 9 km upstream of the upper end of Slapy Reservoir, the mean retention time and volume are nearly three times greater than those of Slapy, and consequently Orlík has a significant impact on Slapy. The Lipno is on the upper section of the river and its main influence on both downstream reservoirs is in buffering of the peak floods.

5.2 SAMPLING PROGRAMME AND METHODS OF ANALYSIS

The analysis of the data gathered over 35 years is based on regular, three-week sampling (usually 17 per year) of the surface layer at the Živohošt Bridge. In some years longitudinal transects were sampled and for several years regular vertical profiles were taken. Evaluation of these data has shown that the surface-water samples give a reasonable representation of the trends for the whole reservoir with the exception of total phosphorus (TP) and some other determinands described elsewhere [1,3]. The reason for these exceptions is the distinct stratification in the reservoir during summer and the depletion of phosphorus from the surface layers. Consequently mean winter TP concentrations are used in this study to define the long-term TP trend.

The methods of analysis were described by Procházková and Blažka [1] but since 1991 new procedures have been introduced for the analysis of TP [4] and ammonia [5].

5.3 RESULTS AND DISCUSSION

The principal changes in the basin during the period of the study are:

- substantial changes in the use of agricultural fertilizers;
- an increasing number of inhabitants linked to sewerage systems;

International River Water Quality. Edited by Gerry Best, Teresa Bogacka and Elżbieta Niemirycz. Published in 1997 by E & FN Spon, London, ISBN 0419215409

- an increasing use of detergents containing phosphorus [6];
- changes in the atmospheric inputs.

The change in the input of nitrogen (N), phosphorus (P) and potassium (K) originating from fertilizers is given in Figure 5.1. Fertilizers, in addition, contain some bulk constituents, particularly sulphate and chloride. Their concentrations in fertilizers are comparable to the active fertilizing elements. Their application rates are not as well known as those of N, P and K. In the 1980s, the estimates of mean inputs were about 50 kg/ha per year and 120 kg/ha per year for chloride and sulphate respectively [2,7]. Figure 5.1. suggests three periods of different fertilization rates:

1. a nearly straight line increase from 1959 to 1982. The amount of applied N increased six times during this period;
2. a levelling off in the period 1983–9 (N at around 100 kg/ha per year); and
3. a gradual fall (following the political and economic changes in 1989) down to 40 kg/ha per year for N and 6 kg/ha per year for both P and K in 1993.

An examination of Figure 5.2 shows that nitrate is the main nitrogenous constituent in the Vltava River. Its contribution to the total nitrogen (TN)

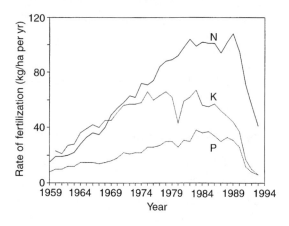

Figure 5.1 Fertilization rates per ha of agricultural land 1959–93. (Sources: Directorate of Agricultural Purchase and Supplies and the Regional Statistical Board, Prague.)

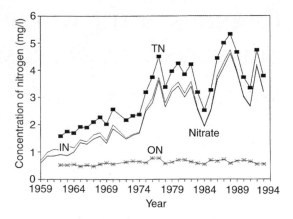

Figure 5.2 Annual mean concentrations of the different forms of nitrogen (ON – organic nitrogen; IN – inorganic nitrogen (nitrate, nitrite and ammonium); TN – total nitrogen).

concentration increased gradually from 60% in 1962 to 90% in 1993. The contribution of organic nitrogen (ON) to TN in the same period decreased from 30% to 10%, although its concentration remained about the same. These data should be compared with the concentration of nitrate in other rivers. For example, the average concentration of nitrate in the middle Rhine increased from 1.2 to 2.7 mg/l from 1965 to 1979 [8], whilst in the Rhur, it increased from 3 to 3.8 mg/l from 1961 to 1978 [9]. Concentrations of nitrite and ammonium were generally low (nitrite < 0.1 mg/l; ammonium < 0.15 mg/l) with the exception of winter 1963, when the severe weather reduced the flow almost completely and anaerobic conditions occurred. This resulted in increased ammonium concentrations and a fish kill in Slapy Reservoir.

Figure 5.3 shows that, in the early years of data collection, the principal variable contributing to the rise in nitrate concentrations in the Vltava River was the increasing rate of fertilizer application. As a result of the use of constant nitrogenous fertilization in the 1980s the flow rate became the main source of variability affecting nitrate concentrations, as shown in Figure 5.4. The influences of fertilization rate and discharge on the concentrations of nitrate cannot be considered separately. High precipitation resulting in an increased flow may give rise to high nitrate levels providing this is available in the soil.

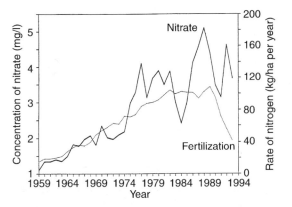

Figure 5.3 Annual mean nitrate nitrogen concentrations and fertilization rates.

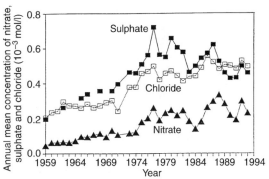

Figure 5.5 Annual mean concentration of sulphate, chloride and nitrate ions v. annual mean discharge.

However, with low nitrate levels in the soil, an increased discharge is not an important factor. For example, the maximum mean discharge in 1965 did not significantly increase the concentration of nitrate. The sudden reduction of fertilizer use after 1989 was followed by only a slight decrease in nitrate in the river because the previous five years had been exceptionally dry and only a part of the nitrate had been eluted from the soil.

During the 1980s, the contribution of nitrogen from the atmosphere was about a quarter of that coming from fertilizer application. After the reduction of fertilization, the contribution from the atmosphere doubled, though its absolute value decreased slightly.

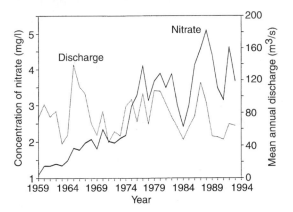

Figure 5.4 Annual means of nitrate nitrogen concentrations v. discharge.

Diffuse sources also determine the concentrations of sulphate and chloride. In Figure 5.5, the fluctuations in the concentrations of sulphate, chloride and nitrate are shown and the lines are roughly parallel. A decrease in sulphate concentrations in recent years has been caused by both a decrease in the application rate of sulphur-containing fertilizers (primarily superphosphate) and a reduction in the deposition of sulphur in Europe (by 37% between 1980 and 1992 [10]). On the other hand, chloride remained nearly constant but for this ion there are significant contributions from human and animal excretion in the catchment.

The concentration of total mineral salts has doubled since 1959, as shown by specific conductivity (1959: 139 μS/cm; 1993: 280 μS/cm at 25 °C). Comparable values for other rivers are: River Clyde (Scotland) 350 μS/cm; Vistula River (Poland) 380–1360 μS/cm. The equivalent ratios of the major cations (Table 5.1) have remained practically the same throughout the period.

In contrast to the cations, the equivalent ratios of the major anions changed markedly (Figure 5.6). The most pronounced increase was observed for nitrate (seven times). Its maximum equivalent ratio was, however, only 12% of the total. As the equivalent ratio of chloride changed only slightly, the biggest fluctuations were observed for sulphate and bicarbonate.

Phosphorus is the element which most frequently limits primary production in fresh water [11, 12,

Table 5.1 Equivalent ratios of major cations 1959–93

Cation	Mean equivalent ratio (%)	Confidence limit (95%) of the mean
Ca^{2+}	51.8	0.87
Mg^{2+}	23.9	0.68
Na^+	18.1	0.57
K^+	6.1	0.35

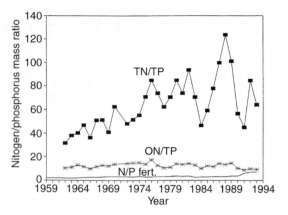

Figure 5.7 Annual means of N:P mass ratios in water and fertilizers for total nitrogen (TN), organic nitrogen (ON) and total phosphorus (TP).

13]. Despite the high fertilization rates, most of the applied phosphorus is retained in the catchment [7] and the relatively high phosphorus concentration in water originates from sewage. The Redfield ratio [14] is the ratio of biogenic elements in the biomass of phytoplankton. For nitrogen and phosphorus, this ratio, based on their atomic masses is N : P = 7 : 3. A ratio significantly higher than this value suggests that phosphorus rather than nitrogen is limiting, and *vice versa*. The total nitrogen (TN) – total phosphorus ratio (TP) calculated from TN and TP annual means, was never lower than 30 (Figure 5.7). On the other hand, the ratio was less than 3.6 in the fertilizers used, and this has increased slightly in the last three years. Figure 5.8 demonstrates the seasonal fluctuations of this ratio in surface water in five selected years. For most years, there is an apparent peak in summer, which is caused by the uptake of phosphorus by phytoplankton, its accumulation in the biomass of consumers and its sedimentation in faeces and dead organisms.

The average annual TP concentrations do not characterize the trends for the whole water column. In Figure 5.9, both annual and winter means are given as well as the maximum and minimum values. The winter mean values represent more truly the long-term trend. Both of them, however, suggest somewhat variable but essentially steady concentrations until about 1980 and a significant increase subsequently. The reason for this might be the result of an increasing TP input due to the increasing use of phosphorus in detergents after 1985 [6]. Accelerated eutrophication resulted in hypoxic and occasionally anoxic conditions in the second part of the growing season. For such situations release of phosphorus from sediments may be assumed. In the Slapy Reservoir, the differences between the annual minimum and maximum concentrations of TP have doubled as a result of increased uptake by phytoplankton. The concentration of chlorophyll, which is a measure of phyto-

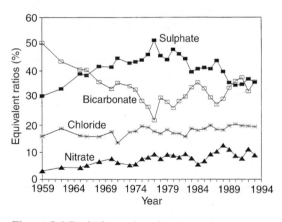

Figure 5.6 Equivalent ratios of major anions.

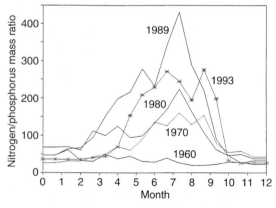

Figure 5.8 Seasonal course of N:P mass ratios in water in selected years.

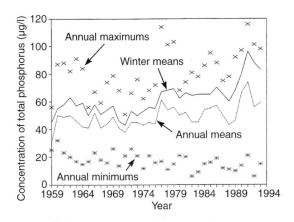

Figure 5.9 Total phosphorus concentrations 1959–93.

plankton biomass, also increased twofold during the period under review. The Vltava River is modified by human activity in terms of the concentration of not only nutrients but also of major ions, which are generally regarded as being determined by the geology of the catchment.

In the near future it may be expected that the fertilization of good agricultural land, which is relatively scarce in the catchment, will increase whilst fertilization of poor-quality land will cease. It is probable, therefore, that concentrations in the Vltava River will decrease.

However, the eutrophication of the Vltava River may be controlled only by reducing the input of phosphorus from point sources. The first and least expensive step is to decrease or eliminate phosphorus in detergents. However, it has been estimated that the contribution of detergent phosphorus to the total phosphorus load in untreated sewage is about 40% [15]. In the long term, facilities for removing phosphorus will be introduced into sewage treatment plants. This will not only improve the Vltava River, but also improve conditions in the Labe (Elbe) River and the North Sea.

ACKNOWLEDGEMENT

This contribution was partly supported by the grant No. 204/93/1310 from The Granting Agency of the Czech Republic. Sampling and analyses were enabled by participation of a number of staff members of what is now The Hydrobiological Institute.

REFERENCES

1. Procházková, L., Blažka, P. (1986) Long-term trends in water chemistry of the Vltava River (Czechoslovakia). *Limnologica* 17, 263–71.
2. Procházková, L., Blažka, P. (1989) Ionic composition of reservoir water in Bohemia: long-term trends and relationships. *Arch. Hydrobiol. Beih. Ergebn. Limnologica* 33, 323–30.
3. Brandl, Z. and Procházková, L. (1985) Changes of some parameters of water quality in reservoirs in space and time. In: *Studies and Qualitative and Quantitative Evaluation of Aquatic Ecosystems.* Proceedings of the 7th Conference of the Czechoslovak Limnological Society, Nitra, (in Czech), 53–6.
4. Kopáček, J. and Hejzlar, J. (1993) Semi-microdetermination of total phosphorus in fresh waters with perchloric acid digestion. *Intern. J. Environ. Anal. Chem.* 53, 173–83.
5. Kopáček, J. and Procházková, L. (1993) Semi-microdetermination of ammonia in water by the rubazoic acid method. *Intern. J. Environ. Anal. Chem.* 53, 243–8.
6. Hejzlar, J. (1993) Pollution of surface waters by P and N compounds, in *Legislative and Technological Development in Treatment of Sewage.* Seminar at the EKO exhibition, Prague 93 (In Czech.)
7. Procházková, L and Brink, N. (1991) Relations between element concentrations and water discharge in agricultural basins. *Catena* 18, 355–66.
8. Hellmann, H. (1987) *Analysis of Surface Waters.* Ellis Horwood, Chichester.
9. Kornatzki, K.H. and Koppe, P. (1981) The content of nitrates in flowing waters taking the Ruhr as an example. *Vom Wasser* 56, 76–83.
10. Agren, C. (1994) *Acid News* 5, 14–15.
11. Vollenweider, R.S. (1970) Scientific fundamentals of the eutrophication of lakes and flowing waters, with particular reference to nitrogen and phosphorus as factors in eutrophication, OECD, Paris.
12. Komárková, J. (1974) Limitation of phytoplankton growth by the lack of nutrients in two reservoirs in Czechoslovakia. *Arch. Hydrobiol. Suppl.* 46, *Algological Studies* 18, 55–89.
13. Nedoma, J., Porclová, P., Komárková, J. and Vyhnálek, V. (1993) A seasonal study of phosphorus deficiency in a eutrophic reservoir. *Freshwater Biology* 30, 369–76.
14. Redfield, A.C. (1958) The biological control of chemical factors in the environment. *Am. Sci.* 46, 205–21.
15. United Kingdom Soap and Detergent Industry Association (1989) Detergent phosphates and water quality in the UK, SDIA, Hayes.

AN INTEGRATED APPROACH: CATCHMENT MANAGEMENT PLANNING FOR THE RIVER TYNE

6

by T. Warburton

6.1 INTRODUCTION

6.1.1 THE CONCEPT OF CATCHMENT MANAGEMENT PLANNING

Since the earliest centres of civilization, society has used river catchments in a variety of ways. The key to making modern society's use of the water environment sustainable lies in a recognition of the need to integrate the management of demands placed on it and the need to treat the river basin as a single environmental management unit. Many countries have now realized the importance of this philosophy.

The rivers, lakes, estuaries and coastal waters of England and Wales, like those around the world, have never before been subject to such large and rapidly increasing demands from the users of water. Many different users interact or compete for water and will inevitably come into conflict with one another. The National Rivers Authority (NRA) has the major responsibility for managing the water environment in England and Wales and for reconciling conflicts between water users. The NRA's mission statement [1] expresses the following principles:

> We will protect and improve the water environment by the effective management of water resources and by substantial reductions in pollution. We will aim to provide effective defence for people and property against flooding from rivers and the sea. In discharging our duties we will operate openly and balance the interests of all who benefit from and use rivers, groundwater, estuaries and coastal waters. We will be business-like, efficient and caring towards our employees.

The National Rivers Authority has developed the philosophy of integrated river basin management into a process called catchment management planning (CMP). Catchment management planning is the process by which the NRA develops catchment management plans through public consultation. Catchment management plans will provide the principal management framework for applying the NRA's powers and in assisting in its role in influencing the actions of others towards the water environment.

This chapter will discuss the NRA's approach to catchment management plans by reference to the River Tyne in the north-east of England and will make particular reference to CMP links to land use and local authorities' development plans.

6.1.2 THE ROLE OF THE NATIONAL RIVERS AUTHORITY

The NRA was created in 1989 as an independent environmental watchdog. Its prime purpose is to

---

International River Water Quality. Edited by Gerry Best, Teresa Bogacka and Elżbieta Niemirycz. Published in 1997 by E & FN Spon, London, ISBN 0419215409

protect and improve the water environment in England and Wales and regulate the use of water by industry, agriculture and the private water and sewerage companies.

The NRA looks after both inland and coastal waters including rivers, lakes and canals as well as groundwaters. It has statutory responsibilities for water quality, water resources, flood defence, fisheries, conservation, navigation and recreation. It also protects people and property from flooding caused by rivers and the sea.

The NRA's core activities cover environmental, regulatory and operational work. It has an important role in environmental emergencies and in influencing land use through the Town and Country Planning process. The NRA is a non-departmental public body sponsored by the Department of the Environment (DoE). The Ministry of Agriculture, Fisheries and Food (MAFF) also has important policy responsibilities in relation to flood defence and fisheries.

Reference is made in this chapter to the National Rivers Authority. In 1996 the National Rivers Authority, Her Majesty's Inspectorate of Pollution and the waste regulation authorities combined to form a new organization called the Environment Agency, which has responsibility for the environmental protection and regulation of water, land and air in England and Wales.

6.2 DESCRIPTION OF THE TYNE CATCHMENT

The River Tyne catchment is one of the largest river catchments in the north-east of England, covering approximately 2935 km^2 (Figure 6.1).

Today, the river provides a convenient and attractive setting for commercial and residential development and is becoming increasingly important for water-based activities and recreation. It is also regarded as one of the country's leading salmon rivers. All of which is a vastly different picture from the Tyne of 20 years ago when sewage and trade discharges into the estuary led to the Tyne's reputation as a grossly polluted river.

The history of the Tyne is similar to that of many great rivers affected by the Industrial Revolution and the associated growth of cities. Historically no effort was made to control the industrial and sewage effluents which discharged directly into the Tyne estuary. At one time, 180 separate major sewage outlets discharged into the river and the results were disastrous. As the river was used as a source of drinking water, numerous outbreaks of cholera are recorded on 19th century Tyneside.

By the middle of the present century little had been done to improve the situation. In the 1960s, the estuary was receiving crude sewage from almost a million people together with large volumes of trade effluent. As a result, in warm weather, the middle reaches of the Tyne estuary were completely dead with no oxygen present in the water at all over several kilometres. The problem was so bad that in 1969 a public enquiry, taking place in Newcastle's Moot Hall on the city's quayside, had to be adjourned because of the stench from the Tyne.

Although moves had been made in the past to solve the problem, no real progress was made until 1966 with the formation of the Tyneside Joint Sewerage Board who laid plans to construct a comprehensive sewerage system to serve Tyneside and its surrounding districts. The scheme, taken over by the Northumbrian Water Authority in 1974, was to become one of the most important and complex ever undertaken in this country.

Miles of interceptor sewers have been installed along both banks of the river and adjacent coastlines. A treatment plant was constructed at Howdon to provide primary treatment and grit removal for the sewage collected by the interceptor sewers.

This work has led to significant improvements in the amenity value and biological status of the estuary. It now supports a range of estuarine animals, and the numbers of salmonid fish returning to the river has increased dramatically (Figure 6.2).

The North and South Tyne rivers drain catchment areas which are sparsely populated and only small volumes of treated sewage and trade effluents are discharged into them. This means that the water quality is generally of a high standard and is abstracted for use as drinking water supply.

The River Derwent is impounded in its upper reaches to form the Derwent Reservoir. In the

Description of the Tyne catchment 47

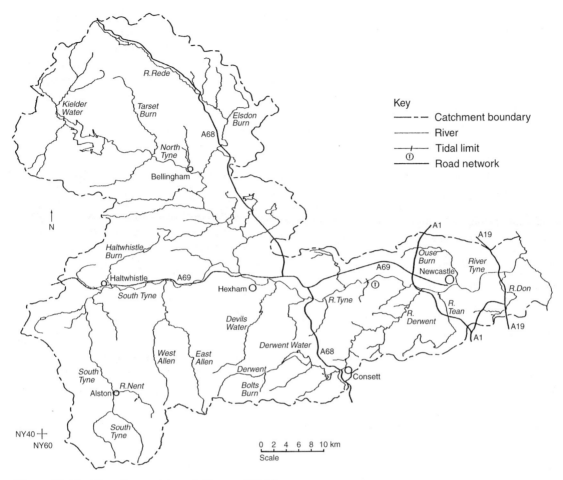

Figure 6.1 The River Tyne catchment. (Source: NRA.)

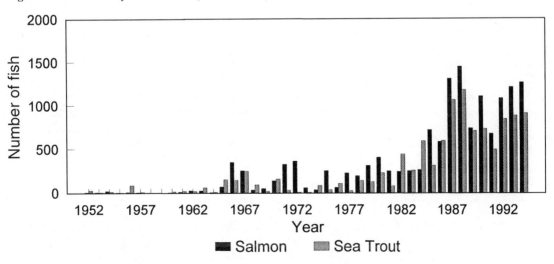

Figure 6.2 Declared rod catch River Tyne 1952–94.

Consett area, discharges from the British Steel Corporation plant, together with effluent from the sewage treatment works, used to cause severe pollution. The closure of the steelworks and extensions to the local sewage works have helped the river to recover and it is now a valuable fishery.

Another major tributary, the River Team, is a heavily industrialized urban stream. It lacks a substantial supply of clean upland water to dilute the large volumes of sewage–trade effluent and urban surface run-off it receives.

Although it is a relatively small river, the Ouseburn is important in that it flows through Jesmond Dene, a popular amenity park in Newcastle. The catchment is heavily populated and the river has suffered from the effects of sewage discharges and contaminated run-off. Some of these discharges have been eliminated and water quality has improved. However, the area still has water quality problems and is under considerable development pressure.

6.3 AN INTEGRATED APPROACH TO CATCHMENT MANAGEMENT PLANNING

The NRA has been introducing the CMP process to the River Tyne catchment [2,3] to assist in applying its powers and responsibilities at a local level in a coordinated way. The River Tyne catchment management plan will provide the principal management framework for applying the NRA's powers and in assisting in its role in influencing the actions of others towards the water environment. The purpose of the River Tyne catchment management plan is described as: 'To define a strategy for realizing the environmental potential of River Tyne catchment within the prevailing economic and political constraints.' This will be achieved by:

- focusing attention on the River Tyne catchment;
- involving all interested parties in planning for its future; and
- establishing an action plan for managing the catchment over the next five years.

The CMP process [4] has been designed to create a framework by which the NRA's diverse responsibilities can be applied in a coordinated and cost-effective manner. The NRA wishes to encourage other bodies to participate in the formation of the action plan. Many improvements in and/or enhancements to the catchment require involvement and commitment from other individuals, groups or organizations. In particular the NRA welcomes the involvement of statutory bodies and key groups, such as planning authorities, landowners and managers, conservation organizations, developers, industries and recreational groups.

The process of CMP can be summarized as follows.

1. Production of consultation report
2. Three month period of public consultation
3. Action plan
4. Action
5. Monitor
6. Review.

The consultation report and action plan together:

- assess catchment resources, uses and activities;
- are used to consult widely on issues to be tackled;
- establish a long-term vision for individual catchments;
- use effective and proactive planning to prevent further environmental problems.

These and the other stages of the River Tyne CMP are outlined below.

6.3.1 THE CONSULTATION REPORT

The NRA, through internal discussion and informal liaison with a range of outside bodies, produced a consultation report for the River Tyne in April 1994. This report described the catchment, the uses of the catchment and activities within it, reviewed the state of the catchment against targets and identified issues which needed to be addressed. Proposals for actions to address these issues were made.

6.3.2 PUBLIC CONSULTATION

The NRA considers public consultation and open discussion to be an essential feature of catchment management planning. The report was therefore widely circulated to statutory bodies and key groups such as the development control authorities, landowners and managers, developers, industries, conservation organizations, and recreational groups. In commenting on the consultation report the NRA encouraged consultees to:

- confirm the range and extent of the resources, uses and activities described;
- express their views on the issues identified in the report (have all the issues been included and have they been fairly assessed?), and provide opinions on them and the proposals being made;
- comment on how the development of strategies and action plans should be progressed;
- identify any actions by other bodies or organizations which may benefit from inclusion in the catchment management plan.

The consultation report was published and interested parties asked for comment during a three-month public consultation period. To encourage formal responses to the consultation report the NRA:

- formally launched the consultation report in April 1994 at the University of Newcastle upon Tyne;
- distributed 600 copies of the consultation report to over 300 organizations, groups and individuals;
- held a public open meeting in June 1994 at Hexham;
- advertised the plan and the meeting in newspapers;
- placed display boards in local libraries;
- conducted press interviews for newspaper, radio and television;
- attended several meetings with interested parties to discuss the plan.

6.3.3 ACTION PLAN

The next phase of the catchment planning process, the action plan, incorporates views, plans and actions arising from consultation and represents a working document for the achievement of improvements to the river environment. It also provides the means of promoting two key aspects of environmental management: land-use planning and water quality objectives. The action plan not only includes commitment to actions from the NRA but from other organizations, statutory bodies and individuals.

6.3.4 ACTION

The most important part of any management system is taking action, the NRA is ensuring that the published action plan becomes its primary business planning tool. Projects identified in the action plan must lead to real improvements for the users of the River Tyne catchment. Examples of some issues in the River Tyne catchment and the projects set up to tackle them are given later in this chapter.

6.3.5 MONITORING AND REVIEW

It is envisaged that the NRA will monitor the implementation of the plan through regular consultations both internally and with committed parties. The plan will be reviewed internally on an annual basis with a full revision on a five-yearly basis.

6.3.6 LIMITATIONS OF THE CATCHMENT MANAGEMENT PLAN

The finished catchment management plan will inevitably be subject to limitations. The plan is non-statutory and therefore not enforceable (beyond the NRA's statutory powers); the aim is to provide a framework for the integrated management of the catchment.

6.4 REVIEW OF RESOURCES, USES AND ACTIVITIES WITHIN THE TYNE CATCHMENT

The resources, uses and activities within the Tyne catchment are described in full in the consultation report [2]. The report also presents issues and proposals for actions to address those issues. Sections 6.4.1–6.4.15 summarize this information including, in some cases, details of the projects included in the action plan [3].

6.4.1 CLIMATE

Average annual rainfall of over 2000 mm in the upper South Tyne and over 1800 mm in the North Tyne decreases to less than 650 mm in the Newcastle area.

6.4.2 SURFACE WATER ABSTRACTION

The River Tyne catchment is extensively used for surface water supply; there are 79 licensed surface water abstractions, which are authorized to abstract up to a total of 694 million cubic metres per annum. A major water resource infrastructure exists to support abstractors. The main elements of this system are Kielder Water – holding 200 million cubic metres in a reservoir of 10 km^2 area held back by a 59 m high dam; and a transfer pumping station at Riding Mill which allows transfer of water south to the River Wear and River Tees. The Kielder Transfer Scheme is of regional importance for water supply and river regulation. The CMP action plan has set up a project, to start in 1996, to evaluate the success of the current transfer system management policy in balancing the needs of the river and reservoir users and the ecology of the river.

6.4.3 GROUNDWATER ABSTRACTION

Major aquifers are the magnesian limestone and the fell sandstone. The magnesian limestone is used for water supply; minor aquifers are important for local use.

6.4.4 HYDROELECTRIC POWER GENERATION

Water in Kielder Water, in addition to supporting regional water demands, is used to generate hydroelectric power.

6.4.5 EFFLUENT DISPOSAL

There are 179 sewage treatment works, a number of which have an adverse impact on the receiving watercourses. There are 140 licensed industrial discharges, only a few are significant. Several tributaries are affected by combined sewerage overflows causing pollution problems. The streams most affected are in the River Team, Ouseburn and River Don systems. The action plan gives details of water quality issues stretch by stretch and the actions needed to deal with those issues; in some cases investigative projects will be set up to clarify the issues and actions needed.

6.4.6 SOLID WASTE DISPOSAL

There are 243 licensed waste disposal facilities. Most of the sites do not pose a threat to the water environment. However, there are sites located close to the River Team which give cause for concern.

6.4.7 MINERAL EXTRACTION AND MINING

There is a long history of mineral extraction; lead, zinc, deep and opencast coal (Great Whin Sill), sand and gravel are, or have been, extracted within the catchment. Problems exist with contaminated land and water quality problems associated with abandoned mining sites. The cessation of mine-water pumping in the Durham Coalfield is also an issue of relevance to the River Tyne. A large multi-functional project has been set up to deal with this issue and the NRA's policy towards it. (The problems associated with the abandonment of coal mines are described by Hammerton in Chapter 9.)

6.4.8 FLOOD DEFENCE

Land and property are at risk from flooding in various locations within the catchment. Developments within the natural flood plains of the rivers are at particular risk.

6.4.9 DEVELOPMENT – HOUSING, INDUSTRY AND INFRASTRUCTURE

Urban areas are concentrated in the lower reaches of the catchment. Future urbanization of previously undeveloped land areas or redevelopment of land has implications for water supply, effluent disposal, solid-waste disposal, flood defence, landscape and ecology. Local authorities are therefore key players in protecting and enhancing the water environment through their development plans. The NRA is putting a large amount of effort into working with the planning authorities to ensure that any future developments do not lead to deteriorations in the water environment. Details of the NRA's planning liaison work are given later in this chapter.

6.4.10 AGRICULTURE AND FORESTRY

The majority of the area of the Tyne catchment is subject to agricultural or forestry land use, including Kielder Forest and large areas of moorland, as well as land put to more intensive agricultural use in the lower valleys. Several projects have been set up to deal with rural issues, including a farm pollution prevention campaign and a habitat enhancement project, in conjunction with the Farming and Wildlife Advisory Group (FWAG).

6.4.11 ECOLOGY AND NATURE CONSERVATION

The Tyne catchment supports a wide diversity of habitat: there are 77 Sites of Special Scientific Interest (SSSI), including three National Nature Reserves (NNRs). Fifty-two of these SSSI are directly influenced by the water environment. There is a small but important population of European otters (*Lutra lutra*) as well as other wildlife.

6.4.12 FISHERIES

The Tyne is of regional importance for migratory salmonids. Brown trout and coarse fish populations are also important. Fish populations are reduced or absent from some tributaries due to water quality problems: these include the River Team, Ouseburn and River Don. Coarse fish and brown trout populations may have decreased in recent years. Several fisheries related projects and investigations are included in the CMP action plan.

6.4.13 AMENITY AND RECREATION

There are considerable demands for many types of water-based recreation within the catchment. In particular the improved water quality of the estuary has resulted in increasing demands for facilities in the Tyneside area. Canoeing is popular throughout the catchment and an access agreement between the British Canoe Union (BCU) and River Tyne Riparian Owners and Occupiers Association ensures that canoeists have access to the river for canoeing at times of year when they will not conflict with anglers. This agreement is considered to be successful by the canoeists and is used as a model for other rivers.

6.4.14 ANGLING AND COMMERCIAL FISHING

Angling on the Tyne is primarily for migratory salmonids and the river is now classed as one of the country's top salmon angling rivers (in 1988, 2500 salmon and sea trout were caught on rod and line). Commercial fishing is limited to eel fishing. The offshore east coast drift net fishery potentially affects migratory salmonid stocks.

6.4.15 LANDSCAPE AND HERITAGE

A large part of the North Tyne catchment is within the Northumberland National Park and the Borders Forest Park. The upper reaches of the South Tyne, the Allens and the Derwent are in the North Pennines Area of Oustanding National Beauty (AONB). Areas of the catchment in the former County of Tyne and Wear have been designated as green belt. The Great North Forest includes part of the green belt. The catchment also contains approximately 230 Scheduled Ancient Monuments (SAMs).

6.5 EXAMPLES OF CMP ISSUES AND PROJECTS

6.5.1 MANAGEMENT OF THE KIELDER SYSTEM

Releases from Kielder Water have a profound effect on the River North Tyne. The reservoir release policy ensures that releases are primarily made to maintain compensation flows or to support abstractions; releases are also used to generate hydroelectric power (HEP). Prior to 1993, the release policy was geared to maximize the revenue from HEP. The introduction of a non-fossil fuel tariff by the Government has meant that a new release policy has been formulated to create a flow regime with a more natural pattern of flows. The previous release policy meant that the temperature profiles in the river were generally colder. This temperature effect may have impacted on the ecology of the downstream reaches and one theory suggests it may also have been responsible for the decline in the dace population which has been noted in the river. The CMP action plan will target resources to examine the current release policy and to make recommendations on how to progress towards a more sympathetic policy geared towards the needs of the aquatic environment.

6.5.2 RIVERSIDE HABITAT ENHANCEMENT

Another project proposed in the CMP is aimed at improving riverside habitats and encourage sensitive habitat management to improve the fishery and the current small otter population. Waterside habitats are important for supporting a wide range of insects, birds and mammals as well as providing cover and food source for fish. Opportunities exist for the enhancement of these habitats through the use of grants to landowners and farmers. The action plan will target NRA resources at encouraging landowners to take up such grant opportunities and coordinate the construction of otter holts to maximize the population's spread. This project will be undertaken in conjunction with the Farming and Wildlife Advisory Group (FWAG).

6.6 CMP AND ITS LINKS TO LAND-USE PLANNING

Urbanization of previously undeveloped land areas, or redevelopment of land, whether for industrial or domestic purposes, has implications for water supply, effluent disposal, solid-waste disposal, flood defence, landscape and ecology (Figure 6.3). The development of some sites may also disturb contaminated land, releasing potential pollutants into the water environment.

The majority of developments are controlled through development plans, published by local planning authorities under Town and Country Planning legislation. These strategic plans take the form of county structure plans, district wide local plans and unitary development plans, which set the context for development in land use, planning and transportation. They identify areas for future residential, commercial and industrial development, and set out the policies against which planning authorities consider development proposals and land uses. The NRA is a statutory consultee for these plans.

Individual development proposals are considered in relation to the approved development plan. The NRA is a statutory consultee for some proposals, but has requested consultation on a more extensive range of proposals. The final decision on planning matters rests with the planning authority. However, if the development includes abstraction, impoundment or discharge, entails work on or near a watercourse, or the introduction of fish, then appropriate consents or licences may be required from the NRA.

Some types of development are further regulated by EC legislation, requiring an Environmental Impact Assessment (EIA) to be submitted as part of the planning application. Developments can also be constrained via other statutory designations such as Sites of Special Scientific Interest (SSSI), National Parks (NPs), Areas of Outstanding Natural Beauty (AONB) and Scheduled Ancient Monuments (SAMs).

Many potential problems or areas of concern between new development and the water environment can be dealt with by careful planning and implementation. The local planning authorities thus have a very important role to play in the long-term protection and enhancement of the water environment. To assist them in this role, and to make developers aware of how they should protect the water environment, the NRA seeks to inform interested parties through (Figure 6.4).

- publication of Guidance Notes [5];
- more detailed localized information given in a land-use statement in the CMP action plan;
- comments on individual planning applications at the time of application.

6.7 CONCLUSIONS

The River Tyne catchment in the 1990s is subject to a variety of uses. It is of strategic importance to the domestic and industrial water supply of north-east England; its waters are used to produce sus-

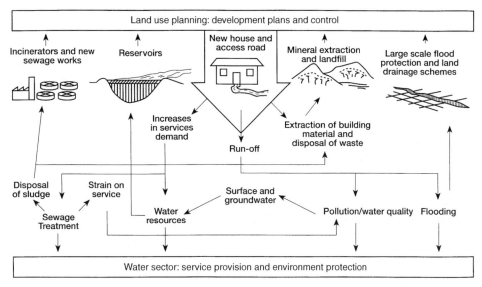

Figure 6.3 Interactions between land use and the water environment. (Reproduced from Slater *et al*, Land Use Planning and the water sector: a review of development plans and catchment management plans; University of Newcastle, 1993.)

tainable energy from HEP generation and provide recreational opportunities for the region's 2.3 million people. Following water quality improvements in the estuary, the river is now one of England's foremost migratory salmonid fisheries. It also supports important wildlife populations including small but significant numbers of European otter (*Lutra lutra*).

The NRA has the major responsibility for managing the water environment in England and Wales and for reconciling conflicts between water users. The NRA is introducing an initiative to coordinate and plan for the future of the catchment through a process called catchment management planning (CMP). This initiative involves open public consultation to assess the future needs of the catchment and identify issues which need to be addressed, for example balancing HEP release policies against the recreational and ecological uses of the river and managing the increasing demand for recreational activities.

The CMP process benefits NRA actions by making them:

- coordinated
- inter-disciplinary
- issue led
- cost effective
- preventative
- proactive.

It also:

- provides a balanced view
- raises awareness
- assists statutory land-use planners.

Figure 6.4 The NRA's role in land-use planning.

Catchment Management Planning will allow, for the first time, catchment-wide coordination of planning for the future of the Tyne river basin. Links between CMP and land-use planning are particularly important. The CMP process will involve statutory bodies and key groups such as the development control authorities, landowners, land managers, developers, industries, conservation organizations and recreational groups.

The views expressed in this chapter are those of the author and do not necessarily reflect the views of the National Rivers Authority.

REFERENCES

1. National Rivers Authority (1990) Corporate Plan 1990–91.
2. National Rivers Authority (1993) River Tyne Catchment Management Plan – Consultation Report.
3. National Rivers Authority (1994) River Tyne Catchment Management Plan – Action Plan.
4. National Rivers Authority (1995) Catchment Management Planning Guidelines. Volume 028 Version 1.
5. National Rivers Authority (1994) Guidance Notes for Local Planning Authorities on Methods of protecting the Water Environment through Development Plans.

WATER QUALITY VARIABILITY IN THE AGÜERA STREAM WATERSHED AT DIFFERENT SPATIAL AND TEMPORAL SCALES

by A. Elósegui, E. González, A. Basaguren and J. Pozo

7.1 INTRODUCTION

The ever-increasing human activities have produced worldwide deterioration of rivers and, more recently, concerns over the assessment and restoration of their quality [1, 2]. Many methods have been developed to determine water and habitat quality, but they are difficult to apply because of the high variability of lotic systems. Spatial and temporal variability also create great difficulties for water quality management [3]. One of the main problems in limnology is knowing the variability of rivers from different ecoregions and under different degrees of disturbance [4]. The problem of the observation scale is still important in limnology [5, 6, 7] and can play a key role in the development of water quality models [8]

The northern Basque Country is a densely populated, highly industrialized area, with severe pollution and depletion of water resources [9]. Thus, the problems of Basque streams and rivers are similar to those elsewhere. Their particular features (e.g., short, steep channels) and management techniques make it difficult to apply the mathematical models used successfully elsewhere. Moreover, there are few studies addressing functional aspects, such as the dynamics of elements of stream–basin relationships in Basque streams.

The Stream Ecology Group of the University of the Basque Country has studied the Agüera Stream since 1988; its aim is to understand the function of a stream in its relatively clean state, in order to gain insight into the critical processes of Basque streams, and to find monitoring guidelines for managing these systems properly. Its first concern was to characterize the stream water; as this implied identifying the relevant spatial and temporal scales of variability, different complementary sampling strategies were devised.

The multiple temporal and spatial scales considered most relevant have been studied. Seasonal variability, driven mainly by changes in temperature and irradiance, is often large [10]; it is usually reflected in a cyclical pattern during the year. Sometimes, there can be broad day–night changes superimposed on this variability [11,12]. Additionally, the changes in flow, which are rather unpredictable in the Basque Country, markedly affect the water chemistry. From a spatial perspective, the tributaries can show broad differences, but their influence in the main channel usually decreases downstream [13]. The effects of more intense inputs, particularly those associated with urban sewage, can extend further downstream; the magnitude, duration and frequency of pollution episodes must be assessed to determine the overall

---

International River Water Quality. Edited by Gerry Best, Teresa Bogacka and Elżbieta Niemirycz. Published in 1997 by E & FN Spon, London, ISBN 0419215409

impact on the aquatic ecosystems [14]. All these aspects, as well as others (e.g. changes over a geological time scale), interact to make the stream quite an unpredictable system.

In this chapter, the results from several research programmes, are reported and the importance of different scales of variability on parameters usually measured when assessing water quality is discussed. Previous work [15] deals with the spatio-temporal variations of seston in the Agüera Stream.

7.2 STUDY SITE

The Agüera Stream (Figure 7.1) is located in northern Spain, an area with a temperate climate, abundant rainfall, and of relatively little change throughout the year. The vegetation is a mosaic of natural forests, pine and eucalyptus plantations, croplands and rangeland. Lithology includes siliceous and calcareous sedimentary rocks and Quaternary alluvials. Guriezo (pop. 1830), Trucios (pop. 520) and Villaverde (pop. 490) are the main villages, but most of the houses are scattered. Industrial activities are of little importance, as the economy is based on forestry and livestock farming. No water treatment plant exists in the catchment. A more detailed description of the study site has been published elsewhere [15, 16].

7.3 SEASONAL VARIATIONS IN THE MAIN CHANNEL

In order to determine the seasonal variations occurring in the main channel, ten sites from the source to the mouth were sampled every two weeks for half a year (November 1988–May 1989) and then for a whole year (December 1989–January 1991). The variables measured were flow, dissolved oxygen concentration, oxygen saturation, pH, water temperature, conductivity, alkalinity, hardness, magnesium, sodium, potassium, silicate, nitrate, nitrite, ammonia, reactive phosphate and sulphate. The period 1988–90 was unusually dry: in Bilbao, the biggest city in the area, the water supply was subjected to restrictions from 1989 to 1991.

Figure 7.2 plots the spatio-temporal dynamics of conductivity. There are steep spatial gradients of mineralization explained mainly by the geology of the basin, with its siliceous headwaters and calcareous central region. There is no clear seasonality, most of the changes detected being due to more or less unpredictable rain events. During low-discharge periods, spatial differences are highest, as is usual in streams [17]. On a larger temporal scale, there are certainly differences between years; comparing the water mineralization in the same month and under similar hydrological regimes, the spatial patterns were stronger during the second year, a fact difficult to explain.

Apart from sampling sites downstream from the main villages, the phosphate concentrations (Figure 7.3) are moderate to low. They recover very rapidly after the sewage inputs have ceased; this is due in part to the retention by the benthos, since nutrient loads, as well as concentrations, decrease between sites 5 and 8 [18]. The temporal variations respond mainly to the flow rates through

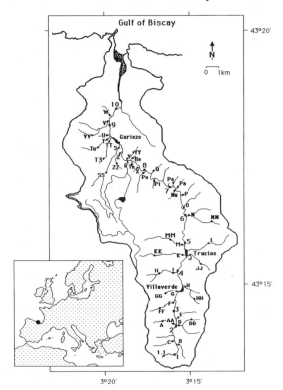

Figure 7.1 The Agüera stream watershed: geographic situation and location of the main villages and sampling sites.

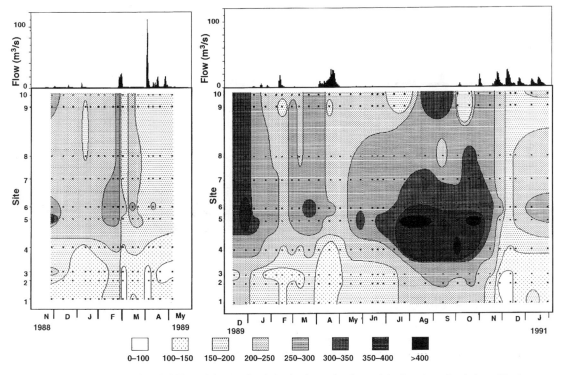

Figure 7.2 Spatio-temporal variability of the conductivity in the main channel (units of conductivity μS/cm).

the dilution of inputs; the differences in phosphate concentration between years mirror the differences in conductivity.

The relative importance of spatial v. temporal variability in the water quality of the Agüera Stream has been explored by means of multivariate analyses [19]. The results show the magnitude of spatial variability along the main reach of the stream, and the dominance of floods on the temporal variability. The spatial differences reflect, first, the geology of the basin and, second, the eutrophication due to urban sewage. A principal component analysis performed to depict temporal differences only, showed the incidence of hydrology on the temporal variations, but low seasonality. Seasonal variations are confined almost exclusively to temperature and oxygen concentration. Variability is highest at sites 4 and 5, downstream from Villaverde and Trucios respectively.

Thus, for most variables measured there is no clear seasonal variability; the hydrological conditions regulate the water chemistry; however, variations between years can also be large.

7.4 VARIATIONS PRODUCED BY FLOW CHANGES

The relationship between flow and concentration can be very complex; inter-site and inter-storm variations occur [20, 21], and it can also be affected by the antecedent conditions [22]. A knowledge of this relationship is necessary to estimate non-point source nutrient or pollutant inputs [23].

Figure 7.4 plots conductivity v. flow in the ten stations along the Agüera stream. There is a clear trend showing conductivity decreasing with flow asymptotically. This pattern is most marked at sites 1 to 5, and could be explained by changes in flow paths of the water through the soil [24]. It becomes less distinct downstream, as reported by others [25]. These authors explained the changing patterns by the areal variations of soil solute chemistry, plus the

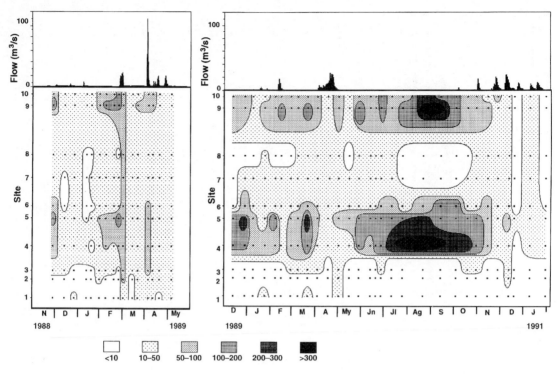

Figure 7.3 Spatio-temporal variability of the phosphate concentration in the main channel (units of μg/l).

effect of tributaries. The geology, vegetation and human activities in the Agüera Stream catchment are more complex than in the catchments others have studied; this would explain the greater variability at the Agüera's downstream sites.

Phosphate (Figure 7.5) shows a rather different pattern in the headwaters, the highest concentrations being measured in medium to high flows. Other authors [26] report concentrations of phosphate (especially particulate) increasing with the discharge in a manner dependent on the basin vegetation [27]. In the Agüera Stream, downstream from Villaverde and Trucios (sites 4 and 5 respectively) there is an abrupt change in the relationship between flow and phosphate concentration. Highest concentrations correspond to base flow, and there is a clear decrease with increasing discharge, as the stream dilutes the urban inputs. Additionally, the increased residence time of the water in the soil during low flows can produce an increase in stream water concentration [28]. As with conductivity, the relationship becomes more complex downstream, a reflection of the more complex mechanism of control that includes hydrological and biological components.

The hydrology is thus a source of variation of the water quality to be taken into account; its effects must be assessed in different reaches, especially when the basin shows large differences in geology or soil uses.

7.5 VARIABILITY IN THE TRIBUTARIES

The ten sites in the main channel plus 52 additional sites in all the tributaries were sampled four times during 1990 (24 January, 16 May, 10 July and 14 November). The flow rates at the stream mouth were 1.342, 1.06, 0.62, and 0.946 m^3/s respectively. The variables measured were the same as in the previous sampling programmes.

Figure 7.6 plots the variations in conductivity; these reflect strongly the geology of the different sub-basins studied. Some tributaries in the central part of the catchment show more than 400 μS/cm;

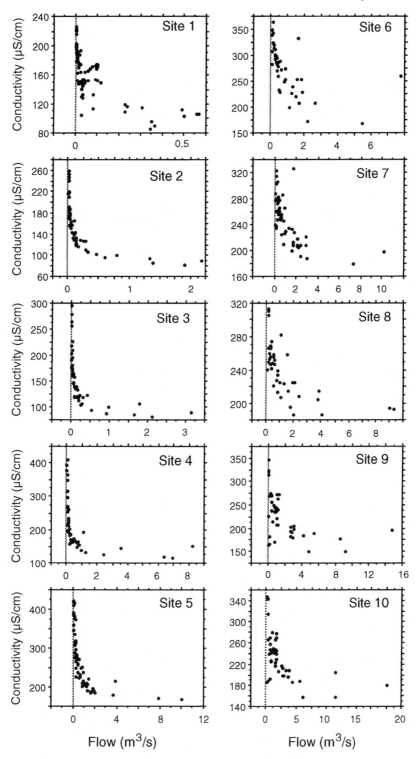

Figure 7.4 Relationship between conductivity and flow at sites 1 to 10.

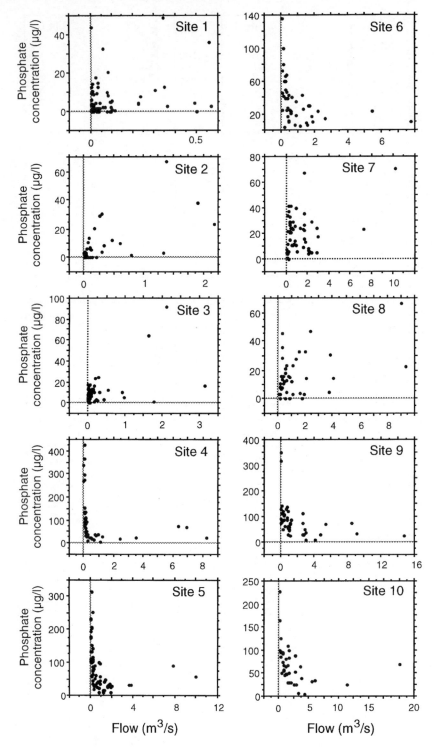

Figure 7.5 Relationship between phosphate concentration and flow in sites 1 to 10.

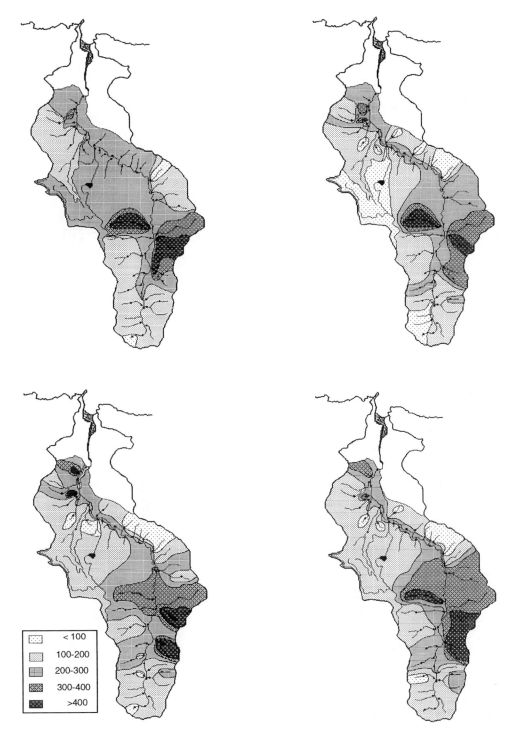

Figure 7.6 Spatial variations of the conductivity in the whole watershed (a) January; (b) May; (c) July; (d) November (units μS/cm).

62 *Water quality variability in the Agüera watershed*

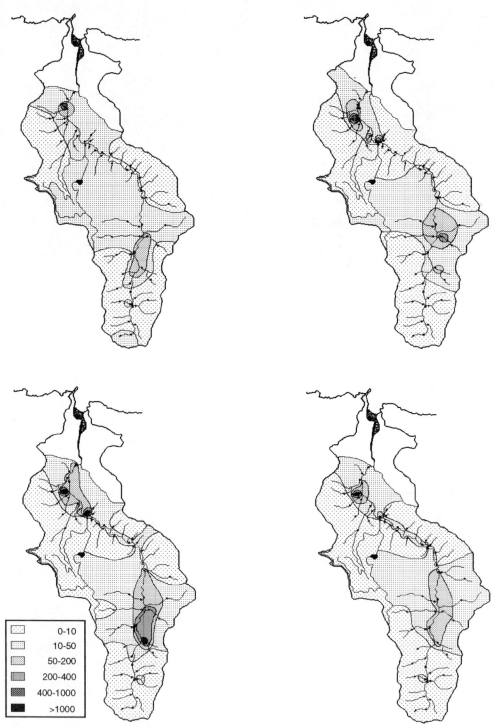

Figure 7.7 Spatial variations of the phosphate concentration in the whole watershed (a) January; (b) May; (c) July; (d) November (units μg/l).

spatial variations are largest during periods of low flow. As has been pointed for nearby streams [29, 30], in some cases the high conductivity can be due to the eutrophication associated with domestic sewage. This is certainly the case in small, siliceous tributaries that pass through villages where the changes though the year are most marked.

The phosphate concentrations (Figure 7.7) also show much broader variations in the tributaries than in the main channel, with up to a tenfold increase in concentration at some points of high local population. In these eutrophic reaches the nutrient load is usually small and, in spite of the high concentration in the tributaries, the main channel is only slightly affected. In streams elsewhere, both the agricultural and urban soil used produce increases in the stream nutrient concentrations [31, 32]. The kind of agriculture practised in the Agüera Stream has little impact because of the small ploughed areas and the low use of fertilizers. Thus, the number of houses or farms in a sub-basin and the flow rate of the stream seem to be the main factors producing eutrophic conditions. This relationship is very complex: the distance from the houses to the stream, the location and type of sewer, and even the kind and extent of riparian vegetation [33, 34] can play a key role. In order to explain the water chemistry it is necessary to determine the proportion of the basin covered by different soil types and the spatial pattern of land use [8].

Although they occur infrequently in the catchment, local black spots are ecologically important: small, shallow tributaries are spawning sites for fishes [35] and refuges for other organisms [36], and highly polluted reaches can act as barriers for the dispersal and migrations of these organisms. Thus, an intense spatial scale should be taken into account when assessing the stream quality from a biological perspective.

7.6 DIURNAL VARIATIONS

Three sites (1, 5 and 7) were sampled for diurnal variations four times during one year. Samples for

Table 7.1 Physico-chemical variables measured hourly during 24 hours in July 1990

Site	Temperature (°C)		pH		Conductivity (µS/cm)		DO (mg/l)		O_2 sat (%)		Alkalinity (meq/l)[a]	
	min	max	min	max	min	max	min	max	min	max	min	max
1	17.1	18.7	7.44	7.68	179	208	8.1	8.4	90	99	0.60	0.76
5	18.2	25.3	7.51	9.17	356	383	2.4	14.9	29	185	2.22	2.82
7	19.6	24.7	7.80	9.17	229	285	6.6	12.3	77	148	1.58	2.08

Site	Hardness (mg/l)		Magnesium (mg/l)		Calcium (mg/l)		Sodium (mg/l)		Potassium (mg/l)		Silicate (mg/l)	
	min	max	min	max	min	max	min	max	min	max	min	max
1	65	81	0.94	2.96	23.9	29.4	1.35	7.84	0.31	1.37	2.21	3.22
5	143	176	0.65	3.12	53.1	67.9	2.03	8.37	1.34	7.88	1.41	2.21
7	101	131	0.32	1.69	37.8	51.2	1.31	5.77	0.29	2.07	0	0.27

Site	Nitrate (mg/l)		Nitrite (µg/l)		Ammonia (µg/l)		Phosphate (µ/l)		Sulphate (mg/l)	
	min	max	min	max	min	max	min	max	min	max
1	0.56	1.52	1	6	18	311	1	44	39.4	45.3
5	0.99	1.75	72	173	78	682	181	326	27.4	40.0
7	0.66	2.74	0	3	16	431	0	12	14.7	19.5

[a] meq/l = molar equivalent per litre

physico-chemical variables were taken hourly during a 24-hour period. Table 7.1 shows, the diurnal ranges of some of the variables for July. Temperature variation, low at the headwaters, is greatest at site 5, which has no riparian vegetation. The high irradiance and nutrient levels of this reach allow for a very active photosynthesis which is usually reflected in diurnal variation of oxygen saturation and pH [37]. However, diurnal ranges as wide as the Agüera's have not been found in the literature. The dissolved oxygen variation has been found to be greatest at warm sites of high productivity [38].

Nutrient dynamics, affected by benthic metabolism and human activities, are highly site-specific [39]. The nitrate concentrations remain relatively constant in the headwaters and change at sites 5 and 7, but in different directions. The sources of nitrate in the Agüera Stream include both the leaching of soils and urban sewage, which results in complex diurnal patterns. Downstream from Trucios, nitrate concentrations show broad diurnal ranges, reflecting mainly the cyclical nature of the sewage inputs [40]. The dynamics of nutrients at sites 5 and 7 are different as a result of the instream processes. Seston also shows broad diurnal variations but the patterns are highly variable between seasons [15].

Diurnal variations are thus greatest in the middle reaches subjected to urban inputs and affects mostly oxygen saturation, nitrite, ammonia and phosphate concentrations. For these variables the range in one day can be greater than that detected during a whole year with conventional sampling schedules [30]. The variables least affected by diurnal variations are conductivity, hardness, calcium and silica, as these are mostly controlled by the flow rate.

The diurnal variations of pH are especially important in the Agüera Stream. The pH changes little in the headwaters, but much more in middle reaches, as a result of photosynthesis [41]. Ranges up to 2 units, and values as high as 9.8 are detected at site 5 under low flows. According to most water quality standards, the whole of the Agüera downstream from site 4, would be classified as poor. Nevertheless, the benthic communities show no adverse effects from pollution between sites 6 and 8 [42]. So, pH seems a poor indicator of water quality in Basque streams [18]. In Denmark wide diurnal pH variability has been reported [37] and this scale of variation should always be taken into account.

The metabolism of the community and the photosynthesis/respiration (P/R) ratios were calculated from the diurnal variations of dissolved oxygen concentration (Table 7.2). The headwaters of the Agüera Stream are always heterotrophic; in the middle reaches, even in winter, metabolism is very active and usually autotrophic provided that there is enough hydrological stability. In places affected by sewage, respiration increases during summer and the community becomes heterotrophic. The low P/R ratio in December was due to the sloughing of most of the benthic community caused by a flood.

The high productivity rates in such a shallow stream result in the withdrawal of most of the inorganic nutrients. Mass balances performed with the nutrient loads in the tributaries and in the main reach show strong nutrient retention (up to 90% of the load) between sites 5 and 8, downstream from Trucios. So, the recovery of water quality is not merely a result of dilution, but also self-purification [18]. The benthic community plays a key role in controlling the characteristics of the stream water [12]. Basque streams are very nutrient-retentive, probably because of their high re-aeration results and strong metabolic activity.

7.7 INFLUENCE OF THE VILLAGES

Water samples were taken upstream and downstream from the three main villages, on four dates (13 February, 23 May, 12 July and 15 November 1990). Three samples per site were obtained for each occasion: early morning, noon and late after-

Table 7.2 Photosynthesis/respiration ratios calculated from the diurnal oxygen saturation curves in 1990

Site	7 March	2 May	25 July	18 December
1	0.356	–	–	–
5	1.840	1.379	0.638	0.049
7	3.108	2.560	1.131	0.085

Table 7.3 Impact of the three main villages on conductivity, nitrate and nitrite levels in 1990

	Conductivity (% increase)				Nitrate (% increase)				Nitrite (% increase)			
	Feb	May	July	Nov	Feb	May	July	Nov	Feb	May	July	Nov
Villaverde	36.7	31.8	31.9	30.6	−1.6	18.6	56.9	16.4	220.7	896.7	2340.9	1282.2
Trucios	29.3	21.8	16.5	21.6	0.9	20.4	51.0	15.6	9.0	0.1	257.5	−16.8
Guriezo	−5.0	16.9	−3.3	19.5	3.6	−0.4	−22.4	9.3	24.2	−17.5	−16.6	38.1

	Ammonia (% increase)				Phosphate (% increase)			
	Feb	May	July	Nov	Feb	May	July	Nov
Villaverde	19.6	208.5	327.6	1237.5	917.0	7777.8	1061.2	541.3
Trucios	23.3	−15.2	−36.7	26.0	23.1	50.9	6.7	9.6
Guriezo	39.1	−35.6	−31.7	35.1	89.0	145.8	45.4	44.1

noon. The data (Table 7.3) show the large impact produced by Villaverde, despite its small size. It mainly affects the nitrite, ammonia and phosphate concentrations. This impact is due to its upstream position, and because the village is adjacent to water of highest quality [40]. It is difficult, however, to assess the impact of other villages on water quality because the main river also receives some tributaries and these can be significant sources of nutrients. The nutrient loads from tributaries change mainly with the flow rate, while urban loads are more constant through the year. To distinguish between the nutrient inputs coming from tributaries and those coming from the sewers, the relationship between water flow increments and nutrient loads has been studied [43]. No significant fit has been found for Guriezo, but for the other two villages it has been shown that the sewage mainly affects the phosphate; nitrate being more affected by the tributaries.

7.8 SYNTHESIS

Water quality in the Agüera Stream changes in a complex way, affected by different sources of variability. With its mild, oceanic climate, seasonality is of minor importance; most of the changes detected through the year are due to variations in the water flow rate. On a shorter time-scale, diurnal variations are greatest in the middle reaches of the stream on receipt of sewage and reflect the metabolism of the fluvial community. This activity can be high through the year, provided that the benthic community can recover from the last flood, which requires a period of hydrological stability. In this case, the stream shows a great self-purifying capacity, and the inputs of nutrients from villages do not seem to adversely affect the aquatic community. There are changes in the trophic structure of the benthic macroinvertebrate community, but these are not reflected either in the taxa richness [42] or in the biotic indices. A good deal of caution is necessary, however, since diurnal variations of pH and oxygen saturation could harm the community, and affect the self-purification capacity of the stream.

Table 7.4 summarizes the most significant observations for the different variables in the Agüera Stream and is probably representative of most other Basque streams. Temperature and pH depend mostly on diurnal and seasonal variations. In contrast to other regions, sampling under different hydrological states gives little information about the pH of Basque streams [44]. Although the necessity of taking diurnal variability of temperature and pH into account has been repeatedly stated [45] this aspect is often neglected; the network of water quality surveillance in the Basque Country relies mostly on low-frequency, fixed-interval sampling [9].

The dissolved oxygen concentration changes seasonally as a result of the effect of water temperature on the gas solubility. The oxygen saturation, however, varies little through the year. Its diurnal

Table 7.4 Most relevant temporal scales of variability in the Agüera Stream

Temporal scale	T	pH	Con	DO	O_2 sat	Alk	Mg	Ca	K	Na	Si	NO_3	NO_2	NH_4	PO_4	SO_4
Diurnal	+	+		+	+						+	+	+	+	+	
Seasonal	+	+		+			+					+				
Other (e.g. discharge)			+				+	+	+	+	+	+	+	+	+	+

variations can be very important, especially in medium and low reaches. Oxygen depletion during floods has not been found, as reported elsewhere [46].

For variables associated with mineralization, the flow regime is the main source of variation, but in a site-dependent way. The magnesium ion is the only cation studied that shows seasonal variability; it is not clear why there is a difference in behaviour between this ion and, for instance, calcium.

Inorganic nutrients are mostly affected by flow rate and diurnal variability. Nitrate is the only nutrient that shows significant seasonality, probably as a result of the differences in soil leaching through the year. When monitoring for water quality standards, samples should be taken when and where peak concentrations are expected [45]. For most nutrients this means sampling mainly during low flows, and downstream from the urban areas, but diurnal variability should also be measured in these situations. When monitoring for total load, it is important to recognize the different discharge–concentration patterns.

These kinds of results are important when designing monitoring guidelines for streams [47], and should be taken into account in sampling protocols. Of course, the relevant scales of observation will change from site to site, making it necessary to carry out similar studies for each geographical area.

REFERENCES

1. Heinonen, P. and Herve, S. (1994) The development of a new water quality classification system for Finland. *Water Science and Technology*, **30**(10), 21–4.
2. Seager, J. (1994) Developments in water quality standards and classification schemes in England and Wales. *Water Science and Technology*, **30**(10), 11–19.
3. Johengen, T.H. and Beeton, A.M. (1992) The effects of temporal and spatial variability on monitoring agricultural non-point source pollution. Nat. Rural Clean Water Symp. Proc., EPA/625/R-92/006.
4. Gore, J.H. and Milner, A.M. (1990) Island biogeographical theory: can it be used to predict lotic recovery rates? *Environ. Manag.*, **14**, 737–53.
5. Minshall, G.W. (1988) Stream ecosystem theory: a global perspective. *J. North Amer. Benthol. Soc.*, **7**, 263–88.
6. Schumm, S.A. (1988) Variability of the fluvial systems in space and time, in *Scales and Global Change* (eds T. Rosswall, R.G. Woodmansee, and P.G. Risser), John Wiley & Sons Ltd, New York, pp 225–50.
7. Boon, P.J. (1992) Channeling scientific information for conservation and management of rivers. *Aquatic Conservation: Marine and freshwater ecosystems* **2**, 115–23.
8. Hunsaker, C.T. and Levine, D.A. (1995) Hierarchical approaches to the study of water quality in rivers. *Bioscience* **45** (3), 193–203.
9. Gobierno Vasco (1995) *Red de vigilancia de la calidad de las aguas y del estado ambiental de los rios de la CAPV*. Servicio Central de Publicaciones del Gobierno Vasco. Vitoria. 178p.
10. Vidal-Abarca, M.R., Suarez, M.L. and Ramirez Diaz, L. (1992) Ecology of Spanish semi-arid streams, in *Limnology in Spain* (eds C. Montes, C. Duarte and J. García-Avilés) Asociación Española de Limnologia, Granada, pp 150–60.
11. Lavandier, P. (1974) Ecologie d'un torrent Pyrénéen de haute montagne. 1. Caractéristiques physiques. *Annals. Limnol.*, **10**, 173–219.
12. Wyer, M.D. and Hill, A.R. (1984) Nitrate transformations in southern Ontario stream sediments. *Wat. Resour. Bull.* **20**, 729–37.
13. Sabater, F. and Armegol, J. (1986) Chemical characterisation of the Ter river. *Limnetica*, **2**, 75–84.
14. Milne, I., Mallett, M.J., Clarke, S.J. *et al.* (1992) Intermittent pollution. Combined sewer overflows,

ecotoxicology and water quality standards. R&D Note 123. National Rivers Authority, Bristol.
15. Pozo, J., Elósegui, A. and Basaguren, A. (1994) Seston transport variability at different spatial and temporal scales in the Agüera watershed (North Spain). *Wat. Res.* **28**, 125–36.
16. Elósegui, A. and Pozo, J. (1992) Physico-chemical characteristics of the Agüera river (Spain) during an unusual hydrological period. *Annals. Limnol.*, **28**, 85–96.
17. Cooper, A.B. (1990) Nitrate depletion in the riparian zone and stream channel of a small headwater catchment. *Hydrobiologia*, **202**, 13–26.
18. Elósegui, A., Arana, X., Basaguren, A. and Pozo, J. (1995) Self-purification processes in a medium sized stream. *Environ. Manag.*, **19**(6), 931–9.
19. Elósegui, A. and Pozo, J. (1994) Spatial versus temporal variability in the physico-chemical characteristics of the Agüera stream (North Spain). *Acta Oecol.* **15** (5), 543–59.
20. Silsbee, D.G. and Larson, G.L. (1982) Water quality of streams in the Great Smoky Mountains National Park. *Hydrobiologia*, **89**, 97–115.
21. McDiffett, W.F., Breidler, A.W., Dominick, T.F. and McCrea, K.D. (1989) Nutrient concentration–stream discharge relationships during storm events in first order streams. *Hydrobiologia* **179**, 97–102.
22. Britton, D.L., Day, J.A. and Henshall-Howard, M.P. (1993) Hydrochemical response during storm events in a South African mountain catchment: the influence of antecedent conditions. *Hydrobiologia*, **250**, 143–57.
23. Reinelt, L.E. and Grimvall, A. (1992) Estimation of non-point source loadings with data obtained from limited sampling programs. *Environ. Monit. Assessment*, **21**, 173.
24. Mulholland, P.J. (1993) Hydrometric and stream chemistry evidence of three storm flowpaths in Walker Branch Watershed. *J. of Hydrol.* **151**, 291–316.
25. Lawrence, G.B. and Driscoll, C.T. (1990) Longitudinal patterns of concentration–discharge relationships in stream water draining from the Hubbard Brook experimental forest, New Hampshire. *J. of Hydrol.*, **116**, 147–65.
26. Cosser, P.R. (1989) Water quality, sediments and the macroinvertebrate community of residential canal estates in South-East Queensland, Australia: a multivariant analysis. *Wat. Res.*, **23**, 1087–97.
27. Prairie, Y.T. and Kalff, J. (1988) Particulate phosphorus dynamics in headwater streams. *Can. J. Fish. Aquat. Sci.* **45**, 210–15.
28. Webb, M.D. and Walling, D.E. (1985) Nitrate behaviour in streamflow from a grassland catchment in Devon, UK. *Wat. Res.* **19**, 1005–16.
29. Orive, E. and Basaguren, A. (1989) Cambios espacio temporales en la composición química de los ríos Cadagua, Butron, Oka, Lea y Artibai. *Ecologia*, **3**, 63–74.
30. Docampo, L. and G. de Bikuña, B. (1991) Analysis of the physico-chemical variables of the stream water of Vizcaya (Basque Country): mathematical model of conductivity. *Archiv. fur Hydrobiologie*, **28**, 351–72.
31. Lenat, D.R. and Crawford, J.K. (1994) Effects of land use on water quality and aquatic biota of three North Carolina piedmont streams. *Hydrobiologia*, **294**, 185–99.
32. Benzie, J.A.H., Pugh, K.B. and Davidson, M.B. (1991) The rivers of North East Scotland: physico-chemical characteristics. *Hydrobiologia*, **218**, 93–106.
33. Owens, L.B., Edwards, W.M. and Van Keuren, R.W. (1991) Baseflow and stormflow transport of nutrients from mixed agricultural watersheds. *J. Environ. Qual.* **20**, 407–14.
34. Correll, D.L., Jordan, T.E. and Weller, D.E. (1992) Nutrient flux in a landscape: effects of coastal land use and terrestrial community mosaic on nutrient transport to coastal waters. *Estuaries*, **15** (4), 431–42.
35. Hall, C.A.S. (1972) Migration and metabolism in a temperate stream ecosystem. *Ecology*, **53**, 585–604.
36. Sedell, J.R., Reeves, G.H., Hauer, F.R. *et al.* (1990) Role of refugia in recovery of disturbance: modern fragmented and disconnected river systems. *Environ. Manag.* **14**, 711–24.
37. Rebsdorf, A., Thyssen, N. and Erlandsen, M. (1991) Regional and temporal variation in pH, alkalinity and carbon dioxide in a Danish stream, related to soil type and land use. *Freshwat. Biol.* **25**, 419–35.
38. Greb, S.R. and Graczyck, D.J. (1993) Dissolved oxygen characteristics of a Wisconsin stream during summer runoff. *J. Environ. Qual.* **20**, 445–51.
39. Elósegui, A. and Pozo J. Variaciones nictemerales de las características físico-químicas de un rio cantábrico. *Limnetica* (in press).
40. González, E., Elósegui, A. and Pozo, J. (1993) Influencia de los núcleos urbanos en la variabilidad físico-química del rio Agüera. *Kobie* **21**, 5–15.
41. Pozo, J., Elósegui, A. and Basaguren, A. (1994) Aproximación sistémica al análisis de la cuenca del rio Agüera. *Limnetica*, **10** (1), 83–91.
42. Riaño, P. Basaguren, A. and Pozo, J. (1993) Variaciones espaciales en las comunidades de macroinvertebrados del rio Agüera (Pais Vasco-Cantabria) en dos epócas condiferentes condiciones de régimen hidrológico. *Limnetica*, **9**, 19–28.
43. Gonzalez, E., Elósegui, A. and Pozo, J. (1994) Changes in the physico-chemical characteristics of the Agüera streamwater associated with human set-

tlements. *Verh. Internat. Verein. Theor. Limnol.* **25**, 1733–8.
44. Burkholder, J.M. and Sheath, R.G. (1985) Characteristics of soft water streams in Rhode Island. 1. A comparative analysis of physical and chemical variables. *Hydrobiologia*, **202**, 143–57.
45. MacDonald, L.H., Smart, A.W. and Wissmar, R.C. (1991) *Monitoring Guidelines to Evaluate Effects of Forestry Activities on Streams in the Pacific Northwest and Alaska*. EPA CSS, Seattle.
46. Graczyk, D.J. and Sonzogni, W.C. (1991) Reduction of dissolved oxygen concentration in Wisconsin stream during summer runoff. *J. Environ. Qual.* **20**, 445–51.
47. Line, D.E. Arnold, J.A., Osmond, D.L. *et al.* (1993) *Water Environ. Res.* **65** (4), 558–71.

SOURCES OF POLLUTANTS AND CURRENT STATUS OF WATER QUALITY IN THE LOWER REACHES OF THE NEMAN AND THE PREGEL RIVERS

by E.V. Krasnov and D.M. Zaporozhsky

8.1 SOURCES OF AIR POLLUTION AND ACID RAIN

There are more than 7000 discharges in the Kaliningrad region near the mouths of the Neman and the Pregel rivers, including their estuaries. Only 60% of these are treated. In Kaliningrad, Sovetsk, Neman and other cities concentrations of dust, gaseous ammonia, nitrogen oxides, carbon dioxide, hydrogen sulphide, carbon sulphide and sulphates are measured at special monitoring stations. Harmful emissions to the air are composed mainly of sulphur dioxide (50%), dust and carbon dioxide (20%), nitrogen oxides (4%), phenols, hydrogen sulphide and hydrocarbons (1% in total). Air pollution is greatly influenced by atmospheric transport of aerosols from western European countries. Acid rains can be mentioned here as a source. No systematic data are presented concerning this phenomenon. Annual harmful emissions from land-based sources account for 100 t (including 20 t of solid substances). Pulp and paper mills, power and heating plants and heating and boiler houses are mostly responsible for the airborne pollution.

8.2 SOURCES OF RIVER POLLUTANTS

The Neman and Pregel rivers catchments are of great importance for the region. The majority of the population have settled in this area and, as a result, elevated concentration levels of ammonia and nitrates are observed at all water quality monitoring stations. At some sites bacteriological pollution exceeds the maximum permissible level. Annually up to 12 t of suspended matter, 11 t of acid and 35 t of tannins are discharged into the mouth of the Pregel River.

The rivers are polluted by petroleum hydrocarbons, phenols, fats and heavy metals. Organic compounds and pesticides in the water originate from cattle-breeding farms and fields. Annual wastewater discharge is estimated at 250 million cubic metres including 40 million cubic metres of untreated water. The quality of the Pregel River is the most threatened by pollution. Near Kaliningrad, in the period from June to August, hydrogen sulphide pollution exceeds the maximum permissible level by 12–15 times; more than 70 000 people are exposed to dangerous levels of hydrogen sulphide contamination.

The Neman River is less polluted due to its greater flow. However, all watercourses in the Kaliningrad region have high concentrations of mineral and organic matter which cause excessive algal growth and reduced oxygen levels.

The most typical contaminants of surface and groundwaters are nitrites, nitrates and pesticides. High concentrations of iron, chlorides and sul-

Table 8.1 Results of water quality sampling and analysis at Niva, 20–26 June 1991 (Source: Nordconsult International A.S. and Jaakko Poyry joint venture. Pre-feasibility study of the Kaliningrad region and the Pregel River basin, 1992)

	pH	Sodium	Chloride	Sulphate	Ammonium	Total nitrogen	Total phosphorus	Total organic carbon	Aluminium	Copper	Zinc	Cadmium	Chromium	Nickel	Lead	Iron	Manganese	Mercury
		(mg/l)	(mg/l)	(mg/l)	(mg/l)	(g/l)	(mg/l)	(mg/l)	µg/l	µg/l	µg/l	µg/l	µg/l	µg/l	µg/l	µg/l	µg/l	µg/l
Lake Vistytis	8.55	2.98	6	13	18	0.5	18	6.45	10	0.8	10	0.1	0.5	5	0.5	100	20.7	–
Instruch	7.99	25.3	22	31	280	2.8	229	15.10	16	1.6	10	0.1	0.5	5	0.8	410	48.4	–
Angrapa	8.24	9.5	13	19	45	1.6	262	9.93	10	1.5	10	0.1	0.5	5	0.8	260	170	–
Pregel downstream Chernyakhovsk	7.98	18.2	23	25	590	2.4	406	10.14	10	4.9	10	0.1	0.6	5	0.8	250	190	2.5
Lava	8.20	13.8	20	33	49	2.6	282	9.64	10	1.1	10	0.1	0.5	5	0.7	200	49.9	2.5
Pregel upstream Kaliningrad	7.85	16.5	23	31	320	2	264	10.33	10	1.6	10	0.1	0.5	5	1.4	330	200	2.5
Pregel downstream Kaliningrad	7.32	180	300	100	320	1.7	358	36.3	73	1.1	10	0.1	0.5	5	2.0	400	260	2.5
Neman upstream Neman city	8.58	10.6	15	21	111	1.5	301	8.84	32	2	10	0.1	1.0	5	0.7	230	120	2.5
Neman downstream Neman city	8.59	10.9	15	18	102	1.4	417	9.12	19	2	10	0.1	0.5	5	0.9	340	45.7	3
Neman downstream Sovetsk	8.34	11.4	16	20	170	1.4	124	9.84	27	1.9	10	0.1	0.9	5	0.8	240	130	3

Table 8.2 Nutrient budget for the input of the Vistula Lagoon and Kurshian Lagoon from sources in the Kaliningrad region and Polish part of the Pregel River catchment area (Source: Krasnov, Proceedings of 3rd International Baltic Ecological Forum; 1991)

Source	Phosphorus (t per year)	Nitrogen (t per year)	BOD (t per year)
VISTULA LAGOON			
Sewage	460	2 200	11 000
Agriculture	970	14 580	not applicable
Industries	50	2 000	25 000
Service	70	500	4 000
Tributaries from Poland	160	1 400	not applicable
Total load	1710	20 680	55 000
KURSHIAN LAGOON			
Sewage	110	500	3 000
Agriculture	710	8 200	not applicable
Industries	30	7 000	41 000
Service	20	75	1 000
Total load	870	15 780	45 000
Total load from Kaliningrad and Pregel River catchment into the lagoons	2580	36 460	85 000

phates are found in the coastal areas. Total nitrogen, from organic and mineral fertilizers, enters the river catchment areas via groundwater and surface water flows (5% and 15% respectively). In the estuaries, oxygen deficiency and pH values in the range of 8.2 to 9.1 can be observed. In the mouths of the Pregel and Neman rivers pH decreases to about 7.4 (Table 8.1). Hydrogen sulphide has been found in these areas (up to 0.8 mg/l) and dissolved oxygen is absent. Table 8.2 provides information on the loadings of nutrients – nitrogen and phosphorus – and BOD_5 from various sources.

High concentrations of organic phosphorus (0.15–0.28 mg/l) have been found in the estuaries. Recent nitrite and nitrate concentrations have been at normal levels, but increasing ammonia nitrogen concentrations have occurred in the Vistula Lagoon near Pregel mouth (on average up to 0.47 mg/l). Detergent pollution is insignificant.

No mercury has been detected in the Vistula Lagoon water in the last two–three years, although the average mercury concentration in the sediments amounted to 0.4 µg/kg.

In the Kurshian Lagoon near the mouth of the Neman the average mercury concentration in the surface water is 0.09 µg/l. In summer there is an increase of blue–green algae (eight–ten times) and algal blooms are very frequent. During recent years, catches of bream, eel, perch, pike and some other species have decreased.

With the recent reduction in industrial production (for example, cellulose–paper production has decreased by at least 50% between 1987 and 1993 the discharge of wastewater decreased by 49%; BOD_5 by 45%; and suspended matter by 70% (Table 8.3).

8.3 BIOLOGICAL INDICATORS

Negative morphological, physiological and genetic changes to the aquatic organisms are evident. The fish in the rivers are mostly of small size and different skeletal deformities are found. Fish are also subject to parasite infestations.

Elevated bacterial contamination of faecal origin results in the temporary closure of public beaches, even in recreational areas. Toxicological studies of aquatic organisms have been carried out. For example, bivalve mussels have elevated concentrations of cadmium and lead in their flesh and skeletal parts [1].

Table 8.3 Dynamics of main pollutants' release with wastewater from cellulose–paper combine in Kaliningrad town (Source: Nordconsult International A.S. and Jaakko Poyry joint venture. Pre-feasibility study of the Kaliningrad region and the Pregel River basin, 1992)

Years	Wastewater release $10^3\ m^3$	BOD_5 (t)	Suspended mattter (t)	Ammonia (t)	Nitrate (t)
1987	26 911	11 989	3070	244	49
1988	25 671	11 966	3033	233	50
1989	25 530	11 921	3023	242	49
1990	23 088	10 879	2547	224	44
1993	13 662	6 549	927	126	24

Table 8.4 Changes in the number of planktonic algae in the Kurshian Lagoon (Source: NIRO, Kaliningrad, 1990)

Species	Number of cells per month in summer ($\times 10^9$)			
	1940	1960	1980	1990
Diatoms				
Melosira	0.2	0.2	0	0
Fragillaria	0.2	0	0	0
Cyclotalla	0.5	0.1	0.1	0
	0	0.3	0	0
Blue-greens:				
Aphanisomenon	0	0.4	30	24
Anabaena	0	0.3	4.8	0
Microcystis	0.2	0.1	4	0
Gomnosphaeria	0	0.1	3	0
Lyngbya	0.2	0.1	0.2	0
Greens:				
Scenedesmus	0.5	0.2	0	0
Ankistrodesmus	0	0	0.3	0

New results show significant changes in the benthic and planktonic communities; there are decreases in the diversity of species, the amount of biomass and the density of the populations. This is due to the deterioration in the conditions of the aquatic environment (Table 8.4).

In order to study the pathological and other responses to the deteriorated water quality, it is necessary to carry out long-term studies on the effects of pollutants on different organisms, from unicellular algae and bacteria to humans at different stages of their life cycles – from larvae to adult forms.

REFERENCES

1. Krasnov, E.V. (1991) Natural water quality and adaptive responses of organisms in Proceeding of the 3rd Intern. Baltic Ecological Forum, Gdansk, pp. 59–64.

ACID MINE-WATER POLLUTION FROM AN ABANDONED MINE

by D. Hammerton OBE

9.1 INTRODUCTION

It has been recognized since the Middle Ages that the extraction of minerals from the Earth can cause serious environmental damage. The first textbook on mining published in 1556 [1] described the impact of mining in Germany in the following vivid terms:

> ... the strongest argument of the detractors is that the fields are devastated by mining operations ... the woods and groves are cut down, for there is need for an endless amount of wood for timbers, machines and the smelting of metals. And when the woods and groves are felled, there are exterminated the beasts and birds. Further, when the ores are washed, the water which has been used poisons the brooks and streams, and either destroys the fish or drives them away ... Thus it is said, it is clear to all that there is greater detriment from mining than the value of the metals which the mining produces.

In the last 100 years the mining industry, particularly coal mining on account of its scale, has produced some of the most severe and intractable water pollution problems experienced in the UK. Moreover, the wholesale closure of coal mines in the last three decades has not resolved all the problems; in fact the very act of closing a mine and removing the pumps, which used to keep the underground galleries dry, has led, in some instances, to more serious pollution than when the mine was in production.

In the central lowlands of Scotland (Figure 9.1) a survey published in 1983 [2] revealed that there were 62 significant discharges of ferruginous waters of which eight were from active mines, two from opencast workings, 25 from abandoned mines, 17 indirectly associated with mining, seven from colliery waste heaps and three from rubbish tips. The scale of the problem can be seen from the fact that daily flows ranged from 10 m^3 to 31 400 m^3 while the iron content ranged from 2 to 500 mg/l. The total flow was 230 000 m^3 per day containing 6139 kg iron (equivalent to 2200 t per annum). At that time it was estimated that 50–60 km of rivers were moderately or severely affected. While the worst pollution was associated with recent closures, problems were recorded for 15 discharges from mines which had been abandoned 80 to 100 years previously! Today only two deep mines remain in production in Scotland and further problems have occurred. The latest survey published in 1994 [3] suggests that there are 176 discharges from abandoned mines which currently pollute 134 km of watercourses.

The River Purification Boards are concerned that, under present legislation, they have no powers to control discharges from abandoned mines – indeed Section 31(2) of the Control of Pollution Act 1974, Part II, seems specifically to protect the act of permitting water from an abandoned mine to enter controlled waters. The prosecution referred to

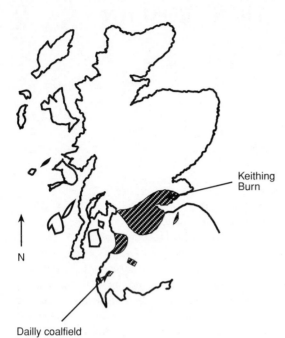

Figure 9.1 Scotland's main coal-working areas. (Redrawn from Carter and Reid, *Pollution Problems with Minewaters in Scotland*; 1988.)

in this chapter was taken under former legislation which was repealed when the Control of Pollution Act was implemented.

9.2 FORMATION AND IMPACT OF ACID, FERRUGINOUS MINE-WATERS

The main cause of acid, ferruginous discharges from mineworkings is the presence of pyrite (FeS) in the coal-bearing strata. It is especially common within the coal seams and within the overlying marine mudstone strata. The three conditions essential to the process are [4]:

1. the presence of iron pyrite (FeS) exposed to a supply of oxygen and water;
2. a sufficient flow of water to elute the oxidation products;
3. insufficient alkaline material in the surrounding strata to neutralize the acidity produced.

It would seem that the network of underground mineworkings and abandoned galleries at Dalquharran colliery, Dailly, Ayrshire (Figures 9.1 and 9.2) provided the ideal conditions for the chemical and biochemical processes which resulted in the formation of ferrous sulphate, ferric sulphate, sulphuric acid and hydrated ferric oxide. The main steps in the complex series of interactions are as follows [5].

1. $2FeS_2 + 7O_2 + 2H_2O \rightarrow 2(Fe^{2+} SO_4^{2-}) + 2(H_2^+ SO_4^{2-})$

2. $2(Fe^{2+} SO_4^{2-}) + H_2^+ SO_4^{2-} + \frac{1}{2} O_2 \rightarrow Fe_2^{2+} (SO_4^{2-})_3 + H_2O$

3. $Fe_2^{3+} (SO_4^{2-})_3 + 6H_2O \rightarrow 2Fe(OH)_3 + 3(H_2^+ SO_4^{2-})$

These reactions can be rewritten:

$2FeS_2 + 7\frac{1}{2}O_2 + 7H_2O \rightarrow 2Fe(OH)_3 + 4(H_2^+ SO_4^{2-})$

In many mines, and certainly at Dalquharran, the presence of the acidophilic iron-oxidizing bacterium, *Thiobacillus ferro-oxidans*, acts as a catalyst which speeds up the oxidation by five or six orders of magnitude over the abiotic rate [6].

In most mines the pumping out of water from the lowest levels keeps the strata dry so that oxidation cannot take place. However, at Dalquharran the problems of acid iron-bearing discharges continued to increase until the final closure and flooding, which produced the disastrous outbreak in 1979.

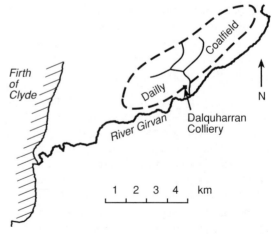

Figure 9.2 Location of Dalquharran Colliery. (Redrawn from Carter and Reid, *Pollution Problems with Minewaters in Scotland*; 1988.)

The mine-water discharge at Dalquharran, both during the working life of the mine and after its subsequent break-out following closure, was always crystal clear as it emerged; when it reached the receiving watercourse it became cloudy and bright-orange in colour. What was happening here was that mixing with the river water produced a rapid rise in pH (in this case from 4 to 7 or above) and the presence of dissolved oxygen brought about hydrolysis, with the production of the typical orange-yellow precipitate of ferric hydroxide ($Fe(OH)_3$), with variable amounts of goethite ($FeO(OH)$) and jarosite ($KFe(SO_4)_2(OH)_4$).

Biochemical processes are often involved, and the sheath bacterium *Leptothrix ochraceae*, a growth form of the commonly known sewage fungus *Sphearotilus natans*, has often been found in rivers suffering from iron pollution. *Leptothrix* obtains energy from the synthesis of organic matter from carbon dioxide and dissolved salts by the biochemical oxidation of ferrous bicarbonate and ferrous sulphate to ferric hydroxide. The ferric hydroxide is formed within the gelatinous coat of the bacterium and in this way a thick coating is built up over the stones or whatever else the bacterium happens to be growing on [6].

9.3 HISTORY OF DALQUHARRAN COLLIERY

Coal had been worked in the Dailly area on the north side of the River Girvan since at least the 14th century. As successively deeper seams of coal were extracted the land underlying Quarrelhill became a veritable honeycomb of abandoned galleries and inter-connecting underground roads. There was a long history of underground fires which is reflected in the local name of Burning Hills.

Dalquharran Colliery was established by the National Coal Board (NCB) at about the same time as the creation of the Ayrshire and Clyde River Purification Boards (in 1956). The entrance was by means of an inclined shaft leading into the hillside from the river valley and coal was extracted from seams at six different depths (Figure 9.3).

In May 1970 the Ayrshire River Purification Board (ARPB) issued a consent to the NCB for the discharge of pumped mine-water which imposed the following conditions [7].

1. (1) Flow not to exceed 350 000 gallons (1590 m^3) per day;
2. The pH to be within range 5–9;
3. Suspended solids max. 60 mg/l when neutralized to pH 7–8;
4. Discharge to contain no substance in such concentration as to cause the receiving stream to become injurious to aquatic fauna and/or flora.

From this time onwards samples of the discharge taken by the ARPB showed a stead increase in pollution. A series of letters were sent in the period 1970–3 asking for remedial measures but to no avail. In October 1974 the ARPB warned the NCB of possible prosecution. A meeting was held with the NCB in November 1974 at which it was pointed out that samples over a five-year period had not only failed to meet the required conditions but had become steadily worse. However, as the NCB still refused to take any steps to comply, a further formal sample was taken and the results reported to the procurator fiscal (public prosecutor) at Ayr sheriff court. At the subsequent court hearing in January 1975 the NCB pleaded guilty and were fined £20 (maximum possible fine £100!). Following reports of further unsatisfactory samples, the NCB was informed that, in the event of another prosecution, the mine would be closed making some 200 miners redundant.

On May 1975 the ARPB was amalgamated with the Clyde River Purification Board (CRPB). Further correspondence between the CRPB and the NCB from May to October 1975 established that,

1. the NCB took the view that there was no practicable method at reasonable cost to resolve the problem; and
2. they, in any case, intended to close the mine by May 1977 when all the machinery, including the pumps, would be removed.

The CRPB recognized that further action through the courts was not viable in view of the imminent closure of the mine and therefore it was important to take action to protect the river from pollution following abandonment of the Dalquharran

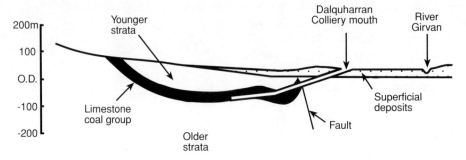

Figure 9.3 Geological section through the Dailly Coalfield. (Redrawn from Carter and Reid, *Pollution Problems with Minewaters in Scotland*.)

Colliery. However, the NCB took the view that the cessation of pumping would resolve the problem completely, and that, taking into account flow rates and the directional flow of water within the mines, no water would escape from the mine mouth.

Based on their own hydrogeological reports the CRPB were convinced that, eventually, following rebound of the water table, a discharge would take place. With this in mind, the CRPB applied to Ayr sheriff court for an order requiring that any discharge of water from the land occupied by the NCB at Dalquharran should not, when neutralized to pH 7–8, contain more than 60 mg/l of suspended solids. The sheriff's order was unopposed by the NCB who stated that they did not expect any pollution after closure. The order was granted in January 1976 but was not to take effect until 1 June 1977 (i.e. after the date by which the NCB had stated that the mine would close).

9.4 THE DALQUHARRAN DISASTER

The colliery was finally closed in May 1977 when the pumps were removed and the discharge of minewater ceased. At about this time the CRPB's installed equipment in an adjacent shaft which allowed them to monitor the water level in the mine as it began to rise. About 15 months later, in September 1979, the CRPB's hydrologist announced that he expected water to break out through the mine mouth on about 1 December.

However, due to heavy rainfall, the expected discharge took place over a month earlier on 21 October in spectacular fashion and was far worse than expected. The initial flow, at a rate of 2270 m^3 per day, contained amounts of iron and aluminium which were individually lethal to most of the stream flora and fauna while the acidity of the discharge was also sufficient to kill all the fish (Table 9.1). In fact, overnight, all fish in the river 16 km downstream were killed, along with most of the other aquatic life.

The CRPB reacted immediately by taking formal samples, the results of which were sent with a report to the procurator fiscal showing that the terms of the sheriff's order had been severely infringed. A letter was sent to the NCB requesting immediate action to control the discharge. At the same time the CRPB had a meeting with the Scottish Development Agency (SDA) asking for assistance in preparing an engineering report on the best methods of solving the problem.

Table 9.1 Mean annual results for significant parameters in the Dalquharran discharge

Parameter	Year			
	1979	1980	1981	1994
pH	4	4.3	5.3	6
Acidity (mg/l)	–	1766	1081	–
Conductivity (μS/cm)	4259	4583	3506	3477
Sulphate (mg/l)	5491	5145	3405	1660
Iron (mg/l)	1093	1044	598	193
Aluminium (mg/l)	92	32	5.6	4
Manganese (mg/l)	50	45	26	–

Within a week the NCB replied to say that, according to their legal advisers, they had not only closed the mine over two years previously but had terminated the lease of the land and therefore could not be held responsible for the discharge or its effects. Meanwhile the SDA, within 24 hours, appointed a firm of consultants, Messrs Babtie, Shaw and Morton, with instructions to investigate the outbreak and identify possible remedial measures.

In July 1980 the case came before Ayr sheriff court. To the great disappointment of the CRPB, the sheriff found the NCB not guilty. However, the Crown appealed against this verdict to the justiciary appeal court in Edinburgh. In February 1981, the court made legal history by reversing the earlier decision and directed the sheriff to find the NCB guilty. This case was the first time that a conviction had been brought for polluting a watercourse by means of a discharge from an abandoned mine.

There can be no doubt that this verdict was of crucial importance and led to agreement by the NCB to carry out remedial measures to meet the CRPB's requirements. The NCB agreed to the CRPB's request to set up a joint working party to ensure the closest collaboration between the two parties. By this time the report from Babtie, Shaw and Morton was available and made a useful starting point for the NCB's approach. This report suggested two alternative solutions:

1. Chemical treatment by means of an aeration cascade, lime dosing and large settlement lagoons together with a pipe from Dalquharran to the treatment works. Capital costs (at 1980 prices) were calculated at £375 000 and annual costs at £65 000. A notable problem with this proposal was the disposal of 30 t per day of sludge.
2. A sea outfall to the Firth of Clyde which required inlet works at Dalquharran, a pumping main, header tank and gravity main to the coast and an outfall 1 km in length. At 1980 prices this worked out at £1.75 million in capital expenditure with annual costs of £50 000.

The NCB considered that both proposals were extremely expensive and, because of the need for planning permission, could incur considerable delays. They believed that their own scientists could produce less expensive but satisfactory solutions within 12 months. The CRPB stated that their main requirement was that the total iron concentration in the River Girvan should not at any time exceed 1 mg/l including the iron already in solution upstream. They also insisted that the NCB should produce their proposals within six months.

In October 1981 the NCB presented their proposals which comprised three main elements:

1. the sealing of all points of ingress of water to the mine including, where necessary, the diversion of streams.
2. the injection of alkali (lime and sodium hydroxide) through a separate shaft to precipitate iron deep within the mine workings.
3. construction of a dam at the mine mouth with controls to regulate the volume of discharge.

A subsequent and valuable modification was proposed by the NCB. Their studies showed that the most severely polluted water was in the deepest part of the mine and water in the uppermost gallery close to the surface on the hillside was almost unpolluted (pH above 6 and iron below 20 mg/l). It was proposed to drill into the gallery and divert large quantities of this clean water into a tributary of the Girvan. The feasibility of this procedure was proved by pumping trials before drilling through solid rock to establish a permanent gravity drain.

These works were completed by the autumn of 1983 and were so successful that the River Girvan met the CRPB's requirements throughout 1984 by which time the salmon fishery was well on the way to full recovery, aided by a donation from the NCB of £1000 to the Girvan Fishery Board for purchase of salmon fry.

The novel feature of the NCB's remedial measures was the attempt to reduce the iron content underground by injection of lime. However, although hundreds of tonnes were put down the shaft, there was no evidence from the CRPB analyses that the lime reached the main flow of polluted minewater since there was no effect on the pH or iron concentration in the discharge. The success of the scheme was due:

1. to the reduction in inflow;
2. to the gravity-flow scheme which removed over 90% of the inflow in the form of clean water; and
3. the control dam at the mine mouth which allowed NCB (later British Coal) to regulate the outflow in accordance with the flow in the River Girvan.

However, because staff were not on site, the control system had obvious weaknesses and, in 1988, a more sophisticated system was introduced in which automatic valves at the mine mouth were actuated by sensors in the main river. This system had teething troubles in that the first surge of minewater (e.g. following heavy rain) caused a short-term visible plume of iron hydroxide, which produced a spate of complaints from the public. As a result, a new valve system was installed in 1993 which operates in a series of steps in accordance with a new form of consent condition (Appendix).

The introduction of the fully automated system, while producing obvious benefits to the river, introduced another problem; the rebound time of the minewater appeared to have been reduced, which, in turn, caused an overflow of about 5 l/sec at the mine mouth during low river flow conditions. Although it had no significant effect on water quality, when fully mixed with the River Girvan, it was a breach of the CRPB consent conditions and created a visible localized deposit of iron oxide. The CRPB took a case against British Coal. In pleading guilty they informed the court of their intention to further improve effluent quality by means of a reed bed treatment plant. The sheriff took these proposals into account and imposed a fine of £250.

The reed bed treatment plant was completed in June 1995 but weather conditions have been such that reed growth has been delayed and it is too early yet to comment on the efficacy of this form of treatment. However, reed beds have been used in Scotland at a site in Lothian region and early results have been encouraging. The reeds used at Dalquharran comprise *Phragmites* (50%), *Typha* (25%) and a mixture of *Iris*, *Phalaris* and *Eleocharis* (25%).

9.5 IMPACT ON THE RIVER GIRVAN

There is no doubt that this pollution incident was probably the most severe yet experienced in Scotland. It caused severe damage to one of the best salmon rivers in Ayrshire over a period of four years which not only affected the fishing community but also gave rise to many complaints from the general public because of the severe discoloration of the river. Its effects were very noticeable in the harbour at Girvan 16 km downstream and at times it caused discoloration of the sea within a radius of at least 1 km. More seriously it had a damaging impact on a large alginate factory which employed some 500 staff living in and around Girvan. The firm was forced to find an alternative source for 9000 m^3/day of freshwater previously abstracted from the River Girvan. On some occasions the firm also had to close its sea water intake because of the high iron content.

However, it is pleasing to record that after the main remedial measures were completed in 1983 the river recovered remarkably quickly and has been in essentially excellent condition in the 12 ensuing years.

9.6 FUTURE OUTLOOK FOR CONTROL OF DISCHARGES FROM ABANDONED MINES

Unfortunately the position today has hardly changed since 1986 when the author stated [6]: 'the River Purification Boards are now very well placed in terms of their legal powers to ensure that never again will mining be allowed to damage the aquatic environment . . . [but] the only area of serious concern is that of pollution from abandoned mines.'

Under the Control of Pollution Act the regulatory authorities are empowered to carry out remedial works themselves. However, they have restricted budgets and the Act does not allow the cost of remedial works to be recovered from the polluters. It is, therefore, extremely unlikely that discharges from abandoned mines will be treated in the foreseeable future unless the Government introduces the necessary legislation. The only satisfactory solution, in the author's opinion, is for the Government to implement a recommendation of the Commission on Energy and the Environment [8]: 'discharges from mines abandoned by the predecessors of the NCB or even the NCB some long time ago, should be regarded as a form of dereliction comparable to abandoned pit-heaps . . . provision should be made through central government to meet the costs of any essential work, as is the case with derelict land schemes.'

APPENDIX

Summary of consent conditions on the discharge from the former Dalquharran Colliery to the Quarrelhill Burn

1. The outlet shall be a 0.45 m × 0.2 m rectangular outfall discharging at National Grid Reference NS 2668 0180 to the Quarrelhill Burn
2. (a) The discharge shall consist of mine-water either separately or in combination with surface water.
   (b) The discharge shall have a pH value between 5 and 9
   (c) The discharge shall not contain more than 350 mg/l of total iron
3. The volume of the discharge shall be related to the flow in the Girvan as measured at Dailly Road Bridge, in accordance with the following requirements, unless otherwise agreed to by the CRPB:
   (a) When the flow in the river is at or greater than 3.17 m^3/s and less than 5.2 m^3/s the maximum rate of discharge shall be 5 l/s.
   (b) When the flow in the river is between 5.2 and 9.09 m^3/s the maximum rate of discharge shall be 17 l/s.
   (c) When the flow in the river is between 9.09 and 15.78 m^3/s the maximum rate of discharge shall be 50 l/s.
   (d) When the flow in the river is greater than 15.78 m^3/s the maximum rate of discharge shall be 76 l/s.
4. British Coal Opencast shall provide and maintain equipment for the continuous measurement of the daily rate and volume of the discharge and accurate records shall be kept of such data. The flow data relating to the discharge shall be made available to the CRPB annually or at any other time following a request in writing.

REFERENCES

1. Agricola, Georgius (1556) *Re de Metallica* translated by H.C. Hoover and L.H. Hoover (1950), Dover Publications, New York.
2. Scottish Development Department (1983) Water Pollution Control in Scotland: Recent Developments.
3. Ross, S.L. (1995) Minewater Problems Experienced by the Clyde River Purification Board Conference on The Environmental Effects of Mining Wastes, IBC.
4. Lothians River Purification Board: Annual Report 1964.
5. Carter, P.G. and Reid, J.M. Pollution problems with minewaters in Scotland. Proc. 2nd Conf. on Construction in Areas of Abandoned Mineworkings, Edinburgh, 1988.
6. Hammerton, D. (1986) Mineral extraction and water quality in Scotland, in *Effects of Land Use on Freshwaters*. WRC and Ellis Horwood, Chichester 1986
7. Loneskie, J (1980) Historical resume and statement of chronological sequence. Internal report, Clyde River Purification Board.
8. Commission on Energy and the Environment (1981) *Coal and the Environment*, Her Majesty's Stationery Office, London.

THE EFFECT OF POLLUTANTS FROM THE VISTULA RIVER ON THE WATER QUALITY OF THE BAY OF GDANSK

M.P. van der Vat and M. Robakiewicz

10.1 INTRODUCTION

A study has been conducted by Delft Hydraulics and IBW-PAN on the water quality of the Bay of Gdansk. The project is entitled 'Vistula River and Gdansk Bay water quality, phase 2' and has been funded by the Programme for Co-operation with Central and Eastern European Countries (PSO) of the Dutch Government and by the Voivodeship of Gdansk. Phase 1 of the project covered the transfer of software and hardware and the set-up of the water quality model. In phase 2 the water quality model has been extended and calibrated and a decision support system (DSS) has been constructed. In phase 3 the model will be extended with the sediment quality and the DSS will be incorporated into an integrated data management and decision support system (IDSS). The project has two objectives:

- incorporation of quantitative tools in policy support in Poland; and
- support of Polish institutes with their development of recent water quality modelling techniques.

The project is carried out in close co-operation with the regional scientific community:

- Centre for Marine Biology, Polish Academy of Sciences, Gdynia (CBM–PAN);
- Institute of Environmental Protection, Gdynia (IOS);
- Institute of Hydroengineering, Polish Academy of Sciences, Gdansk (IBW-PAN);
- Institute of Marine and Tropical Medicine, Gdynia;
- Institute of Meteorology and Water Management, Marine Branch, Gdynia (IMGW);
- Institute of Meteorology and Water Management, Marine Branch, Department of Water Protection, Gdansk-Wrzeszcz (IMGW);
- Marine Fisheries Institute, Gdynia (MIR); and
- University of Gdansk, Institute of Oceanography, Gdynia.

The following authorities have contributed to the project and are now working with the DSS:

- Maritime Office, Gdynia
- Municipality of Gdansk
- Municipality of Gdynia
- Municipality of Sopot
- Sanitary Inspectorate
- Voivodeship of Gdansk; and
- Voivodeship of Gdansk Inspectorate of Environmental Protection.

Water quality modelling as a part of water quality management serves two objectives. Firstly it increases knowledge about the present situation.

International River Water Quality. Edited by Gerry Best, Teresa Bogacka and Elżbieta Niemirycz. Published in 1997 by E & FN Spon, London, ISBN 0419215409

The results of the water quality model for the present situation provide for interpolation and extrapolation from existing data. The calibration of the model, furthermore, provides an insight into the dynamics and causes of the present situation, and the relationship between the actual water quality and the loading of pollutants.

The second objective is to analyse the consequences of possible changes in the input of pollutants to the water system. After calibration of the model, the information on pollutant loading can be changed to represent a possible future situation incorporating autonomous developments as well as management alternatives. The results of the present implementation of the model are estimations of the long-term effects of the changes of the water quality. In this way, the model may be used to assess the long-term development of the water quality under prescribed conditions. The analysis of several different situations is of value for decision-makers in the evaluation of different management alternatives.

The water movement in the Bay of Gdansk has been described by an application of the three-dimensional hydrodynamic model TRISULA [1]. The water quality has been modelled by application of the general purpose Delft water quality model (DELWAQ) [2, 3].

10.2 THE WATER QUALITY MODEL

The water quality model describes the transport of toxic substances and also processes such as eutrophication. Only the general outline and some specific points of the model will be discussed here. A complete description is given in Van der Vat et al.[4].

The calculation of the change of concentrations is based on the advection–diffusion equation:

$$\frac{\delta C}{\delta t} = -V_x \frac{\delta C}{\delta x} + D_x \frac{\delta^2 C}{\delta x^2} - V_y \frac{\delta C}{\delta y} \\ + D_y \frac{\delta^2 C}{\delta y^2} - V_z \frac{\delta C}{\delta z} + D_z \frac{\delta^2 C}{\delta z^2} + L + P \quad (1)$$

with:

C concentration (mg/l)
t time (s)
V_x velocity in x-direction (m/s)
V_y velocity in y-direction (m/s)
V_z velocity in z-direction (m/s)
x distance in x-direction (m)
y distance in y-direction (m)
z distance in z-direction (m)
D_x dispersion coefficient in x-direction (m²/s)
D_y dispersion coefficient in y-direction (m²/s)
D_z dispersion coefficient in z-direction (m²/s)
L loads (mg/l/s)
P processes (mg/l/s).

The velocities are defined by the results of the hydrodynamic model; the vertical dispersion, based on the density gradient, is calculated using Richardson's number; the loads are based on measurements; and the water quality processes are calculated separately. A time step of two days is used with (1) to arrive at an empirical solution.

For application of the model the following data are required:

- water movement;
- pollution loads;
- boundary conditions;
- meteorological conditions, for the calculation of the water quality processes; and
- the actual water quality, for verification of the model results.

Monthly data on the actual water quality have been made available by IMGW (Gdynia) for 15 stations in the Bay of Gdansk. These data have been used for the calibration of the model. In this chapter only the results for station P110 are reported.

1.2.1 TRANSPORT

The water quality has been modelled for the year 1991. For each month a three-dimensional flow field has been calculated with the validated TRISULA application by defining characteristic conditions [5]:

- water level at the boundary with the Baltic Sea;
- vertical temperature distribution at the border with the Baltic Sea;
- vertical salinity distribution at the border with the Baltic Sea;
- velocity and direction of the wind; and
- discharge and temperature of the Vistula.

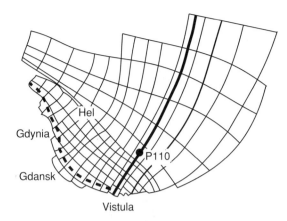

Figure 10.1 The horizontal schematization of Gdansk Bay (solid line: south–west profile; broken line: east–west profile).

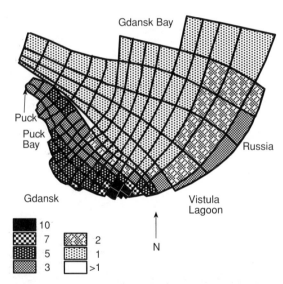

Figure 10.2 The average distribution, both vertically and annually, of water from the Vistula (showing proportions of total water in %).

The hydrodynamic model uses a curvilinear orthogonal horizontal schematization. The water column is divided into 20 layers according to the sigma transformation. The schematization for the water quality model has been generalized, resulting in a horizontal grid of 153 segments for Gdansk Bay (Figure 10.1) and five vertical layers. A dispersion coefficient of 10 m²/s is used in TRISULA.

Figure 10.2 presents the resulting average distribution of water from the Vistula in the bay and provides a good indication of the dilution of the river water: only near the mouth does the contribution of the Vistula to the total amount of water exceed 5%.

It is possible to verify the results of the model with respect to transport by comparing the model results with the measurements of the chloride distribution. Figure 10.3 presents these data for station P110. The model results have been calculated with different values for the horizontal dispersion coefficient (D_x and D_y in (1)): 0, 10, 100 and 200 m²/s. Both the model results and the measurements are averages for the whole water column. It appears that there is no significant difference between the results for the different coefficients. The water movement seems to be dominated by the advective transport as described by the hydrodynamic model and therefore no additional dispersion is used in the water quality model.

10.2.2 LOADS

The types of loads considered in the water quality model are the inflow of water from rivers, streams and point sources (e.g. wastewater treatment plants and factories), the substances contained in the water and atmospheric deposition. The loading of diffuse sources entering the bay directly from the land have been ignored as they are regarded as insignificant.

The major sources of total nitrogen are presented in Figure 10.4. For other substances the division between the different sources is similar, except for atmospheric deposition, which only applies to nitrogen. It is clear that the Vistula River is by far the major source of pollutants in the Bay of Gdansk. The contribution of the other point sources is in the order of 1% of the contribution of the Vistula. These data are comparable to the estimated 52 600 t of nitrogen which entered the Baltic Sea from agricultural sources in Germany in 1989 [6].

The accuracy of the data can be verified by comparing the model results with the measurements for total nitrogen and total phosphorus, which are not highly affected by processes. Figure 10.5 presents the vertically averaged total nitrogen concentration for station P110. This shows, that the data on pollutant loading are quite accurate and that no important sources are neglected or overestimated. The short-term dynamics of the measurements are not covered

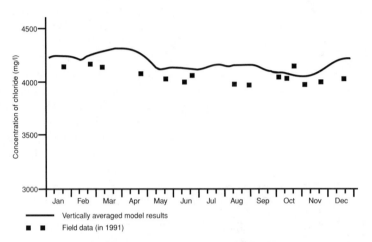

Figure 10.3 Vertically averaged model results and measurements of chloride for station P110.

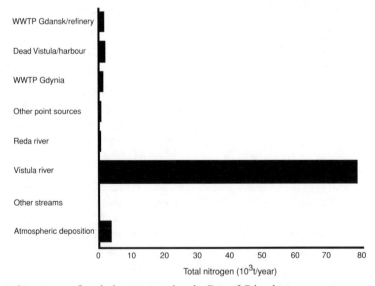

Figure 10.4 The major sources of total nitrogen entering the Bay of Gdansk.

by the model because of the use of monthly rather than daily data. For the purpose of this study long-term developments of water quality, transport and loads are sufficiently accurate.

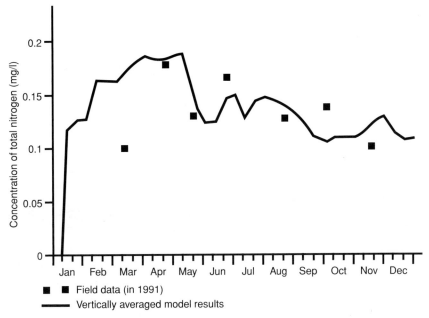

Figure 10.5 Vertically averaged model results and measurements of total nitrogen for station P110.

1.2.3 WATER QUALITY PROCESSES

Figure 10.6 presents the distribution of nutrients into different compartments and the interacting processes. The nutrients carbon, nitrogen, phosphorus and silica were considered. The dissolved inorganic carbon pool was not considered as it was assumed that the availability of carbon does not limit the production of algae in the Bay of Gdansk. For nitrogen, two dissolved inorganic forms are defined – ammonium and nitrate, and the interacting process is nitrification. Nitrogen can also be removed from the system by the process of denitrification. Inorganic phosphates are considered to be partly dissolved in the water and partly adsorbed by suspended matter. The main processes are:

- the production of algae by consumption of nutrients and the associated production of oxygen;
- the mortality of algae, resulting in dead organic material and the autolysis of dissolved inorganic nutrients;
- the mineralization of dead organic material and the associated consumption of oxygen;
- the sedimentation and resuspension of dead organic material, algae and adsorbed phosphates;
- the release of nutrients from the sediment by mineralization of dead organic material and the associated consumption of oxygen; and
- the burial of nutrients from the upper sediment into the lower inactive part of the sediment.

Process coefficients define the transformation rate of mass by the processes. The DELWAQ model provides default values for the different coefficients based on values mentioned in literature and on experience with the application of this water quality model in different circumstances [3]. During the calibration of the model, some parameters were adjusted to the local situation.

Figure 10.7 presents the results for nitrate at station P110. In winter, nitrate is the main pool of nitrogen, while in summer most nitrate is consumed by the algae. The curve reflects the dynamics of algae growth and mortality, and the sedimentation and decay of dead organic matter illustrated in Figure 10.6. From this and other evidence it is concluded that the model describes

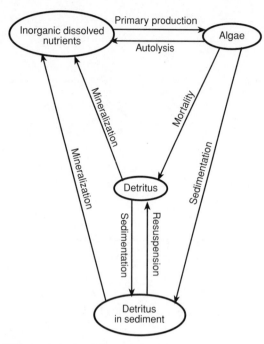

Figure 10.6 The different pools of nutrients and the interacting processes.

these processes in accordance with the available data.

10.3 RESULTS AND CONCLUSIONS

Figure 10.8 presents the calibration results with respect to the average vertical and annual total phosphorus concentrations. Other parameters for which concentrations are calculated include orthophosphate (dissolved as well as particulate), nitrate, ammonium, total nitrogen, silicates, chlorophyll-*a*, biological oxygen demand and dissolved oxygen. Within the decision support system the user can prepare scenarios by varying the input of pollutants, thereby representing the effect of possible developments and management alternatives. It is possible to compare the results of those scenarios with the results of the calibration, which provides a quantitative evaluation of the effect of the scenario on the water quality.

Three scenarios were analysed to evaluate the effect of the Vistula on the water quality of Gdansk Bay:

1. in the year 2000, if no measures are taken. Based on the development of the population and

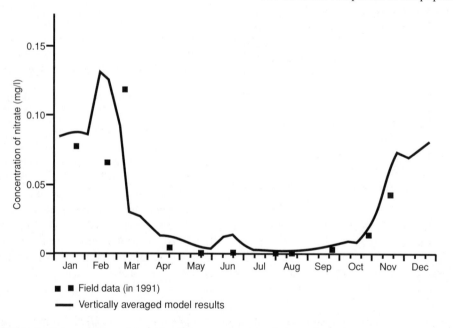

Figure 10.7 Vertically averaged model results and measurements of nitrate for station P110.

Results and conclusions 87

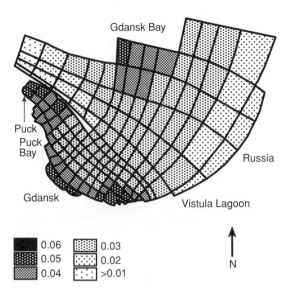

Figure 10.8 Averaged vertical and annual simulated total phosphorus concentrations (showing total phosphate concentrations in mg/l).

Figure 10.9 presents the average vertical and annual total phosphorus for the base case and the three scenarios for the south–north profile shown in Figure 10.1. The mouth of the Vistula is located at 0 km. It appears that the influence of the Vistula, as represented by the different scenarios, does not extend more than approximately 12 km from the mouth. Figure 10.10 presents the same information for the east–west profile shown in (Figure 10.1). Here the influence of the Vistula appears to be significant for the whole 50 km stretch of the coast. Figure 10.11 presents a synoptic view of the relative difference between the situation in 1991 (the base case) and scenario 3. Here again it appears that the influence of the Vistula extends only a limited distance to the north, whilst the pollution in Gdansk Bay from human activities contributes some 10% to 20% of the pollution of the urban coastal zone around the city of Gdansk.

The following conclusions are drawn with respect to the influence of the Vistula River on the water quality of Gdansk Bay based on these results.

- The Vistula River is the major source of pollutants in Gdansk Bay;
- The polluted water of the Vistula is diluted with relatively clean water from the Baltic Sea;
- Dilution limits the significant influence of the Vistula to the north of the mouth to a distance of 10–15 km;
- The dilution is less to the west of the mouth, where the influence of the Vistula River in the coastal zone remains significant as far as Puck Bay, some 50 km away;

the economy, an average increase of pollution of 10–15% is expected;
2. pollutant loads from the Vistula set at 50% of 1991 levels.
3. pollutant loads from the Vistula set at estimated natural background levels for undisturbed rivers [7].

In each scenario only the loads of pollutants from the Vistula were adapted. The other sources were the same as 1991.

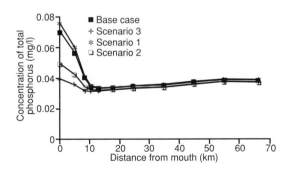

Figure 10.9 South–north profile of the average vertical and annual total phosphorus concentrations.

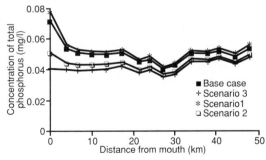

Figure 10.10 East–west profile of the average vertical and annual total phosphorus concentrations.

88 The effect of pollutants on the Bay of Gdansk

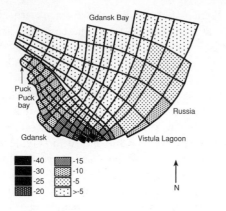

Figure 10.11 Relative difference in annual average total phosphorus concentrations between the actual situation and the situation with the natural background contribution of the Vistula.

- Although the Vistula is the major contributor to the pollution of the Bay of Gdansk, local sources of pollution are of primary importance for Puck Bay and the coastal zone west of the mouth of the Vistula.

REFERENCES

1. Delft Hydraulics, (1993). *TRISULA User's Manual*, Release 2.0.
2. Postma, L. (1984) A two-dimensional water quality model application for Hong Kong coastal waters, *Water Science and Technology*, 16.
3. Delft Hydraulics, (1994). *DELWAQ User's Manual*, Version 4.0.
4. Van der Vat, M.P., Jasińska, E. and Robakiewicz, M. (1994) Gdansk Bay water quality: water quality modelling. Project note prepared for: Rijkswaterstaat, Hoofddirectie. Delft Hydraulics and IBW–PAN, T1029.
5. Robakiewicz, M. (1994) Hydrodynamics of Gdansk Bay by 3D model. Sensitivity study and validation. Project report. Delft Hydraulics, Z672.
6. Werner, W. and Wodsak, H.P. (1994) Germany, the Baltic Sea and its agricultural environmental status, *Mar. Poll. Bull.* **29** 6/12, 471–6.
7. Laane, R.W.P.M. (ed.) (1992). Background concentrations of natural compounds (in rivers, sea water, atmosphere and mussels). DGW Report 92.033.

METAL LOADINGS TO THE BALTIC SEA FROM POLAND'S RIVERS

E. Heybowicz and R. Ceglarski

11.1 INTRODUCTION

The assessment of the total outflow of metals from Poland's river catchments is of great importance for two reasons:

1. it gives an indication of the relative amounts of loadings of the different metals which originate from Polish inland surface waters;
2. it determines the discharge of metals from the territory of Poland into the Baltic Sea.

This study is based on the results of river monitoring of the Vistula, Odra and ten smaller Pomeranian rivers. The drainage areas of these rivers represent 89.7% of Poland's total land area. At the same time, 90.3% of the drainage areas of the rivers within Poland's borders flow into the Baltic Sea. The area constitutes 20.2% of the total area of the Baltic basin. Twelve per cent of the drainage areas of the Vistula and Odra rivers are located outside Poland.

11.2 METHOD

The river monitoring was carried out by the Department of Water Pollution Control of the Institute of Meteorology and Water Management in Gdansk. Data for the Vistula and Odra rivers were for the period 1987–92, whilst for the remaining rivers the data were for the period 1988–92. The data set consisted of about 100 results of measurements made each year for the Vistula and Odra rivers, and about 50 results per year for the remaining rivers. Data on chromium and nickel were based on the results of measurements taken during 1990. The following ten metals were analysed: sodium, potassium, manganese, iron, zinc, copper, lead, chromium, nickel and cadmium. All were analysed by atomic absorption spectrometry (flame and electro-thermal method) either on the sample directly or after mineralization.

11.3 RESULTS

The mean concentrations of individual metals found in the river waters vary by a factor of five (Figures 11.1, 11.2 and 11.3). The biggest ranges of concentrations occur in the Pomeranian rivers and the smallest in the Odra River. The variation of potassium is small whilst the variation of zinc is large. When the results of the mean concentrations of the metals are compared to the standards set for Polish surface waters and drinking water (Table 11.1), the greatest exceedence of the standards is for manganese. Of all the elements measured, manganese is the only one to exceed the permissible limits for water of class I quality. Surface waters that are used as sources of drinking water and in the food industry have to meet the requirements of class I. This suggests that the current standard for manganese is too strict when compared with the natural levels present in Poland's rivers.

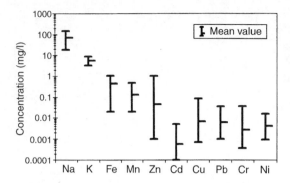

Figure 11.1 Range of metals' concentrations in the Vistula River.

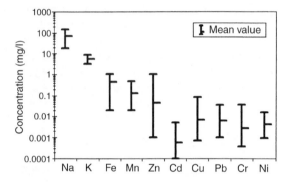

Figure 11.2 Range of metals' concentrations in the Odra River.

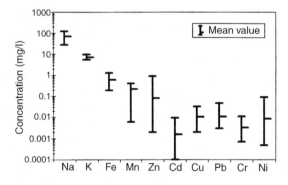

Figure 11.3 Range of metals' concentrations in ten Pomeranian rivers.

From Figures 11.1, 11.2 and 11.3 it can also be seen that occasionally the concentrations of iron, zinc, copper and lead in the river waters exceed the limits for class III. This is because the same standard is used for each of the three classes of water quality (except for iron). There are few water resources in Poland and so it is important to prevent river waters from being polluted by metals because the water may be required as a source of drinking water in the near future. For this reason, it was decided to introduce the same standards for these metals for all classes of water quality.

The investigations also show that occasionally the chloride concentrations in the Vistula and Odra rivers exceed the permissible levels for surface waters. The chloride originates from discharges of industrial brines and saline waters from mines. Significant excess salinity occurs mainly in the upper parts of the Vistula and Odra rivers.

There are also limits on the concentrations of metals in discharges of waste-waters entering surface waters. The standards for permissible metal concentrations in wastewaters are given in Table 11.1.

The high frequency of the measurements carried out on the concentrations of metals in riverine waters and water flow rates enabled metal load outflows to be calculated accurately. The instantaneous loads are the product of the water flow and the concentration of metal in the random sample. It is assumed that there is a linear variability between two adjacent measurements. The mean or total loads for the selected time intervals are derived from the equation:

$$L = \frac{m}{n} \sum_{i=1}^{n} C_i q_i$$

where:
L = load
C = measured concentration
q = measured flow
n = number of measurements
m = conversion factor for units.

The irregularity of a load is given by its coefficient of variability, i.e. the standard deviation expressed as a percentage of the mean concentration. The metal with the lowest variability coefficient is sodium in the Odra River (Table 11.2). The

Table 11.1 The national permissible limits of metals in different types of waters (Source: Rybiński *et al.*, Metals, in *Environmental Conditions in the Polish Zone of the Southern Baltic Sea during 1992*, IMWM, 1993; and Legislation Gazette 35, 205, Ministry of Health and Social Welfare, 1990)

Metal	Waste-water (mg/l)	Surface waters (mg/l)			Drinking water (mg/l)
		Class I	Class II	Class III	
Sodium (Na)	800	100	120	150	200
Potassium (K)	80	10	12	15	–
Iron (Fe)	10	1	1.5	2	0.5
Zinc (Zn)	2	0.2	0.2	0.2	5
Chromium (Cr^{3+})	0.5	0.05	0.1	0.1	–
Chromium (Cr^{6+})	0.2	0.05	0.05	0.05	0.01
Cadmium (Cd)	0.1	0.005	0.03	0.1	0.005
Manganese (Mn)	–	0.1	0.3	0.8	0.1
Copper (Cu)	0.5	0.05	0.05	0.05	0.05
Nickel (Ni)	2	1	1	1	0.03
Lead (Pb)	0.5	0.05	0.05	0.05	0.05

Table 11.2 Variability coefficients of monthly metal loads (%)

Metal	Vistula River	Odra River	Pomeranian rivers
Sodium (Na)	35.4	26.1	53.7–102.4
Potassium (k)	38.8	38.1	35.8–74.4
Iron (Fe)	100	42.6	48.0–141.0
Manganese (Mn)	74.5	32.4	41.7–196.1
Zinc (Zn)	113.5	97.6	89.7–248.7
Cadmium (Cd)	142.3	79.8	109.7–390.8
Copper (Cu)	85.7	55.8	75.1–166.7
Lead (Pb)	80.2	66.9	80.5–272.7
Chromium (Cr)	158.8	70.9	81.8–245.2
Nickel (Ni)	75.5	88.4	85.7–223.2

highest coefficient of variability is for cadmium and for chromium in the Pomeranian rivers. Here, at low mean loads, the discharges of wastes containing metals caused a considerable increase in deviation of the samples. In general the metal load data appear to fit a log–normal population distribution.

Analysis of the seasonal distribution of the riverine outflow of metals shows that, in all cases, the highest loadings occur from March to May, with a reduction in loadings in September and October. The hydrograph created from mean monthly values indicates that the hydrological conditions are responsible to a large extent for the metal outflow. These conditions cause the increase in loads during spring freshets and the reduction at the time of the autumn low water levels.

11.4 DISCUSSION

The contribution of each river to the total metal load varies considerably but the Vistula and Odra rivers dominate by conveying 90.6%–98.2% of the metal loads (Table 11.3). However when the loads are related to the catchment area, i.e. unit loads, then the relative contributions change considerably (Figure 11.4). Large catchment basins produce significantly lower unit outflows than small ones (except for unit loads of cadmium and nickel in the Odra River). This probably results from the accumulation of metals in the bottom sediment of the reservoirs and stagnant areas. These layers are resuspended during spring freshets and magnify the relationship between metal loading and water outflow.

The variability of metal loads with water flow enables, to a certain extent, estimation of the sources of the load. Total load (outflow) comprises outflows of diffuse (non-point) and surface (point) origin. The latter can be further differentiated into minimal outflow and (by substracting minimal out-

Table 11.3 Contribution of each of the rivers to the total metal load (% of total load)

River	Metal (% of total load)									
	Na	K	Fe	Mn	Zn	Cd	Cu	Pb	Cr	Ni
Odra	31	35.6	33.9	40.3	42.6	51.8	38	40.5	32.4	45.8
Ina	0.3	0.96	1.22	0.84	0.76	0.55	0.74	0.75	0.36	0.36
Rega	0.33	1.18	1.62	1.24	0.68	0.41	0.79	0.63	0.77	0.44
Parsęta	0.41	1.07	1.9	1.3	0.88	0.7	1.2	1.01	1.14	0.63
Grabowa	0.06	0.13	0.3	0.24	0.18	0.12	0.22	0.17	0.2	0.1
Wieprza	0.14	0.39	1.21	0.85	0.49	0.61	0.58	0.78	0.54	0.26
Słupia	0.18	0.46	0.79	0.74	0.55	0.48	1.19	0.95	0.61	0.3
Łupawa	0.06	0.19	0.32	0.31	0.19	0.22	0.25	0.38	0.3	0.09
Łeba	0.11	0.39	0.82	0.65	0.4	0.63	0.88	0.52	0.46	0.22
Reda	0.04	0.14	0.21	0.15	0.15	0.12	0.18	0.23	0.1	0.1
Vistula	67.2	58.6	56.7	51.9	50.3	43.3	54.6	52	61.3	51
Pasłęka	0.2	0.8	0.95	1.42	2.8	1.11	1.39	2.09	1.93	0.67

flow from the surface outflow) incidental outflow. The polluting load in the surface outflow is found to be caused predominantly by the flux of pollutants from point sources, i.e. wastes. Incidental load originates from poorly managed treatment plants resulting in periodic discharges of inadequately treated wastes.

If this division is applied to the total loads in the riverine outflow between 1988 and 1991 (Table 11.4), the metals that originate from point sources most often are sodium, manganese and potassium. The rest of the metals, excluding chromium, come from point sources in 52%–63% of cases, of which 43%–54% are considered as incidental outflows. The very low incidental outflow of sodium and potassium is a characteristic of their general occurrence in point and diffuse outflows. It might be concluded from these values, that 50% of riverine metal loads could be removed by improving the disposal of wastewater from metallurgical plants. This is in agreement with the actual condition of the waste management system in most of these plants.

In order to describe the magnitude and directions of the changes in the riverine metal outflows over time, the data sets were reduced to 90%-ile values in order to eliminate random extreme values. These values were then plotted in a linear regression of load against time. The correlations obtained give an indication of the trend, correlation coefficients and the significance level of non-correlation. Apart from very few exceptions, the change with time seems to be very weak and many cases of low correlation can be observed. Setting a limit on the significance level of non-correlation at $\alpha = 0.05$ eliminates 46.5% of the data, i.e. there is no tendency of any change. Of the 77 remaining, 54 indicate a negative trend, while 23 show a positive trend. The Odra River shows the most significant correlation with negative trend slopes for all the metals, except cadmium. Second place is occupied by the Vistula River, which shows six cases of significant correlation including four negative trends [1].

This lack of visible negative trends appears to be not only a national but also a Baltic problem. This was apparent from the Second Baltic Sea polution load compilation taking zinc, cadmium, copper and lead as examples [2]. The contribution of Poland to the total load of metals discharged into the Baltic Sea in 1990 was as follows: zinc 48%, cadmium 61%, copper 25.5%, lead 23%.

REFERENCES

1. Rybiński J., Heybowicz E., Ceglarski R., Metals, *Environmental Conditions in the Polish Zone of the Southern Baltic Sea during 1992*. Maritime Branch, IMWM, Gdynia, 1993.
2. Helsinki Commission (1993) Second Baltic Sea pollution load compilation, *Baltic Sea Environment Proceedings*, No. 45.

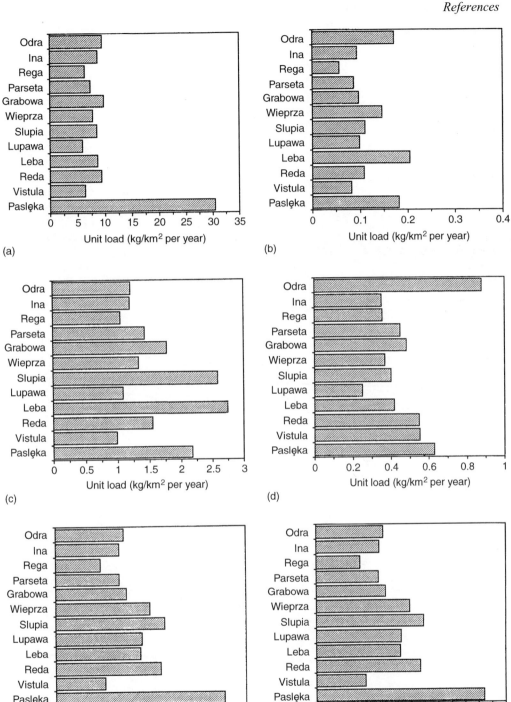

Figure 11.4 Mean unit load of metals in river basins (a) zinc; (b) cadmium; (c) copper; (d) nickel; (e) lead; (f) chromium.

Table 11.4 Load of metals from Polish rivers entering the Baltic Sea

Parameter	Years 1988–91									
	Na	K	Fe	Mn	Zn	Cd	Cu	Pb	Cr	Ni
Mean annual outflow (t/year)	2 820 966	275 675	27 771	7570.1	2757.8	36.2	374.0	366.7	121.4	219
Mean annual unit outflow (kg/km² per year)	8696.9	849.9	85.6	23.3	8.5	0.111	1.153	1.131	0.374	0.675
Proportion of total (%)	100	100	100	100	100	100	100	100	100	100
Minimal outflow (t/year)	1 411 360	138 542	3877.5	1414	217.7	2.43	41.7	55.6	17.2	27.9
Minimal unit outflow (kg/km² per year)	4351.1	427.1	12	4.4	0.67	0.008	0.128	0.171	0.053	0.086
Proportion of total (%)	50	50.3	14	18.7	7.9	6.7	11.1	15.2	14.1	12.7
Underground outflow (t/year)	2 242 463	183 015	14 354.6	5240.4	1729.8	20	201.7	225.1	54.7	135.7
Underground unit outflow (kg/km² per year)	6913.4	564.2	44.3	16.2	5.33	0.062	0.622	0.694	0.169	0.418
Proportion of total (%)	79.5	66.4	51.7	69.2	62.7	55.2	53.9	61.4	45.1	62
Incidental outflow (t/year)	831 103	44 473	10 477.1	3826.4	1512.1	17.5	160	169.5	37.6	107.8
Incidental unit outflow (kg/km² per year)	2562.2	137.1	32.3	11.8	4.66	0.054	0.493	0.523	0.116	0.332
Proportion of total (%)	29.5	16.1	37.7	50.5	54.8	48.5	42.8	46.2	30.9	49.2
Non-point outflow (t/year)	578 503	92 660	13 416.1	2329.8	1028	16.2	172.3	141.6	66.7	83.3
Non-point unit outflow (kg/km² per year)	1783.5	285.7	41.4	7.18	3.17	0.05	0.531	0.437	0.206	0.257
Proportion of total (%)	20.5	33.6	48.3	30.8	37.3	44.8	46.1	38.6	54.9	38

SURFACE WATER POLLUTANTS AND PROBLEMS WITH THEIR ANALYSIS

by B. Zygmunt, W. Wardencki, M. Biziuk and J. Namieśnik

12.1 INTRODUCTION

The degradation of water quality is usually caused by pollution. An accurate, if clumsy, definition of pollution is 'too much of something in the wrong place'. When applied to chemicals in the aquatic environment this definition recognizes that any chemical can be a pollutant if present at a high enough concentration. This includes non-toxic substances such as sugar. Nevertheless some chemicals can be selected as being of higher priority for control in the aquatic environment than others. They are selected because they are frequently found there, they are capable of exerting adverse effects at low concentrations, they remain unchanged in the aquatic environment for long periods or they biomagnify in food chains. Often priority chemicals display more than one of these characteristics.

For chemicals carrying the greatest threat to the aquatic environment, special regulations with respect to their discharges and concentration levels are issued by the regulatory organizations. In the EU those substances selected as being especially dangerous to the aquatic environment are published in the Black List and the Grey List. Chemicals on the Black List are regarded as hazardous and have limit values and environmental quality standards (EQS) agreed at EU level. Grey List chemicals are regarded as less hazardous and are controlled using the EQS approach with quality standards set by each member state.

There is a separate Red List to control dangerous chemicals present in discharges entering the North Sea from surrounding countries. The UK prioritized 23 substances or groups of substances for the Red List, although this was later extended to 36. For Red List chemicals strict environmental quality targets have been set to reduce pollutant loads entering the North Sea within a set timetable. Industrial processes discharging significant amounts of these chemicals must be scheduled. Measures must also be taken, where possible, to reduce inputs of Red List chemicals from diffuse sources.

The most important way to control water pollution is to reduce discharges of dangerous chemicals which enter water bodies directly. However, some pollutants are discharged to other sectors of the environment such as the atmosphere. These pollutants circulate through the environment by interchanging between the biotic, terrestrial, atmospheric and aquatic sectors. In fact, any chemical which is moderately stable will eventually become widely dispersed in the environment, no matter how carefully it is controlled. This is a natural consequence of the second law of thermodynamics.

The quality of water depends on the concentration levels of chemical pollutants and on other factors. In Poland, the degree of degradation of water quality is determined by chemical, physical and microbiological analyses. Using up to 57 parameters, inland surface waters are divided into the following three quality classes [1]:

International River Water Quality. Edited by Gerry Best, Teresa Bogacka and Elżbieta Niemirycz. Published in 1997 by E & FN Spon, London, ISBN 0419215409

- class I – waters suitable for production of drinking water or for supporting a salmonid fish population;
- class II – waters suitable for recreation;
- class III – waters for agriculture and industry.
- waters which do not satisfy the predetermined standards are unclassified.

By this classification, the quality of Polish inland surface waters looks rather poor (Tables 12.1 and 12.2).

12.2 CLASSIFICATION OF ORGANIC WATER POLLUTANTS

The most important classes of organic pollutants in water are:

- polychlorinated dibenzo-*p*-dioxins (PCDD) and dibenzofurans (PCDF)
- petroleum hydrocarbons
- chlorinated hydrocarbons
- polycyclic aromatic hydrocarbons
- chlorinated and other pesticides
- phenols
- polychlorinated biphenyls.

12.2.1 POLYCHLORINATED DIBENZO-*P*-DIOXINS (PCDDS) AND DIBENZOFURANS (PCDFS)

Dioxins and furans are, in fact, chlorinated dibenzo-*p*-dioxins (CDDs) and chlorinated dibenzofurans (CDFs) respectively. On the basic structure shown in Figure 12.1, up to eight chlorine atoms can be placed. This gives rise to 75 dioxin congeners and 135 furan congeners. All 75 dioxins are congeners of one another or members of a like group. A group of dioxin or furan isomers, i.e. compounds having the same molecular formula and hence the same

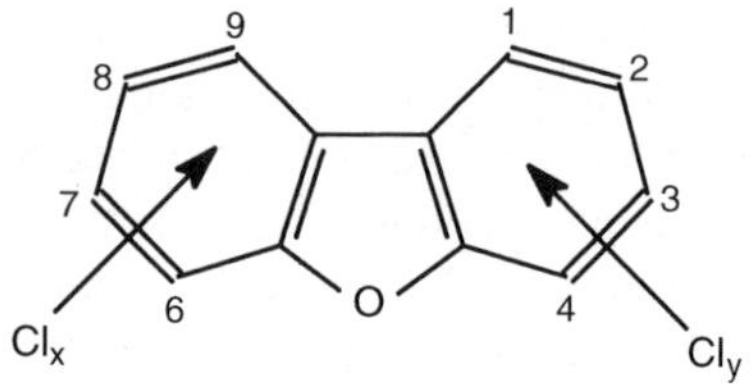

Figure 12.1 The basic structures of (a) chlorinated dibenzofurans and (b) dibenzo–p–dioxins.

Table 12.1 Lake water quality in Poland 1974–90

Class	Number of lakes	Volume ($10^3\ m^3$)
I	26	2 242
I/II	3	249
II	170	6 984
II/III	11	378
III	154	3 531
III/unclassified	2	311
Unclassified	119	2 581
All classes	485	16 276

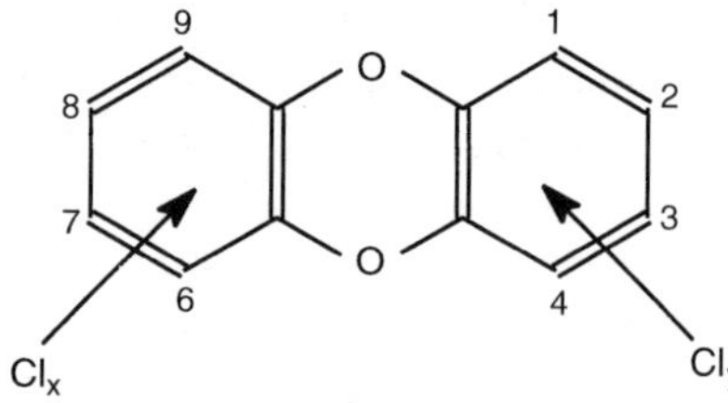

Table 12.2 River water quality in Poland 1978–92

Class	Vistula River 1992	Fraction of rivers length studied (%)				
		1978	1984–8	1990	1991	1992
I	0	6.9	4.8	6	2.3	2.4
II	5.4	28	30.7	27.9	32.7	12.8
III	37.5	30.1	29.7	30.3	30	24.8
Unclassified	57.1	35	34.8	35.8	35	60

number of chlorine atoms, is often referred to as a congener group [2,3].

Dioxins and furans do not occur naturally and are not deliberately manufactured, but this does not mean that they are not present in the environment. In fact their synthesis was first reported by German chemists in 1872 [4]. Scientists from Dow Chemical Company performed extensive studies on dioxins and furans. According to them traces of these substances may be produced by any combustion source [5]. If this trace chemistry of fire is true, dioxins and furans may have been with us since the first forest fire.

These days, the main sources of dioxins and furans include:

- the pulp and paper industry;
- the manufacture and use of chlorophenols and related compounds;
- production of chlorine;
- production of magnesium;
- automobile emissions; and
- incineration of municipal, hospital and industrial wastes.

Polychlorinated dibenzo-p-dioxins enter surface waters primarily through liquid discharges from these industries. Fly ash from incinerators contains considerable amounts of dioxins and furans and this material can find its way to surface waters.

The solubilities of PCDDs and PCDFs in water are low; for example, for TCDD it is only 0.2 µg/l. In the aquatic environment therefore they fractionate predominantly onto suspended solids and sediments. Their concentrations in surface waters are generally below detectable limits, except in cases of extreme ambient pollution. For example, concentrations in Lake Ontario and Lake Erie in Canada are generally < 1 pg/l despite the input of substantial amounts of PCDD-bearing wastes. They have a high degree of chemical stability and are poorly biodegradable, so their concentrations will gradually increase until an equilibrium is reached between their input and removal; they are removed by sorption onto sediments, chemical degradation, biodegradation and accumulation by aquatic life. There are no data on their concentrations in Polish inland surface waters.

These is concern about dioxins and furans and their extreme toxicity. In '*The Pendulum and the Toxic Cloud*' [6]. Thomas Whiteside presents data on the properties and deadly effects of dioxins and furans on human and animal life and shows that humans are exposed to a class of compounds that are toxic at ppm to ppb levels and mutagenic, carcinogenic and teratogenic at ppt levels

The compounds are often considered to be the most toxic chemicals known. The adverse health effects on humans, particularly in the long term, are not clear. There is no doubt that the compounds are very toxic. Their potential threat to humans depends on the different toxicities of the 75 dioxin and 135 furan congeners.

The toxicity depends on the number and position of chlorine substituents. The most toxic dioxin congeners belong to the 2,3,7,8-group. This group includes: 2,3,7,8-tetra-; 1,2,3,7,8-penta-; and 1,2,3,4,7,8-, 1,2,3,6,7,8- and 1,2,3,7,8,9-hexachlorinated dibenzo-p-dioxins. The first listed is the most toxic dioxin. Its LD_{50} for guinea pigs is 0.6 µg/kg body weight and for mice 284 µg/kg body weight [7]. It is clear from these data that toxicological response to CDDs and CDFs are species dependent.

There is much less information with respect to the toxic effects on humans. Most existing data have been derived from occupational exposure or industrial accident victims. The highest tolerable daily intake has been estimated at 1–5 pg/kg body weight of 2,3,7,8-TCDD or other CDD equivalent [8]. The World Health Organization suggests that there are insufficient data to draw final conclusions about safe levels of CDDs and CDFs for humans or their resistance to them.

The dioxins and furans emitted into the environment undergo significant accumulation in the body. Of particular concern is the accumulation of CDDs and CDFs in human breast milk. Accidental exposure to dioxins has shown that acute exposure results in persistent skin acne, neurological disorders and liver disfunction. Of more concern is the linking of increased cancer incidence with long-term exposure to low environmental levels of CDDs and CDFs. Special care need to be taken over dioxins and furans since there is still much that is not known about their effects on biological systems. As part of the care programme, their concentration levels in different waters should be monitored.

12.2.2 PETROLEUM HYDROCARBONS

Petroleum hydrocarbons can be a problem not only to sea water but also to inland surface waters. They can enter soil from leaking subsoil containers and be carried to surface waters by groundwater. In Poland thefts of fuels from pipelines are a particular problem, with oil spills often occurring as well. In cities, a large proportion of hydrocarbons in rivers comes from rainwater run-off from paved roads. Major pollution incidents are usually accidental and it is difficult to evaluate average concentration ranges.

Some data on pollution of various waters in the Gdansk region are presented in Table 12.3.

In the aquatic environment these pollutant chemicals undergo several sink processes. These are sorption on sediments, volatilization and biotransformation. Virtually all crude oils and hydrocarbons are vulnerable to microbial degradation under favourable conditions.

12.2.3. VOLATILE ORGANOHALOGEN COMPOUNDS [2, 3]

In this group the most common are halogenated methanes, ethanes and ethylenes. They are used as solvents, cleaning agents, degreasers, cooling agents etc. They are frequently found in municipal wastewaters and certain types of industrial discharges. In general, chlorination of liquid wastes results in the formation of halo-derivatives that are subsequently discharged into the environment. Chlorinated pulp mill effluents contain a range of substances in this group from chlorinated methanes to butanes.

Chloromethanes or THMs have little or no affinity with suspended solids or sediments and evaporation is the principal route through which they are lost from water. Chemical oxidation is not significant in the fate of THMs in natural waters. They are commonly found in surface waters, but at relatively low concentrations. Concentrations of selected volatile organohalogen compounds in surface waters throughout the world are given in Table 12.4 whilst some data for waters in the Gdansk region are given in Table 12.5.

12.2.4 POLYCYCLIC AROMATIC HYDROCARBONS (PAHS)

PAHs originate from both natural and anthropogenic sources and are distributed in plant and animal tissues, surface waters, sediments, soils and air. There are many sources of PAHs in surface waters and these include municipal and industrial effluents, atmospheric fall-out, fly-ash precipitation, road run-off, and leaching from contaminated

Table 12.3 Results of the determination of volatile hydrocarbons in natural waters in the Gdansk region[a] (Source: Zygmont)

Sampling sites	Sampling date	Heptane (mg/l)	Benzene (mg/l)	Toluene (mg/l)	Nonane + Xylene (mg/l)	Decane (mg/l)	Undecane (mg/l)	Chlorobenzene (mg/l)	Isooctane (mg/l)
Vistula River at Kiezmark	7 March 1994	nd	0.78	2.14	nd	nd	nd	0.89	nd
	10 June 1994	nd	nd	nd	0.35	nd	0.16	nd	0.23
Radunia River at Żukowo	7 March 1994	nd	1.08	1.23	nd	3.14	nd	0.38	nd
	14 March 1994	nd	nd	nd	nd	nd	nd	nd	nd
Motława River at Gdansk	7 March 1994	nd	0.83	1.22	nd	nd	nd	nd	nd
	10 June 1994	nd	nd	nd	0.03	nd	0.18	nd	nd
Kacza River at Orłowo	7 March 1994	nd	0.18	0.93	nd	nd	nd	nd	2.05
	14 March 1994	nd	0.08	0.6	nd	nd	nd	nd	nd
Rozwójka River (near Refinery)	7 March 1994	nd	33.7	0.07	nd	48.8	nd	31.3	nd
	14 March 1994	nd	22.3	nd	nd	22.5	6.7	nd	nd

[a] nd = not detected

Table 12.4 Concentration of selected volatile organohalogen compounds in natural waters worldwide[a] (ng/l) (Source: Biziuk; Technical University of Gdansk, 1994)

Location	Year	$CHCl_3$	CCl_4	$C_2H_3Cl_3$	C_2HCl_3	C_2Cl_4
Industrial wastes Sweden	1978	760 000	–	–	2 800	1 400
Aurajoki River, Finland	1980	1 300	30	nd	40	10
Crawford Lake, Canada	1981	58	3.8	5.9	32	9
Avon River, UK	1984	2.6	1.1	9.7	15.2	14.7
Tama River, Japan	1983	590			140	
Rhine River, 506 km Germany	1982–3		450–44 400		200–5 100	260–4 600
Rhine River, Koblenz	1983	5 900	750		340	940
St Clair River, Canada	1985	5–4 482	2–2 411	5–4 174		2–740
Elbe River, Czech	1989				0–30 000	0–8 600
Straszyn Reservoir, Poland	1990	500	120		340	
Vistula River, Kiezmark, Poland	1990	430	130		730	
Borowo Lake, Poland	1990	220	80		120	

[a] nd = not detected

soils. Aromatics with two- or three-ring structures such as naphthalenes, fluorenes, fenanthrenes and anthracenes are volatile and relatively toxic. Higher molecular weight, four- to seven-ring aromatics are not acutely toxic but have been shown to be carcinogenic. Benzo-*a*-pyrene is regarded as the most dangerous.

PAHs are sorbed by particulate matter and deposited in bottom sediment because of their low aqueous solubility and vapour pressure. Volatilization can also be a significant sink process for two-ring members. They are commonly found in natural waters, particularly in industrialized areas of the world. Total PAH concentrations in the Rhine River range from 0.7 to 1.5 µg/l [9].

Concentrations of selected PAHs in Krokowa in Gdansk district are given in Table 12.6.

12.2.5 PESTICIDES

The introduction of DDT during World War II marked the beginning of a period of very rapid growth in pesticide use. Large quantities of persistent pesticides entered water directly, e.g. through mosquito control, or indirectly, primarily from drainage of agricultural lands. There have been great advances in the nature of pesticides since the first examples were synthesized over 60 years ago. In recent years there has been a move to produce substances that are targeted to the specific pest but

Table 12.5 Results of the determination of volatile organohalogen compounds in natural waters in the Gdansk region. Sampling date 7 March 1994 (units µg/l) (Source: Zygmunt)

Sampling place	$CHCl_3$	$CHBrCl_2 + C_2HCl_3$	$CHBr_2Cl$	$CHBr_3$	C_2Cl_4	CCl_4
Sea water at Orłowo	2.2	1.44	0.55	0.02	nd	0.11
Kacza River	1.28	0.86	0.36	0.01	nd	0.01
Vistula River	1.44	0.08	0.14	0.04	0.03	0.03
Motława River	0.26	0.02	nd	0.02	0.02	0.03
Rozwójka River	0.65	0.06	nd	0.01	0.02	0.28
Radunia River	0.23	0.03	nd	nd	0.02	0.02

[a] nd = not detected

Table 12.6 Concentration of selected polycyclic aromatic hydrocarbons in surface waters in Krokowa

Compound	Concentration (ng/l)				
	Site 1	Site 2	Site 3	Site 4	Site 5
Acetonaphthalene			not detected		
Phenanthrene	7	16	7	9	6
Anthracene			not detected		
Fluoranthrene	7	9	7	7	5
Pyrene			not detected		
Benzo-a-anthracene	12	6	4	2	5
Chrysene	273	73	62	45	183
Benzo-k-fluoranthrene			not detected		
Benzo-a-pyrene			not detected		
Benzo-ghi-perylene			not detected		

do not adversely affect the environment. There are now over 725 pesticides listed in the UK Pesticides Manual and many of these are effective against the target pest at a fraction of the application rate of a decade ago [10].

Major water-pollutant pesticides belong to the classes of chlorinated hydrocarbons, organic phosphates, and carbamates. Their residence times in aquatic environments differ considerably but, in general, those that have the greatest stability in the environment are gradually being removed from the market.

The toxicities of pesticides vary widely, too. Hence the freshwater quality criteria range from 1 ppt to 10 ppb [11]. Pesticide concentrations in surface waters in various places around the world vary widely (Table 12.7). Levels of some chlorinated pesticides in rivers and other bodies of water in Gdansk and its neighbourhood were also studied (Table 12.8) and in general, the concentrations are low.

12.3 SELECTING ANALYTICAL METHODS AND PROCEDURES

Water quality has been examined for several centuries. The first examinations were organoleptic but, as analytical chemistry developed, the methods of water analysis changed and became more sensitive and specific. The targets for reducing the amount of noxious contaminants, especially in waters used for drinking, were considered the most important issues for the regulators and thus their identification and quantification in water are priority tasks.

The identification and determination of contaminants in surface waters has proved to be a difficult task because such waters have been found to be chemically complex, containing hundreds or even thousands of inorganic, organometallic and organic compounds. Moreover, the pollutants are present in wide-ranging concentrations, from 1 to 10^{-10} mol/l or less [12]. In addition, these pollutants must be determined precisely because of the continuing concerns about acceptable trace-level concentrations of many pollutants in the environment.

The most important tasks for the analytical chemist are:

- description of the current status of the quality of water using a range of parameters;
- defining ecotoxicological effects, e.g. biomethylation of metals, such as mercury, tin, lead, etc.;
- mobilization of metals and other pollutants from sediments back into the water column as a result of redox reactions;
- quantitative and qualitative control of emissions;
- determination of the pathway of the pollutants, from source to their interaction with humans;
- consideration of the avoidance or minimization of wastes using new technologies with zero or low discharge rates.

Table 12.7 Concentration of selected pesticides and PCBs in waters in different countries (units μg/l) (Source: Biziuk; Technical University of Gdansk, 1994)

Location	Year	Lindane	DDT	Aldrin	HCB	PCB	Atrazine	Simazine
Kupa River, Croatia	1989	2–20	2–6		<1	2–8		
Park Donana, Spain	1984	23	145	18		665		
Niagara River, Canada	1986		45	15	11	908		
Piemont, Italy, (max. conc.)	1991						>1000	>2100
Ebro estuary, Spain	1983	>0.81	0.84		3.5	0.25		
Danube River, Yugoslavia	1990	57	110			trace		
Klang River, Malaysia	1990		0.2	0.01	0.12			
Adyga River, Italy	1990	1.94					5.05	6.68
Roden River, France	1990						32	18
Saint Laurent River, Canada	1986–8	0.88	1.59	0.19	0.35		do670	do300
Radunia River, Poland	1984–8	0–39	0–139				0–2000	0–2500
Drinking water – Straszyn, Poland	1986–9	0–54	0–43				0–3000	0–10 000

Problems occur with the analysis of water mainly because of the unknown matrix from which a substance has to be determined at trace level. Clean-up procedures and preconcentration steps have to be included in order to enrich the target substance(s) or to remove unwanted matrix components. In metal analysis, care has to be taken to ensure that the volatile elements (arsenic, mercury, etc.) are not lost during the clean-up or pre-concentration.

The criteria that should be considered when choosing the analytical method and the instrumentation for the analysis of different waters are:

- high sensitivity and low detection limit;
- sufficient precision and accuracy;
- good selectivity and resolution;
- speed of analysis and wide application range;
- the possibility of determining several substances;
- high identification potential;
- simple preparation of samples for analysis using small samples volumes;
- easy operation of the instrument including the possibility of automation for 'out-of-working-hours' analysis;
- possibility of working in the field;
- acceptable costs.

To fulfil most of these criteria, often sophisticated instumentation must be used: there are many books on this topic [13–16]. For organic pollu-

Table 12.8 Results of the determination of pesticides after sorption on XAD-4, C-18 and LC-Ph in natural waters (units μg/l)

Sample	Solid sorbent	7 March 1994			14 March 1994		
		Lindane	p,p' DDT	Metoxychlor	Lindane	p,p' DDT	Methoxychlor
Vistula River	XAD-4	48	<10	24	56	<10	32
	C-18	43	nd	29	60	16	30
	LC-Ph	39	nd	nd	43	nd	17
Sea water at Orłowo	XAD-4	26	12	17	33	<10	21
	C-18	31	nd	20	35	12	19
	LC-Ph	27	nd	<10	24	nd	<10

[a] nd = not detected

tants, chromatographic methods such as gas and liquid chromatography are indispensible. Of the methods currently known, they have the highest resolving power. Additionally, the detection systems used in these techniques are very sensitive. Some of them are selective and even specific with respect to selected elements [17]. An atomic emission detector (AED), for example, can be selective to any element except helium [18]. Unfortunately, chromatographic methods have low identification potential. However, when coupled with spectroscopic techniques such as mass spectroscopy (MS) or Fourier transform–infra-red (FT–IR) they are really powerful tools for environmental organic analysis [18,19]. Also, some advanced electroanalytical techniques, such as advanced polarography and voltametry, can provide alternatives to chromatographic and spectroscopic techniques [20]. However, these techniques suffer from the following disadvantages.

- There is still limited availability of such fully automated instrumentation because of the high cost;
- Their operation requires highly trained staff.

One of the early tasks facing the analytical chemist is to choose the most appropriate procedure to obtain the desired information about the particular material of interest. However, a very important first step is the collection of samples and their subsequent treatment and storage to prevent contamination and losses. Natural waters are systems consisting of suspended particles and biological/chemical species in equilibrium and these can change when the sample is placed in contact with air and/or container walls or when physical changes (temperature, pressure) occur [21]. Sediments may cause physico-chemical changes when in contact with air (oxidation, flocculation). Futhermore, microbiological activity may result in changes in concentration and composition, e.g. methylation may occur. Methyl–tin compounds, for instance, can be produced by natural methylation of inorganic tin, whereas butyl–tin arises exclusively from industrial processes.

Often, samples need extensive clean-up, preconcentration and selective isolation of the materials of interest. Otherwise, even with the use of the best instruments, false positive identification can occur. Sample preparation is a very difficult and critical step. Without proper sample pretreatment money spent on expensive instrumentation is useless and can be harmful. Incorrect sample preparation gives erroneous results and interpretation. Money may be spent on treatment processes that are not required, or, alternatively, the quality of the water may be worse than that suggested by the analysis.

In conclusion, it must be emphasized that water analysis is a matter for analytical experts. It is much too important to be performed by dilettantes.

REFERENCES

1. Dojlido, J. (1987) *Chemia Wody*, Arkady, Warsaw.
2. Namieśnik, J., Górecki, T., Wardencki, W. et al. (1992) *Polish J. Environ. Stud.*, **1**, 5.
3. Namieśnik, J., Górecki, T., Wardencki, W. et al. (1993) *Secondary Effects and Pollutants of the Environment*, Technical University of Gdansk.
4. Long, J.R. and Hanson, D.J. (1983) *Chem. Eng. News*, Jun **6**, 25.
5. Bumb, R.R., Crummett, W.B., Cutie, S.S. et al. (1980) *Science*, **210**, 385.
6. Whiteside, T. (1979) *The Pendulum and the Toxic Cloud*.
7. Baker, P.G. (1981) *Anal. Proc.*, **18**, 478.
8. Oakland, D. (1988). *Filtr. Sep.*, **25**, 39.
9. Andelman, J.B. and Snodgrass, J.E. (1974) *CRC Crit. Rev. Environ. Control*, **4**, 69.
10. Wilson, M.F. (1995) Monitoring and adapting to the changes in pesticide use profiles that occur in response to modern pest control and environmental requirements, in *Pesticides–Developments, Impacts and Controls* (eds G.A. and A.D. Ruthven), Royal Society of Chemistry, Cambridge.
11. Holmes, C.W., Slade, A.E. and McLerran, C.J. (1974) *Environ. Sci. Techn.*, **8**, 255.
12. Dojlido, J. and Best, G.A. (1993) *Chemistry of Water and Water Pollution*, Ellis Horwood, Hemel Hemstead.
13. Poole, C.F. and Poole, S.K. (1991) *Chromatography Today*, Elsevier, London.
14. Crompton, T.R. (1991) *Analytical Chemistry for the Water Industry*, Butterworth–Heinemann, Oxford.
15. Smith, R.K. (1993) *Handbook of Environmental Analysis*, Genium Publishing, New York.

16. Baugh, P.J. (Ed.) (1993) *Gas Chromatography – A Practical Approach*, IRL Press, Oxford.
17. Wardencki, W. and Zygmunt, B. (1991) *Anal. Chim. Acta*, **255**, 1.
18. Zygmunt, B. and Wardencki, W. (1993) Proceedings of the Conference on Organic Compounds in the Environment and Methods of their Analysis, Jachranka, 18–21 May 1993, Warsaw, pp. 55–76.
19. Zygmunt, B. and Wardencki, W. (1993) *Polish J. Chem.*, **67**, 369.
20. Bersier, P.M. Howell, J. and Bruntlett, C. (1994) *Analyst*, **119**, 219.
21. Hunt, D.T.F. and Wilson, A.L. (1986) *The Chemical Analysis of Water – 2nd Edition*, Royal Society of Chemistry, London.

MONITORING THE WATER QUALITY OF THE RADUNIA RIVER

by A. Walkowiak and D. Kozak

13.1 INTRODUCTION

About 20% of drinking water for Gdansk is supplied by a surface intake from the Straszyn Reservoir, which is on the Radunia River. The catchment area of the Radunia River contains an administration area of nine government communities – the local self-governments. The quality of water supplying the reservoir does not meet basic requirements because of pollution of the Radunia River.

Between 1986 and 1992 a decision was taken to create a protection zone for the Straszyn intake, comprising almost the whole of the drainage basin of the Radunia River.

This decision was accompanied by a programme of activities and tasks to be undertaken by the individual communities situated in this protection zone, in order to improve the water quality of the Radunia River. The following communities are involved: Gdansk, Żukowo, Kolbudy, Przodkowo, Przywids, Szemud, Pruszcz Gdański, Chmielno, Kartuzy, Somonino and Stężyca. The priority tasks for the years 1993–96 are:

- construction of sewerage system and collectors;
- pumping stations for wastewater;
- sewage treatment plants;
- disposal site for waste.

The Voivodeship Inspectorate of Environmental Protection in Gdansk, which until 1991 was known as the Centre of Environment Inspection and Control has monitored the quality of water of the Radunia River for many years. In addition, the quality of water of the Straszyn Reservoir was tested in 1984, 1985 and 1992. This institution also carried out, in 1991, a detailed assessment of the communities in the drainage basin of the Radunia River. The management of sewage and wastes was investigated, based on the obligations of the communities. Also, the investments made by the larger industries, water companies and local authorities towards water quality protection were evaluated.

This chapter presents selected results of the investigations carried out and their conclusions. It is based on internal analysis and reports of the Voivodeship of Gdansk Inspectorate of Environment at Protection.

13.2 THE MONITORING PROGRAMME

The waters of the Radunia River should be of class I quality from the source to Pruszcz Gdański and class II quality from there to the mouth. The following tributaries should be of class I: Mała Supina, Trzy Rzeki, Strzelenka and Reknica, and the Klasztorna Struga should be class II [1]. The assessment of the water quality was based on a ministerial decree concerning surface water classification [2] and an instruction from the former Central Office of the Water Economy concerning

106 *Monitoring the water quality of the Radunia River*

the execution of control measurements and assessment of the results for estimation of quality for surface waters [3].

The evaluation of the pollution is based on the characteristic concentration of individual pollutants, calculated from the most unfavourable results measured during each year of monitoring [3]. For the general classification the evaluation is carried out according to the rule that exceeding just one one factor of the various standards disqualifies the water for the prescribed use (Figure 13.1).

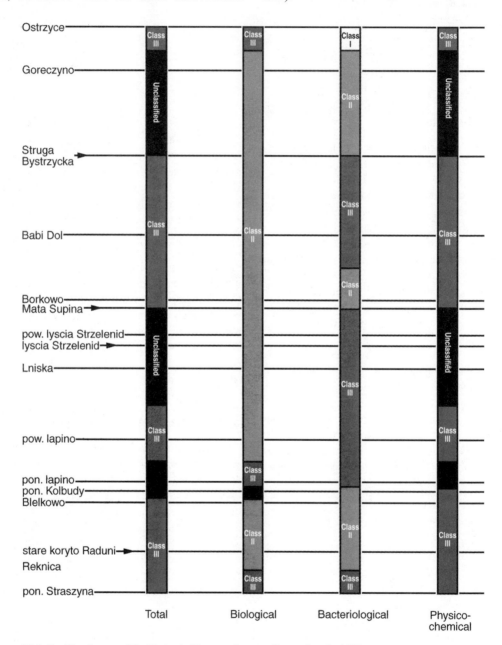

Figure 13.1 Quality classes of the Radunia River at the sampling points in 1992.

The streams were sampled five times a year and the flow of the streams was measured simultaneously with the sampling. The investigations were carried out taking into account the physical, chemical, biological and bacteriological factors. The investigations included the following indicators: temperature, colour, reaction, dissolved oxygen, BOD_5, COD, chlorides, sulphates, dissolved substances, general suspensions, ammonia nitrogen, nitrate nitrogen, nitrite nitrogen, organic nitrogen, phosphate phosphorus, total phosphorus, total iron, chlorophyll-a, faecal coliform numbers, volatile phenol, total hardness, total basicity, substances extracted by petroleum benzine (ether extract), calcium, magnesium, sodium, potassium, zinc, cadmium, copper, lead and mercury.

The waters of the Radunia River catchment area were sampled at the following points – distance from the river mouth in parentheses (Figure 13.2):

1. Radunia River downstream from Ostrzyce Lake (75.5 km);
2. Radunia River at Goręczyno (71.6 km);
3. Radunia River at Babi Dół (55 km);
4. Radunia River at Borkowo (48.7 km);
5. Klasztorna Struga River downstream from Klasztorne Lake (15.3 km);
6. Klasztorna Struga River at Kobysewo downstream from the sewage treatment plant at Kartuzy (11.3 km);
7. Trzy Rzeki River at the mouth of the Klasztorna Struga River in Kczewo (8.3 km);
8. Mała Supina River at Żukowo at the mouth of the Radunia River (48.0 km);
9. Radunia River upstream from the Strzelenka River mouth (45.3 km);
10. Strzelenka River downstream from Tuchomskie Lake (13.6 km);
11. Strzelenka River at the Banino–Firoga road (5.9 km);
12. Strzelenka River at the mouth of the ditch at Bysewo (4.4 km);
13. Strzelenka River above the Pepowo–Leżno road bridge (3 km);
14. Strzelenka River at the mouth of the Radunia River at Lniska (44 km);
15. Radunia River at Lniska (42.6 km);
16. Radunia River up Łapino (35.2 km);
17. Radunia River channel downstream from Łapino, at the mouth of Kolbudy Reservoir (31.3 km);
18. Radunia River downstream from Kolbudy Reservoir (30 km);
19. Reknica River outflow at Ząbrowskie Lake (9.9 km);

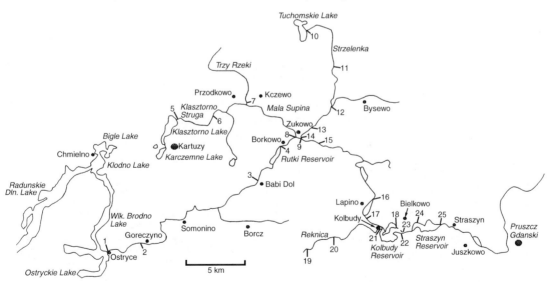

Figure 13.2 Location of the sampling points in the Radunia River catchment.

20. Reknica River upstream from Czapelsko (5.8 km);
21. Reknica River in Kolbudy, at the mouth of the old Radunia River (31.6 km);
22. Old Radunia River in Bielkowo (24.5 km);
23. Radunia Channel in Bielkowo (28.7 km);
24. Tributary from Bąkowo, at the entrance to Straszyn Reservoir (0.1 km);
25. Radunia River downstream from Straszyn Reservoir (20.8 km).

13.3 CHARACTERISTICS OF THE RADUNIA RIVER WATER QUALITY BASED ON SELECTED INDICATORS

The tributary rivers Mała Supina, Klasztorna Struga and Strzelenka, have a distinct influence on the water quality of the Radunia River. The water quality of the Mała Supina and Klasztorna Struga is determined by wastewaters from the sewage treatment plant OPWK in Kartuzy and the high level of pollution in the lakes of Kartuzy region. The water quality of the Strzelenka deteriorates at the mouth of the ditch from Bysewo and also at Lniska where the flow is largely wastewater from the sewage treatment plant in Żukowo (Figure 13.3).

The water quality in the Straszyn Reservoir determines the quality of water in the Radunia River. The water which flows into the Straszyn Reservoir from the Radunia Channel and the Old Radunia River in Bielkówo does not meet the standards of class I for the following reasons.

- There are high concentrations of phosphates in both inputs, organic compounds (BOD_5) in the Radunia Channel and suspensions in the Old Radunia River;
- The permitted levels for bacteriological and the biological parameters of both input are exceeded.

The pollution of the Straszyn Reservoir by phosphate and by BOD_5 is caused mainly by the waters of the Radunia Channel. The Straszyn is a flow-through reservoir with a specific water ecosystem which has an open metabolic cycle. The chemical and biological composition of the supply river is the most important influence on this metabolism, whilst the local soil, geological conditions and climate have less effect because of the rapid flow of water through the reservoir (the average retention time is 170 hours).

The reservoir shows features of eutrophic water and does not meet the standards for class I quality. This is because of high concentrations of phos-

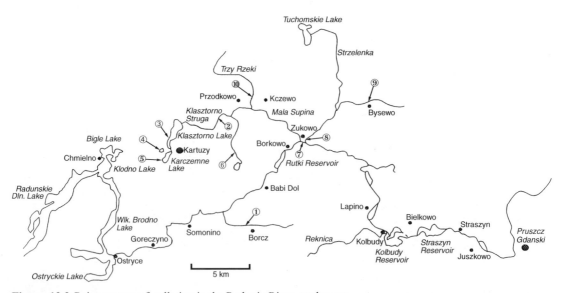

Figure 13.3 Point sources of pollution in the Radunia River catchment.

phates and organic nitrogen and the bacteriological and biological conditions.

13.4 CONTROL OF THE DRAINAGE BASIN

The controls of the Inspectorate of Environmental Protection during 1992 achieved no improvements in the area of the protection zone of the Straszyn water intake. In particular:

- The management of communal and industrial sewage did not improve;
- The uncontrolled inflow of inpurities to surface and groundwaters and the land continued; and
- The supervision of environmental protection of communities did not improve.

The activities undertaken to improve the Radunia River were not very effective and as a result the quality of water in the Straszyn Reservoir used for consumption did not improve. However it should be stressed that not all activities were completed according to plan and some were not started.

There were also no significant changes to the treatment of sewage in any of the river basin communities in the years 1988–92. In the whole area, which comprises 800 km^2 and 100 000 people, only two communal sewage treatment plants were operating with outputs of about 6000 m^3 per day of sewage (in Żukowo and Kartuzy) together with several small plants with outputs up to 20 m^3 per day. Other sewage is led to ground without treatment. The communities of Somonino, Stężyca, Chmielno and Kartuzy started operations to improve the treatment of sewage in 1992.

In recent years several dozen small firms were established (slaughter houses, butcheries and food processing), which operate without correct waste management and sewage treatment. The wastewaters are led to ground without treatment and the solid wastes are stored on temporary dumps.

The communal wastes are stored on several illegal dumps often sited close to sewer outfalls (Figure 13.4). In addition to the dumps used by the community authorities, there exist several dozen 'wild dumps' in forests, pits, water eyes and near roads. Incorrect dumping can pollute surrounding land and groundwaters. There is only one correctly managed dump for communal wastes, in Kaplica village in Somonino.

Those areas attractive for recreation and tourism have been exploited by the construction of many, often illegal, summer houses that are often located by lakes and rivers.

The regulation of construction, regional planning, sewage and waste management is the duty of the community bodies (some activities should be undertaken with the government administration of the region). However, these local administration controls were insufficient and ineffective up to the end of 1992.

13.5 CONCLUSIONS

The waters of the Radunia River and its tributaries do not meet the class I quality standards at any of the sampling points. This is mainly because of the presence of phosphorus compounds, nitrite nitrogen and faecal Coli bacteria.

As a result of additional testing in 1993, the following changes were noted when compared to the previous year:

- The state of water of the Radunia Channel deteriorated, particularly below the Straszyn Reservoir and in Bielkowo, from class III to unclassified; this was caused by high concentrations of nitrite nitrogen;
- The quality of water below Straszyn Reservoir also decreased from class III to unclassified, caused mainly by the presence of nitrite nitrogen and the faecal Coli bacteria.

There was a noticeable decrease in the concentration of total nitrogen and phosphorus in 1993 and a tendency for the inorganic forms of both elements to increase at almost all the sampling points along the Radunia River. The reduction in the total nitrogen and in nitrate nitrogen was noticed mainly downstream from the reservoirs. The quality of water from the bacteriological point of view meets the class III standards of quality at almost all the sampling points along the Radunia River.

The highest pollution in the main river was measured between the mouth of the Mała Supina tributary and the Łapino Reservoir and was similar to

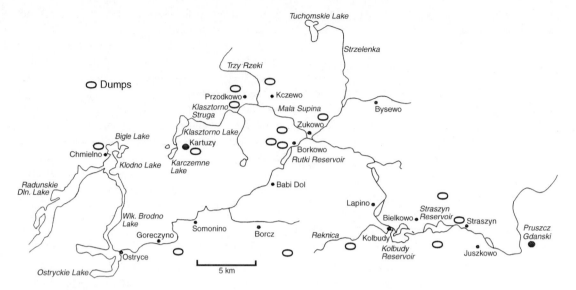

Figure 13.4 Illegal waste dump sites in the Radunia River catchment.

previous years. The highest pollution in the tributaries was measured in the Klasztorna Struga, particularly at Kobysewo below the discharge of the sewage treatment plant. The water quality decreased from class III to unclassified in two tributaries as compared with the previous year:

- the Strzelenka tributary at the outflow from Tuchomskie Lake;
- the Reknica tributary upstream from Czapelsko and at its mouth.

The pollution control of the waters of the Radunia River, which has been carried out for several years, has indicated that the waters do not meet the standards of class I in the section from its spring to the Straszyn Reservoir.

The improvement in water quality requires the elimination of point sources of pollution, surface pollution, agricultural pollution, etc. It is necessary for local governments to prioritize the many problems of building, engineering, regional planning, and sewage and waste management. This work should not exceed their capabilities and some activities should be undertaken with the government administration of the region. The first priority should be to control activities within the designated protection zone according to the timetable scheduled for them.

The good co-operation that exists within the public administration for solving the basic problems in the communities leads one to believe that the decision to create a protection zone on the Radunia River will not be an empty document this time.

REFERENCES

1. PWRN Gdansk (1972) Instruction 6, Official Journal 10.
2. Ministry of Environmental Protection, Natural Resources and Forestry (1991) Legislation Gazette 116, item 503.
3. Central Office of the Water Economy, Instruction 35.

THE INFLUENCE OF INDUSTRIAL AND MUNICIPAL WASTE ON THE QUALITY OF THE ENVIRONMENT IN POLAND, TAKING GDANSK REGION AS AN EXAMPLE

by P.J. Kowalik and H. Obarska-Pempkowiak

14.1 INTRODUCTION

In the 1970s, the Polish natural environment degraded to such an extent that it became a threat to the health of the people. This was largely as a result of an incorrect state investment policy and an indifferent attitude to the principles of the preservation of nature. The degradation involved the atmosphere, water and the soil with its vegetation. It was caused by the activity of industry, progressive urban development, and the lack of adequate legislation to safeguard the natural environment. Until recently, the law was very liberal with respect to the quantity and the quality of wastes discharged to the natural environment.

The ongoing process of integration into Europe makes it necessary to change the Polish standards and technological solutions for environmental protection so as to harmonize them with those of the EU. The degraded natural environment cannot support living to a higher standard. Hence the concept of ecological safety, which is based on the conviction that 'there is no high quality of life without a high quality of the surrounding environment', has more followers.

This chapter is intended to present the degree of the pollution of the natural environment in Poland with particular attention on the region of Gdansk.

14.2 ATMOSPHERIC POLLUTION IN POLAND

The emission of dust and gaseous pollutants in Poland in 1992 amounted to approximately 5530 t: this includes 3950 t of industrial discharges and 1480 t of road transport exhausts. The largest proportion of the gas and dust emissions, however, as much as 2.1×10^6 t in 1992, originates from the power industry. This industry includes power stations which are separate industrial enterprises. They generate the electricity required for both industry and households.

The main raw material used for generating power in Poland, is hard coal. The amount of hard coal as a proportion of the total raw material used for power generation amounts to 63%, while only 17% is brown coal. The principal air pollutants generated in the course of burning coal are finely dispersed dusts, sulphur dioxide, nitrogen oxides, and carbon dioxide. Table 14.1 illustrates the major atmospheric pollutants emitted to the air in Poland.

Nearly 80% of the emitted pollution comes from 96 industrial enterprises. These include the Sendzimir metallurgical works, the Katowice steel mill, the Głogóow plant, the heating power plants of Turow, Bełchatów, Polaniec, Jaworzno III, Kozienice, Siersza, and Łaziska [1]. The number of companies which lack equipment to

International River Water Quality. Edited by Gerry Best, Teresa Bogacka and Elżbieta Niemirycz. Published in 1997 by E & FN Spon, London, ISBN 0419215409

Table 14.1 Emission balance of main atmospheric pollutants in Poland in 1992

Sources of pollution	Sulphur dioxide (10^3 t)	Nitrogen oxides (10^3 t)	Dust (10^3 t)
Commercial power industry	1370	380	410
Local power plants	750	95	520
Industrial power industry	415	115	315
Transport and transport services	95	395	–
Industrial technologies	230	160	312

remove pollutants from the emitted gases amounts to 1165.

In addition to the losses of materials due to the emission of dusts and gases, these pollutants are responsible for reducing sunlight intensity and visibility, increasing the corrosion of metals and increasing the destruction of building construction plasters, facades of historic buildings and infrastructure.

The primary effect on the biosphere is the absorption of solar radiation in the range that is necessary in the carbon dioxide assimilation process. Another effect is the clogging of leaf stomata with dust, which adversely affects photosynthesis. Acidification in precipitation gives rise to a reduction in the activity of the bacteria producing nitrates and the decomposition of soil organic matter.

As a result, the forests are disappearing rapidly even though they are regarded as creating a protective ecosystem for the whole of life on Earth. In Poland the forest stands affected by harmful gases and dusts amounted to 3820 km² in 1980, 5880 km² in 1985, and 14 640 km² in 1992, which represents 21.5% of the total area of forest. International experts are unanimous in the view that, unless all nations take effective measures to protect forests, a chain reaction of climatic variations will take place, initiating the start of an ecological disaster. There will simply be a shortage of oxygen.

It is difficult to estimate the losses suffered as a consequence of atmospheric pollution. A variety of experiments are being carried out on the effect of air pollution upon people and animals, and from these it may be possible to determine the harmful and the acceptable levels of atmospheric pollution.

During 1986–7, the Institute of Environmental Protection in Warsaw prepared some computer programs from different local databases. These included annual data on significant industrial and communal polluters, quantities of emissions and charges imposed, together with an inventory of all the testing points, methods and measuring techniques used. From this study, an efficient national control system for the measurement of quality is envisaged. This would include, the control of the quality of sampling and measurement, by checking the quality of the meters and their accuracy, and inter-laboratory calibration procedures. Such a control-system should be implemented by the Polish Ministry of Environmental Protection [2].

14.3 POLLUTION OF WATER IN POLAND

In Poland, surface waters receive 10.5×10^9 m³ of sewage each year [1]. The large majority of this comes from 'conventionally pure' cooling waters, about 5.59×10^9 m³ yearly, i.e. about 55% of the total. Communal sewage contributes 20.7% and industrial waste about 14% of the total. The remainder comes from small individual discharges. In spite of the purity of cooling water, it is regarded as a threat to receiving waters because it raises the water temperature and this results in a decrease in the amount of dissolved oxygen and speeds up the eutrophication processes. If the cooling water element is ignored then about 3.8×10^9 m³ communal and industrial sewage is produced in Poland each year.

Of the 833 towns in Poland in 1991, 729 were equipped with a sewerage system, although 251 did not possess a sewage treatment plant [1]. The latter group included many large district towns, for instance, Białystok, Koszalin, Bydgoszcz and others. Many of the sewage treatment plants in the former group (160 mechanical and 318 biological) were largely ineffective because they were hydraulically overloaded and outdated. As a result only preliminary treatment of sewage was being achieved.

An improvement to the management of sewage requires the construction of new mechanical–biological treatment plants in towns with no treatment, and an increase in the effectiveness of the existing treatment plants by modernization and expansion. The investment needs, expressed in terms of the sewage treatment plant capacity, are estimated to be 4×10^6 m^3 per day, to be provided for 700 towns.

In 1991, of the 4414 large factories, only 2297 (52%) were provided with a sewage treatment plant. Out of this group, 353 did not possess any equipment at all and discharged their wastewater directly to the surface waters, while 1767 discharged to the municipal sewerage system, or to the ground. A major proportion of the industrial effluent was particularly harmful to the aquatic environment because it came from chemical, metallurgical, or wood and papermaking factories. The sewage from these industries is characterized by a high concentration of pollutants and toxic substances that cannot easily biodegrade [3]. The treatment of sewage from industrial enterprises requires the construction of new, or modernization of existing treatment plants, for an output of approximately 2×10^6 m^3 per day.

Other sources of water contamination are pollutants that originate from agricultural fields, meadows and grazing land, as well as those rural housing settlements with no sewerage systems. A control of the inflow of this kind of pollution is especially difficult. It can be achieved however, by

1. an improvement of the atmospheric purity;
2. raising the level of the sanitary conditions of the villages and recreation centres;
3. construction of buffer vegetation protective zones along streams;
4. construction of simple local sewage treatment plants (very small – up to 50 people, or small – between 50 and 1000 people);
5. dissemination of information to farmers on the use of fertilizers and soil testing. The aims of the latter would be to optimize the amounts of fertilizers and the timing of fertilization to the requirements of the plants, and to rationalize the use of crop protection products.

Liquid manure from industrial animal farms is a great hazard to the water environment. If it is discharged into fields in an uncontrolled way it causes contamination of the surface water and groundwater, and degradation of the soil. Organic pollutants contained in the liquid manure represents about 17% of the national waste load flowing into receiving waters.

A specific Polish problem is the salinity of water caused mainly by coal mines in the Upper Silesia (Górny Śląsk) area. The load of salt pumped out of mines and discharged into rivers amounts to about 9000 t of chloride and sulphate ions per day at present and it is anticipated that by the year 2000 it will have risen to 13 000 t per day. In the Vistula River catchment (45% of the total salt load) is discharged by just three coal mines, Czeczot, Piast, and Ziemowit, and for this reason 57% of its total length fails to conform to requirements [2]. Salt water can be pumped down into the ground below the coal mines and some experimental solutions are being tested.

The chemical fluids used in society also pollute the aquatic ecosystems. Only 50% of the products are used and the remaining part is disposed of to the environment. It is estimated that each year approximately 250 000 t of petroleum solvents, oils, lubricants, coolants, electrolytes, brake fluids, preservatives, and sludges [2] flow into surface waters, groundwaters and municipal sewerage systems. These enormous loads of wastes discharged to the water have a harmful effect upon water quality.

Since 1964, following the introduction of a unified system of measuring and estimating the quality of water, the length of class I rivers has systematically decreased. The lengths of rivers lower than class I, especially class III and unclassified, have increased.

In 1992, 7005.6 km of river were subjected to quality assessments. According to the physicochemical criteria only 2.4% of rivers were class I, 12.8% were class II, 24.8% were class III and 60% were unclassified as they were too polluted. If the microbiological criteria are taken into account, the situation is even worse (class I – 0%, class II – 1.6%, class III – 10.4%, unclassified – 88%). The classification for the Vistula River alone was: class I – 0%, class II – 5.4%, class III – 37.5% and unclassified – 57.1% [1].

Lake waters are also excessively polluted. On the basis of investigations of 161 lakes performed in 1984–8, including 120 large ones representing 26.9% of the total area of lakes above 0.5 km^2, 4 were class I, 64 were class II and 47 were class III. The water in 46 of the lakes was deemed to be unclassified [1].

An improvement of the quality of water in Poland should be attained by, among other things, a change in the administrative arrangements. For this purpose in 1991 there were created some new bodies for the management of water economics based on hydrographical regions, rather than administrative units of the country. They are seven Regional Boards of Water Management (RZGW) and the extent and locations of their headquarters are presented in Figure 14.1. Three of them, centred in Kraków, Warsaw, and Gdansk, manage the Vistula River basin. The next three (in Szczecin, Poznan, and Wrocław) – manage the Odra River basin, while the Katowice includes Górny Śląsk situated on the highest part of the Vistula and the Odra Watershed.

14.4 SOLID WASTES IN POLAND

The quantity of solid industrial waste and municipal refuse is increasing rapidly. From 1980 to 1992 the amount of wastes at disposal sites and dumps rose from 904 × 10^6 t to 1783 × 10^6 t; of this, 392 × 10^6 t accumulated in 1981–5. The quantity of recycled industrial wastes did not increase by any significant amount (it rose from 52.4% to 52.7%).

Only 0.4% of waste is treated, while 46.9% is dumped in controlled landfills. The remainder is disposed of in uncontrolled sites. In 1991 there were about 1800 registered dumping sites larger than 1 ha, totalling approximately 120 km^2 [1]. The municipal solid waste disposed of at dumping sites increased from 13 × 10^6 m^3 per year in 1975 to 46.5 × 10^6 m^3 per year in 1992 [1].

These statistics do not include the enormous amounts of soil displaced by surface mining and various kinds of earth work. Nor do the statistics include unregistered dumping such as construction materials (refuse) dropped at building sites and chemicals discarded by small production plants, service workshops, agricultural farms and households [4].

Figure 14.1 Regional Authorities of Water Management.

The public's awareness of the environmental hazards resulting from the inadequate management and rising production of wastes is increasing. However, the pollution control equipment fails to meet even the basic requirements. Of particular cause for concern are:

- the advanced pollution of the landscape with more and more uncontrolled waste disposal sites and the unsatisfactory conditions of communal waste dumps;
- severe problems connected with the location of the waste dumps, the management and reuse of the municipal wastes and lack of resources for securing technical equipment;
- losses of fertilizers and energy resources contained in liquid manure and in the biomass of the municipal wastewater sludge, and industrial and agricultural wastes.

Table 14.2 Characteristics of the Gdansk region

	The Gdansk Voivodeship
Area	7394 km^2
Total population	1.547×10^6
Urban population	1.032×10^6
Density of population	192/km^2
Arable land	60% of the total area
Forest	23% of the total area

Table 14.3 Annual emissions of sulphur dioxide and dust in the Gdansk region

Parameter	1980		1991	
	10^3 t per year	t/km^2 per year	10^3 t per year	t/km^2 per year
Sulphur dioxide	50.6	6.8	33.3	4.5
Dust	27.9	3.8	15.5	2.1

14.5 THE GDANSK REGION

For the purpose of the case-study, the Gdansk region is defined in Table 14.2 as the Gdansk Voivodeship, a unit in the administrative structure of Poland.

One of the main features of the region is the Vistula River estuary. This estuary has, for hundreds of years, been a centre of urban and economic activity.

The Gdansk region is known for its large sea harbours of Gdansk and Gdynia, for shipping and fishing activities as well as shipbuilding and ship repairing. It is a centre of the oil refining industry, phosphate fertilizer production, electronics, and fish and food processing industries. Several heat and power plants are also located in Gdansk and Gdynia. The towns are centres of university and cultural life.

When compared to other regions of Poland, the region of Gdansk has a rich supply of water from groundwater and from surface water (Radunia River and Reda River). The total length of the rivers in the region is some 1500 km, the major one being the Vistula. Together with tributaries the Vistula drains some 35% of the area. Other rivers draining directly to the Baltic are relatively short. The region has over 590 lakes with areas larger than 1 ha. Altogether the lakes constitute 2.4% of the area. The shoreline is 200 km long, but of this some 70 km constitute the shoreline of the Hel Peninsula.

14.6 SOURCES AND THE AMOUNTS OF POLLUTANTS IN THE GDANSK REGION

Pollutants are brought to the surface waters from the land and the atmosphere, and the sink is the Bay of Gdansk. Most of the pollutants emitted to the atmosphere originate from the phosphate fertilizer industry and combined heat and power plants (CHP) which are fuelled with poor quality coal.

The annual emissions of sulphur dioxide and dust are presented in Table 14.3. Emissions from 1980 and 1991 are presented in the table for comparison and show that the emissions decreased over this period. This is due to the extensive programme aimed at limiting air pollution by the installation of air purification facilities.

The average concentration of dust in air exceeded the acceptable value which is equal to 22 mg/m^3. The highest concentrations of over 100 mg/m^3 sulphur dioxide were found at Tczew, Puck and Kościerzyna.

Tczew 103 mg/m^3
Puck 133 mg/m^3
Kościerzyna 100 mg/m^3

The average annual concentrations of sulphur dioxide are within the acceptable limits of 64 mg/m^3. In some districts of Gdansk the concentration is 60 mg/m^3.

Air pollution causes severe but temporary problems in Gdansk, Gdynia and Sopot. In small cities and in rural areas the situation is better as far as air pollution is concerned but is much worse when the pollution of surface water is considered.

The total annual load of liquid wastes entering surface waters in the region in 1991 was 145.4×10^6 m^3. This comprised 26.2×10^6 m^3 industrial wastes and 119.2×10^6 m^3 municipal wastes.

Considering that 1.5 million people live in the region and assuming that 0.2 m^3 of water is used by each person per day, the total volume of municipal sewage should theoretically be about 110×10^6 m^3

Table 14.4 Pollutant loads from the Gdansk region

Source of pollutants	Loads (10^3 t)		
	BOD_5	Total nitrogen	Total phosphorus
Industry	1.9	0.7	0.1
Municipalities	11.2	3.5	0.9
Rivers from the region	22.5	2.9	1.1
Total from the region	38	7	2.1

per year, which is close to the measured value of 119.2×10^6 m^3 per year. The difference results from the fact that wastewater purification plants receive not only municipal sewage but industrial wastewater as well. On the other hand small towns and villages do not have sewage purification facilities so the wastewater produced by them is directed to surface waters without any treatment.

The annual loads of various pollutants are given in Table 14.4. It should be remembered that a large part of wastewater is discharged without any treatment to the Bay of Gdansk. In the marine environment nutrients are especially hazardous since they support primary productivity and this can result in secondary pollution of the marine environment because of the additional organic matter coming from weed and algal growth.

The load of pollutants in rivers within the boundaries of the Gdansk region is 22.5×10^3 t of BOD_5 and the load brought in by the Vistula from the interior of Poland is about 135.2×10^3 t of BOD_5. The Vistula River therefore brings in more than 75% of the load of organic pollution which enters the sea. This is not suprising considering that the Vistula drains 55% of Poland's land area. Data presented in Table 14.4 indicate that municipal wastewater, even after treatment, is still a significant pollutant load and exceeds that originating from industry. However, industrial wastewaters can contain many toxic substances, for instance heavy metals and/or chlorinated hydrocarbons, and therefore can be much more harmful than municipal liquid wastes. It is estimated than some 40%–75% of the load of pollutants indicated as 'rivers from the region' originate as non-point sources.

There is no doubt that the load of pollutants entering the Baltic is huge. It would be reduced if biological wastewater treatment plants were used in municipalities.

The extent of purification of sewage in the Gdansk region is presented in Table 14.5.

In Table 14.6, the percentages of elimination of pollutants at various stages of treatment are presented. Except for suspended solids, the efficiency of primary treatment is very limited. Despite this, 10% of the wastewater is directed to the surface water without any treatment.

In the Gdansk region there are only 12 plants constructed for the treatment of municipal wastes. Of these, seven are equipped with mechanical purification only; the other five plants have biological treatment facilities.

Solid wastes produced in the region also cause problems. Each year, factories producing phosphate fertilizers dump about 50 000 t of refuse on the environment. Heat and power producing plants dispose of 50 000 t of fly-ash, whilst municipal solid waste amounts to 2×10^6 t annually. Solid wastes are stored in 35 places, 5 of them are for industrial wastes alone.

From the foregoing it is clear that in the Gdansk region the atmosphere, soil and surface waters are contaminated with substantial amounts of liquid and solid wastes. The extent of this damage in the case of surface waters can be estimated on the basis of several quality criteria. When one or more of these criteria exceeds certain limits the water body classification changes. Biochemical oxygen demand, nitrogen and phosphorus are the chemical parameters most often used for water quality assessment, whilst the coliform count is the commonest biological parameter. Changes in the percentage of

Table 14.5 Volume of municipal wastes treated by various methods in the Gdansk region in 1991

Sewage type	Volume ($\times 10^6$ m^3)
Treated sewage	105.7
Stage I (mechanical)	79.7
Stage II (biological)	23.5
Stage III (chemical)	2.5
Untreated sewage	13.5
directed to surface waters	7.3
directed to sewerage	6.2

Table 14.6 Percentages of pollutants removed at various stages of treatment

Type of treatment plant	Efficiency (%)			
	Suspended solids	BOD_5	Total nitrogen	Total phosphorus
Mechanical	90	35	10	15
Biological	90	90	25	25
Chemical	–	–	5	95

rivers attributed to various classes of water quality according to the coliform index are presented in Figure 14.2. The most dramatic changes in the quality of water took place in the 1970s. The situation is as bad when chemical parameters are used. At present few rivers with water fulfilling the criteria for class I can be found in the Gdansk region. In Figure 14.3, the rivers in the Gdansk region are categorized according to chemical parameters. The most polluted rivers in the Gdansk region are the Wierzyca and the Martwa Wisła. Pollution of the Wierzyca is caused by municipal wastewater discharged from Kościerzyna, Starogard Gdanski, Pelplin and Gniew. The Martwa Wisła River is polluted with industrial wastewater from the Gdansk oil refinery, heat and power plants, food industry and shipyards.

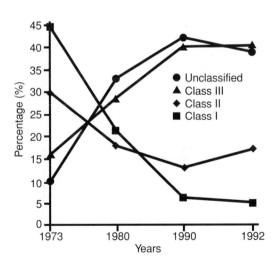

Figure 14.2 Percentages of rivers belonging to different classes in the Gdansk region according to coliform index.

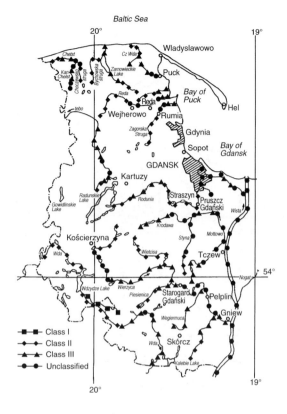

Figure 14.3 River quality classification in the Gdansk region using chemical parameters.

Municipal wastewater from the city of Gdansk is also dumped in this river from time to time.

There are particular difficulties in preserving water quality in the Radunia and Reda rivers. (See Chapter 13 by Walkowiak and Kozac for a description of the Radunia River.) These two rivers are used to supply drinking water to Gdansk and Gdynia.

Most of the polluted rivers discharge into the Bay of Gdansk. Despite the extensive exchange of water with the open sea, in the near-shore areas the effects of these polluted waters are noticeable. This is most clearly seen in the Bay of Puck where the ecological equilibrum deteriorated markedly in the 1970s. Within four years, traditional types of vegetation in an area of 100 km² were replaced with new ones. In Figure 14.4, the shaded area represents parts of the bay covered with the seaweed *Fucus vesiculosus*.

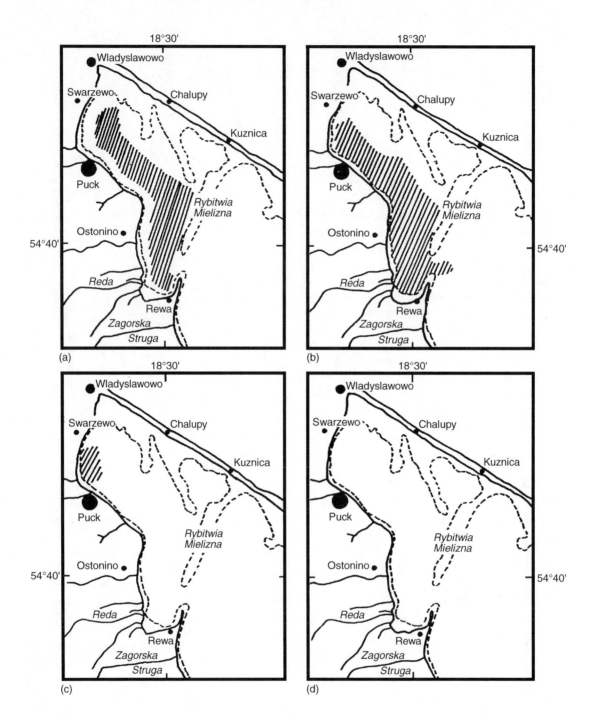

Figure 14.4 Distribution of *Fucus vesiculosus* in the Bay of Puck (a) in 1968; (b) from 1969 to 1971; (c) in 1977; (d) in 1979.

The area of the bay covered with seagrass diminished just as rapidly. If the seaweeds were taken as indicators of the ecological equilibrium than the most rapid changes took place from 1977 to 1979. Recently there are some indications of a recovery of the maritime ecosystem and the situation in 1994 is similar to that shown for 1977.

The deterioration of the quality of the environment in the bay was also measured at an increasing frequency using the coliform index to monitor safe baking levels. In 1993 alone only 20% of the beaches in the Bay of Gdansk were open for bathing (see Chapter 17 by Olańczuk-Neyman).

14.7 PREVENTIVE MEASURES

Local authorities of the Gdansk Voivodeship and municipalities directed 20% of their budgets to environmental protection in 1993. Additional funds were acquired from the Government and the Fund for Ecoconversion of Polish Debts. The total funds allocated to environmental protection in 1993 were some 600 billion Polish zlotys, which is the equivalent of US$ 40 million. The activities are divided into two groups: municipal and non-municipal. In this case 'non-municipal' activities are those financed from funds obtained from outside the Gdansk municipalities.

Most of the Gdansk investments are being directed to development of the wastewater treatment plants:

- purification at Wschód wastewater treatment plant is only mechanical and it will be equipped with a biological stage of purification;
- the construction of local wastewater treatment plants in the districts of Orunia, Górki Zachodnie, Olszynka, Letniewo, Wyspa Sobieszewska, Lipce, Święty Wojciech;
- the construction of the sewerage system in the newly developing districts of Gdansk.

The non-municipal investments are directed to:
- substitution of technology in the sulphur reloading station with a new one in the harbour of Gdansk, recultivation of soil around the reloading station;
- introduction of ecological engineering methods (hydrobotanical methods, soil filters and soil drainage) for small producers of wastewater.

14.8 CONCLUSIONS

The urban development and the industrialization of Poland has resulted in the degradation of the natural environment.

All power-generating stations and most industrial enterprises are primarily responsible for the polluted state of the atmosphere. Municipal sewage causes appalling sanitary conditions of the surface water. The discharge of wastewater caused a breakdown of the ecological balance in numerous rivers and lakes in Poland in the 1970s and the effects of this are evident 25 years later. Chemical degradation of soil and plants due to the effects of industrial pollutants is likely to intensify. This is of particular significance to forest ecosystems, where, although there is resistance to the degrading effects of pollutants, the technical means to counteract them are insufficient. A reduction of the pollution load emitted to the atmosphere will eventually diminish, but is unlikely to prevent the progressive degradation of forests since the immunity of forests' ecosystem against pollutants will also be weakened. The increased land area for the needs of housing, transport, industry, recreation, and reforestation will result in a reduction in the amount of arable land. By the year 2010 it is estimated that productive land will be reduced by approximately 12 000 km^2, whilst if the special reforestation programme is implemented the reduction will amount to 17 000 km^2.

Preventive action is proceeding at present, shown in the form of investment plans. This includes limiting polluting emissions by installing dust removal and absorbing equipment for flue gases, and upgrading existing sewage treatment plants and construction of new ones. The main limiting factors are a lack of capital for new investments and a lack of human resources for managing the country's environmental protection policies.

REFERENCES

1. Environmental Protection (1993) Polish Central Statistical Office (GUS), Warsaw.
2. Nowicki, M. (1990) General outline of Polish system for air pollution measurements. Environment Protection, Institute of Environmental Protection, Warsaw, 1, 13–27.
3. Sobczyk, M. (1993) Major problems of water management in Poland. Scientific – technical conference at Technical University of Koszalin and NOT (Chief Technical Organization) entitled *Actual Problems of Water and Wastewater Management*, Kolobrzeg, May 1993, pp. 71–86.
4. Siuta, J. (1990) Prognosis of degradation and land use in Poland. Environmental Protection. Institute of Environmental Protection, Warsaw, 1, 51–4.

FURTHER READING

1. Niemirycz, E., Taylor, R., Bogacka, T. (1992) Input of pollutants to the Bay of Gdansk, IMWM, Gdansk.
2. Nowacki, J. (1993) Trends of concentration changes of nitrogen and phosphorus compounds in the Bay of Puck, in *Ecological problems of the Puck region – state and preventional measures*, (ed. M. Pliński), Krokowa, pp. 79–108.
3. Environmental Protection (1992) Statistical report of Voivodship Statistical Office in Gdansk, Gdansk.
4. Sobol, Z., Szumilas, T. (1994) Reasons of poor sanitary state of coastal sea water and water of the Bay of Gdansk, in *Proceedings of Seminar Pollution and restoration of the Bay of Gdansk*, Gdynia 1991 (ed. J. Błażejowski and D. Shuller), pp. 104–11.
5. Trzosińska, A., Pliński, M. and Rybiński, J. (1994) Hydrochemical and biological characteristics of the Polish coastal zone of the Baltic Sea. Studies and Oceanological Materials (at present Oceanological Studies), 54, 5–70.
6. Zmudziński, L., Ciszewski, P., Kruk-Dowgiałło, L. *et al.* (1987) Assessment of environmental state of the Bay of Puck. Institute of Environmental Protection in Gdansk, Gdansk.

ACTION TAKEN BY THE GDANSK MUNICIPALITY TO REDUCE THE LOADS OF POLLUTANTS DISCHARGED INTO GDANSK BAY

by J. Kopeć and E. Niemirycz

15.1 INTRODUCTION

Gdansk is a port city located at the mouth of the Vistula River which flows into the Baltic Sea. The Vistula River, which is Poland's largest river, enters the sea through the Przekop Canal which forms the eastern boundary of Gdansk. Part of the southern border of Gdansk is formed by the Dead Vistula. This part of the Vistula River forms a reservoir of almost stagnant water downstream from the Bold Vistula and encloses Sobieszewska Island. The western part of the Vistula, with its small tributaries, forms a shipyard area with many outlets.

Other large rivers in the region include: the Motława River which flows into the Dead Vistula near Sienna Grobla and the Radunia River which splits in the City of Pruszcz into Old Radunia which flows into the Motława River at Niegowo and the Radunia Canal which flows through downtown Gdansk into the Motława River. There are also smaller watercourses including the Jelitkowski (Oliwski) Stream and the Kołobrzeski Canal (a stream presently flowing in an underground culvert) which flow directly into the sea. Also of importance is the Strzyża Stream which flows through the districts of Br'towo and Wrzeszcz eventually leading to the Dead Vistula, and Siedlicki and Oruński streams which discharge into the Radunia Canal.

15.2 SOURCES OF POLLUTION

Officially, these rivers and streams receive only storm waters and treated sewage effluent which are permitted by the water licence. There are some 30 wastewater outlets within the borders of Gdansk controlled by the Voivodeship Inspectorate of Environmental Protection. The largest of these include discharges from industrial plants like Gdansk Refinery, the Gdansk phosphorus fertilizer plant and shipyards, Siarkopol fat processing plant, Maritime Commercial Port, and the heat and power generating plant. These larger discharges are carefully controlled. Three years ago, the Gdansk authorities decided to evaluate the municipal pollutant load. This evaluation included quantifying the following parameters: BOD_5, COD, total suspended solids, total nitrogen and total phosphorus. Future expansion is expected to include oil derivatives. Tables 15.1–15.3 present the flows, concentrations of pollutants in the water and pollutant loads discharged into Gdansk Bay in 1994. Table 15.4 shows the reduction in total pollutant loads discharged into the Baltic Sea between 1988 and

---

International River Water Quality. Edited by Gerry Best, Teresa Bogacka and Elżbieta Niemirycz. Published in 1997 by E & FN Spon, London, ISBN 0419215409

Table 15.1 Pollutant loads discharged into Gdansk Bay in 1994 – sampling and flow data

Item	Sources of pollution	Flow ($\times 10^3$ m^3 per year)	Examination year	Number of examinations
	Water courses			
1	Jelitkowski Stream	356.6	1994	21
2	Bystrzec (Strzyza) Stream	736.6	1994	7
3	Siedlicki Stream	522.1	1994	7
4	Oruński Stream	49.1	1994	7
5	Stara Radunia	9 864.9	1994	7
6	Radunia Canal	4 289.3	1994	7
7	Motława Canal	13 234.1	1994	7
8	Dead Vistula		1994	7
	Total – water courses	29 052.7		
	Wastewater treatment plants			
9	Zaspa wastewater treatment plant	11 622.4	1994	52–208
10	Wschód wastewater treatment plant	45 381.5	1994	48–240
	Total – wastewater treatment plants	57 003.9		
	Industrial Plants			
11	Fat processing plant	1 012.4	1994	24
12	Heat and power generating plant (EC II)			
	Bystrzec Stream	27.5	1994	12
	Dead Vistula	114.1	1994	12
13	Gdansk Shipyard SA		1994	
	Gdansk Shipyard SA (ramp)		1994	
14	Heat and power generating plant (EC I – Ołowianka)	113.9	1994	12
15	Phosphorus fertilizer plant of Gdansk (Fosfory)	48.2	1994	156
16	Maritime Commercial Port SA			
	I Port canal (container wastewater treatment plant)	83.7	1994	156
	II Port canal (outlets)	8.6	1994	4
	III Gdansk Bay	14.4	1994	156
17	Cosmetics producing plant (Pollena)	0	1994	0
18	Siarkopol SA Gdansk	411.4	1994	104
19	Gdansk Refinery SA			
	I Vistula – Przekop	1 085.8	1994	104–1095
	II Rozwójka Canal	2 241.2	1994	104–1095
	Total – industrial plants	5 161.2		
	Total – Gdansk	91 217.807		
20	Vistula Kiezmark[a]	26 410 000	1993	53
	Total – sources of pollution	26 501 217.8		

[a] Data from 1992 and first quarter of 1993.

1993. The data set for 1988 is incomplete so, in some cases, more recent data have been included.

Ninety-five percent of the total pollution load discharged from Gdansk into the bay originates from the Vistula River, however this river carries a lot of water and the concentrations of selected pollutants are such that the quality of the river is class II or III (Chapter 3 by Dojlido). The remaining 5% of the pollutant load originates from:

- Wschód and Zaspa wastewater treatment plants (92.2%);
- watercourses (5.7%);
- industrial plants (2.1%).

Table 15.2 Pollutant loads discharged into Gdansk Bay in 1994 – average concentrations

Item	Sources of pollution	BOD_5 (mg/l)	COD^b (mg/l)	Tot. N (mg/l)	Tot. P (mg/l)	Suspended solids (mg/l)
	Water courses					
1	Jelitkowski Stream	3.3	20.0	0.90	0.07	12.2
2	Bystrzec (Strzyża) Stream	8.6	41.0	2.48	0.45	17.9
3	Siedlicki Stream	3.8	24.0	2.42	0.19	85.5
4	Oruński Stream	2.1	22.0	2.16	0.22	9.4
5	Stara Radunia	2.3	20.0	1.38	0.17	6.8
6	Radunia Canal	3.8	31.0	1.60	0.15	13.3
7	Motława Canal	4.8	22.0	1.63	0.20	10.7
8	Dead Vistula	5.7	27.0	1.01	0.44	8.8
	Total – water courses					
	Wastewater Treatment plants					
9	Zaspa wastewater treatment plant	24.0	49.0	18.80	5.90	13.0
10	Wschód wastewater treatment plant	96.0	161.0	44.80	1.50	85.0
	Total – wastewater treatment plants					
	Industrial plants					
11	Fat processing plant	104.6	246.6			135.5
12	Heat and power generating plant (EC II)					
	Bystrzec Stream	2.1	23.1			17.8
	Dead Vistula	2.2	25.9			25.5
13	Gdansk Shipyard SA					
	Gdansk Shipyard SA (ramp)					
14	Heat and power generating plant (EC I – Ołowianka)	1.1	20.3			21.4
15	Phosphorus fertilizer plant of Gdansk (Fosfory)					27.9
16	Maritime Commercial Port SA					
	I Port canal (container wastewater treatment plant)	9.8	61.2			43.3
	II Port canal (outlets)	27.0	101.0			30.0
	III Gdansk Bay	7.9	41.6			25.0
17	Cosmetics producing plant (Pollena)					
18	Siarkopol SA Gdansk	3.6	35.5			11.9
19	Gdansk Refinery SA					
	I Vistula – Przekop	3.5	45.4			15.7
	II Rozwójka Canal	9.7	67.4			24.0
	Total – industrial plants					
	Total – Gdansk					
20	Vistula Kiezmark[a]	4.6	28.9	2.83	0.26	26.7

[a] Data from 1992 and first quarter of 1993.
[b] Pollutant load calculations based on unadjusted flow values for 1993.

15.3 REMEDIAL ACTIONS TAKEN

During the last two years, the load of pollutants discharged by Wschód waste-water treatment plant has reduced significantly even though the treatment process at the plant is not yet complete.

Removing surplus wastewater from Zaspa wastewater treatment plant and reducing its hydraulic overloading to the design capacity, i.e. approximately 30 000 m³/day, has resulted in the effluent quality remaining (with some exceptions)

Table 15.3 Pollutant loads discharged into Gdansk Bay in 1994 – annual loads

Item	Sources of pollution	BOD_5 (t/year)	COD^b (t/year)	Tot. N (t/year)	Tot. P (t/year)	Suspended solids (t/year)
	Watercourses					
1	Jelitkowski Stream	1.2	7.1	0.32	0.03	4.4
2	Bystrzec (Strzyża) Stream	6.3	30.2	1.82	0.33	13.2
3	Siedlicki Stream	2.0	12.5	1.26	0.10	44.6
4	Oruński Stream	0.1	1.1	0.11	0.01	0.5
5	Stara Radunia	23.1	197.3	13.59	1.68	67.1
6	Radunia Canal	16.4	133	6.86	0.63	57
7	Motława Canal	63.4	291.2	21.55	2.7	141.6
8	Dead Vistula					–
	Total – watercourses	112.5	672.4	45.51	5.48	328.4
	Wastewater treatment plants					
9	Zaspa wastewater treatment plant	278.9	569.5	218.5	68.6	151.1
10	Wschód wastewater treatment plant	4 356.6	7 306.4	2 033.1	68.1	3 857.4
	Total – wastewater treatment plants	4 635.6	7 875.9	2 251.6	136.6	4 008.5
	Industrial plants					
11	Fat processing plant	105.90	249.7			137.2
12	Heat and power generating plant (EC II)					
	I Bystrzec Stream	0.06	0.6			0.5
	II Dead Vistula	0.25	3			2.9
13	Gdansk Shipyard SA					
	Gdansk Shipyard SA (ramp)					
14	Heat and power generating plant (EC I – Ołowianka)	0.12	2.3			2.4
15	Phosphorus fertilizer plant of Gdansk (Fosfory)					1.3
16	Maritime Commercial Port SA					
	I Port canal (container wastewater treatment plant)	0.82	5.1			3.6
	II Port canal (outlets)	0.23	0.9			0.3
	III Gdansk Bay	0.11	0.6			0.4
17	Cosmetics producing plant (Pollena)					
18	Siarkopol SA Gdansk	1.48	14.6			4.9
19	Gdansk Refinery SA					
	I Vistula – Przekop	3.80	49.3			17
	II Rozwójka Canal	21.74	151.1			53.8
	Total – industrial plants	134.5	477.1	0	0	224.3
	Total – Gdansk	4 882.5	9 025.4	2 297.1	142.1	4 561.2
20	Vistula Kiezmark[a]	121 486	763 249	74 740.3	6 866.6	705 147
	Total	126 368.5	772 274.4	77 037.4	7 008.7	709 708.2

[a] Data from 1992 and first quarter of 1993.
[b] Pollutant load calculations based on adjusted flow values for 1993.

within licence limits. At the same time, the introduction of chemical precipitation at Wschód wastewater treatment plant has improved the efficiency of treatment from 40% to 70%.

Table 15.4 Reduction of total pollutant loads discharged into the Baltic Sea from Gdansk between 1988 and 1993 (Source: Institute of Meteorology and Water Management and municipal authorities)

Sources of pollution	Parameter surveyed				
	BOD_5 (t/year)	COD (t/year)	Total nitrogen (t/year)	Total phosphorus (t/year)	Solids (t/year)
Watercourses (Items 1–6 in Tables 15.1–15.3)	1 017	3 232	433	68	2 253.2
Sewage treatment plants	5 777	10 263	no data available	no data available	1 152.7
Industrial plants	–14	237	no data available	–207	17 579.7
Total sewage discharged by the city of Gdansk	6 779	13 731	no data available	no data available	20 985.6
The Vistula River	13 132	95 511	56 236	657	no data available
Total including the Vistula River	19 911	109 242	incomplete	incomplete	incomplete

Industrial plants in general meet the requirements defined in their licences. The only exception is the fat processing plant. By contrast, the phosphorus fertilizer plant has succeeded in reducing the amount of phosphorus discharged from 111 t/year to approximately 340 kg/year. This is primarily attributed to the introduction of recycling of process water. At the same time, fluoride loads were reduced by 99.6%, i.e. from 76 t/year to 1.2 t/year, and total suspended solids by 98.4%, i.e. from 2200 t/year to 2.4 t/year.

Some industrial plants discharge their wastewater into the municipal sewerage system. This waste thus flows directly into municipal wastewater treatment plants. In some cases, industrial contaminants in this waste cannot be removed. Pollutant loads discharged by both industrial plants and watercourses are summarized in Tables 15.2 and 15.3.

For three years efforts were concentrated on reducing pollution in three selected watercourses: the Jelitkowski Stream, the Kołobrzeski Canal and the Radunia Canal. The first two required special attention since they discharge directly to the sea, whilst the Radunia Canal flows through the downtown area of Gdansk. The Jelitkowski Stream has been thoroughly cleaned. It has been canalized, all illegal connections to it cut off and ponds formed by the stream substantially cleaned (five out of ten so far, with four more to be cleaned during 1994). Surface run-off from the zoo flowing into the Rynarzewski Stream (a tributary of the Jelitkowski Stream) is treated by specially designed filters made of plants and sand (Chapter 31 by Obarska-Pempkowiak). At the stream estuary an experimental ecological screen with mussels has been constructed. In addition, the city plans to build a storm water pretreatment basin. At present the chemical parameters of the stream meet class I and class II water quality requirements.

At the Kołobrzeska Canal oil and sludge separators have been constructed and a pretreatment (sedimentation–aeration) basin installed adjacent to the estuary. As for the Radunia Canal, the sewer outlets and sources of pollution connected to them were surveyed in 1990–1. Funding in the municipal budget has been earmarked for providing a wastewater treatment system on the left side of the Canal in the Orunia area. A directive was issued concerning the need to connect sources of sewage to the wastewater sewerage system and eliminate illegal discharges. This has resulted in a reduction in the number of working outlets from 130 to 13. The water quality in the Radunia Canal now meets class II with regard to chemical parameters.

15.4 FUTURE PLANS

The next phase is to improve the Strzyża Stream. This will include the restoration of old detention ponds and construction of new ones; a green buffer zone will be established around the stream.

It is also intended in 1995 to pressurize the authorities of the municipal landfill site in Szadółki to treat the leachate from the site which is presently discharged by a drainage ditch to the Oruński Stream.

The municipal authorities have set themselves the objective of meeting the Helsinki Commission (HELCOM) requirements concerning the protection of the Baltic Sea environment. This goal is to reduce by 1995, the 1988 pollutant loads discharged into the Baltic Sea by 50%.

At a meeting of the Union of the Baltic Cities (UBC) held in Turku, Finland on 7 April 1994, the city of Gdansk in association with the cities of Gdynia and Sopot suggested that UBC members establish a pollutant load compilation and pollutant control database. This activity will start with a list of pollutants discharged by the participating countries adjacent to the Baltic Sea. This work could be done independently of HELCOM. The Helsinki Commission periodically organizes pollutant load assessments of all rivers with flows over 5 m^3/s. In Poland, these include the Vistula and Odra Rivers and ten other rivers along the coast.

Compiling a list of pollutant discharges by the Baltic Sea coastal communities would increase the interest of the authorities and promote better regional co-operation. This suggestion was tentatively accepted at the meeting in Turku and was sent to HELCOM. Poland's cities were asked to elaborate on this idea and present a more detailed programme. The mayor of the city of Gdansk approached other neighbouring districts located in the area of Gdansk Bay to discuss these possibilities. Many other districts showed enthusiasm and tentatively accepted this proposal. A more detailed workplan will be prepared in the near future.

Gdansk municipal authorities believe that joint efforts to support the construction of the wastewater treatment plants in Swarzewo adjacent to Puck Bay and in Jastarnia will create opportunities for biological life, recreation and natural diversity to return to the waters of Gdansk Bay.

The Vistula River has a significant influence on the water quality of Gdansk Bay. Although the amounts of pollutants discharged by this river to the sea are large, the contamination is significantly diluted. Also, the Vistula River has no negative influence on the bacteriological contamination of the waters of Gdansk Bay because bacteria have such a limited life expectancy (Chapter 17 by Olańczuk-Neyman).

According to the city authorities' plans, it will not be possible to complete Gdansk's sewerage system until the year 1998, but in the meantime the modernization and enlargement of the Wschód sewage treatment plant will improve treatment efficiency up to 98%. The desire is to see the 1000-year old city of Gdaask become a city with clean waters.

MONITORING OF VOLATILE ORGANIC COMPOUNDS (VOC's) IN THE SURFACE AND SEA WATERS OF GDANSK DISTRICT

by M. Biziuk, Ż. Polkowska, D. Gorlo, W. Janicki and J. Namieśnik

16.1 INTRODUCTION

Volatile organic compounds are one of the most common groups of anthropogenic water pollutants [1–6]. There are two main groups of them: volatile organohalogen compounds and volatile hydrocarbons. As they are not naturally occurring, the presence of them in surface waters is a measure of their anthropogenic use. Volatile organohalogen compounds are widely used domestically and in industry and are also formed during water disinfection by chlorination [1–6]. They enter the sea and rivers via discharges of municipal and industrial sewage and also can be transported by air. The concentration of volatile organic compounds in surface waters depends markedly on the presence of point sources of pollution, but also on weather: temperature, wind and rain. The investigations concerned the determination of these compounds in the surface waters of Gdansk district.

16.2 EXPERIMENTAL METHOD

Two methods were used to determine the volatile organic compounds. For volatile hydrocarbons (heptane, benzene, iso-octane, nonane, decane, undecane, xylene, toluene, chlorobenzene), a purge and trap technique followed by sorption on a solid sorbent (Tenax TA), then thermal desorption and final determination by gas chromotography – flame ionization detector (GC–FID) was applied [7,8]. The diagram of the apparatus used for this purpose is shown in Figure 15.1. A stream of purified argon was bubbled through an aqueous layer at a flow rate of 0.5–0.67 ml/s for ten minutes. The liberated volatile compounds were dried by the Nafion drier and trapped on the solid sorbent Tenax TA. After switching over the six-port switching valve the compounds to be analysed were desorbed in a stream of helium (20 ml/min) at a temperature of 250°C for two minutes and were quantified by GC–FID. The volume of the purging device and water sample were 20 ml and 10 ml respectively, and the mass of the sorbent layer was 120 mg. The chromatographic conditions were: Hewlett Packard 5830A gas chromatograph; 10% Dexil 300 coated on Chromosorb WAW DMCS (80–100 mesh) packed column (2 m); the detector temperature was 150°C. The detection limit of this method was about 0.1 µg/l.

For the determination of volatile organohalogen compounds (trichloromethane, bromodichloromethane, dibromochloromethane, tribromomethane, tetrachloromethane, trichloroethylene), direct aqueous injection (DAI) onto a capillary col-

International River Water Quality. Edited by Gerry Best, Teresa Bogacka and Elżbieta Niemirycz. Published in 1997 by E & FN Spon, London, ISBN 0419215409

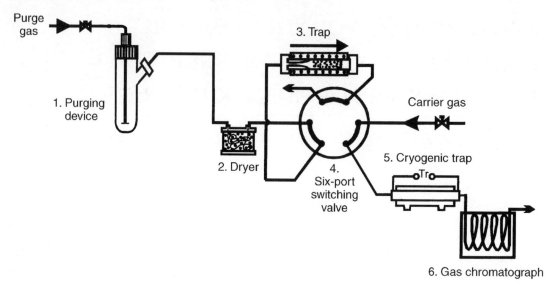

Figure 16.1 Schematic diagram of apparatus for determining volatile organic compounds using purge and trap techniques.

umn in a Carlo Erba (Italy) Vega 6180 gas chromatograph equipped with an electron capture detector (ECD 40/400) [9–13]. The chromatographic conditions were: 30 m × 0.32 mm fused silica capillary column coated with bonded 5 μm apolar DB–1 phase (J&W); 2 m × 0.32 mm fused silica pre-column; temperature programme – 102°C isothermally; injection system – cold on-column with secondary cooling; detector temperature – 350°C, with the pure nitrogen (99.99%) make-up gas (30 ml/min); carrier gas–hydrogen at 0.4 ml/s; injection volume – 1 μl. The detection limits of the DAI–ECD method is dependent on the species being determined but is generally about 0.01 μg and the standard deviation varies between 1.74% and 3.02%.

Samples of water were collected in glass flasks with ground-glass seals in such a way that no air was left in the flasks. The analysis was performed immediately after the samples were transported to the laboratory.

These methods have been used to determine the volatile hydrocarbons and volatile organohalogen compounds in:

- the Vistula river near Kiezmark;
- sea waters of the Bay of Gdansk (Gdynia–Orłowo quay and Sopot quay);
- rivers entering the Bay of Gdansk (Kacza, Radunia, Motława, Rozwójka); and
- rivers of the Krokowa agricultural area Piaśnica River near Zarnowieckie Lake, Czarna River near Sulicice and Karlikowo, Bychowska Struga near Wąglikowice and the agricultural drain near Szary Dwór.

Sampling was carried out in different seasons (autumn, winter, spring and summer).

16.3 RESULTS AND DISCUSSION

The results of the volatile organic compounds determined in the surface waters of the agricultural area of Krokowa are listed in Table 16.1. Very low levels of volatile organohalogen compounds were found and their concentrations did not exceed 0.07 μg/l. The concentrations of volatile hydrocarbons were also very low. The volatile compounds can be transported by air or can originate from vehicles and agricultural equipment. Table 16.2 lists the results of volatile organohalogen compounds determined in the surface and sea waters of Gdansk district. Much higher concentrations of these com-

Results and discussion 129

Table 16.1 Results of the determination of volatile organic compounds in surface waters in Krokowa[a] (units µg/l)[a]

Compound	Drainage ditch (Szary Dwór)	Piaśnica River (Żarnowiec)	Bychowska Struga (Wierzchucino)	Czarna River (Karlikowo)	Czarna River (Sulicice)
$CHCl_3$	0.03	0.07	0.07	0.03	0.04
$CHBrCl_2$	0.01	0.02	0.02	0.01	nd
$CHBr_2Cl$	nd	nd	nd	nd	nd
$CHBr_3$	nd	0.02	0.01	nd	nd
Benzene	0.1	1.5	1.5	1.6	0.24
Toluene	0.41	2.9	3.2	4.3	0.87
Nonane	0.74	4.8	5.3	3.2	1.79
Decane	1.24	7.9	8.1	1.8	2.07
Undecane	0.27	0.9	1.5	1	0.33

[a] nd = not detected

pounds were found at the sampling sites than in the rural areas. Relatively low concentrations were found in the Radunia River and the Moława River and slightly higher values for the Rozwójka River near Gdansk Refinery.

The largest concentrations were found in the Vistula River, the Kacza River and the sea water at Gdynia Orłowo near the mouth of the Kacza River. The Vistula River is Poland's largest river, transporting impurities from half of the country. The

Table 16.2 Results of the determination of volatile organohalogen compounds in natural waters (units µg/l)[a] (Source: Zygmunt)

Sampling site	Sampling date	$CHCl_3$	$CHBrCL_2 + C_2HCl_3$	$CHBr_2Cl$	$CHBr_3$	C_2Cl_4	CCl_4
Sea water at Orłowo	7 March 1994	2.2	1.44	0.55	0.02	nd	0.11
	14 March 1994	0.37	0.12	0.19	0.11	nd	0.01
	22 April 1994	0.34	0.02	nd	0.01	0.01	0.01
Kacza River	7 March 1994	1.28	0.86	0.36	0.01	nd	0.01
	14 March 1994	0.91	0.15	0.26	0.21	nd	nd
	22 April 1994	0.11	0.01	nd	nd	0.02	nd
Vistula River	7 March 1994	1.44	0.08	0.14	0.04	0.03	0.03
	14 March 1994	0.84	0.06	0.05	0.02	0.01	0.01
	22 April 1994	0.30	0.02	nd	nd	0.01	0.01
	10 June 1994	0.88	0.02	nd	nd	0.03	nd
Motława River	7 March 1994	0.26	0.02	nd	0.02	0.02	0.03
	14 March 1994	0.11	0.04	nd	nd	0.01	0.01
	22 April 1994	0.05	0.02	nd	nd	0.01	nd
	10 June 1994	0.03	nd	nd	nd	0.01	nd
Rozwójka River	7 March 1994	0.65	0.06	nd	0.01	0.02	0.28
	14 March 1994	0.37	0.11	nd	nd	0.01	0.01
	22 April 1994	0.06	0.02	nd	nd	0.01	nd
	10 June 1994	0.35	0.03	nd	nd	0.04	nd
Radunia River	7 March 1994	0.23	0.03	nd	nd	0.02	0.02
	14 March 1994	0.16	0.15	nd	nd	0.01	0.01
	22 April 1994	0.05	0.04	nd	nd	nd	0.01
	10 June 1994	0.04	0.12	nd	nd	nd	nd

[a] nd = not detected

Kacza River flows through the densely populated districts of Gdynia which lack a good sewerage system. In the Kacza River, a relatively high level of trichloroethylene, a typical industrial solvent was identified. Trichloroethylene was also found in the underground water supply at Sopot which explains why trichloroethylene was found in the sea waters at Gdynia Orłowo and Sopot.

Table 16.3 gives the results of the determination of volatile hydrocarbons from the same sampling points. High levels of benzene, decane, chlorobenzene and hexane were found in the Rozwójka River flowing near Gdansk Refinery. Other samples of river and sea water did not contain significant amounts of these compounds.

Although two of the sampling dates listed in Tables 16.2 and 16.3 were separated by just one week, very large differences in the concentration of the substances were noted. During the week between the sampling, there was a lot of wind. The results suggest the large influence that weather conditions can have on the concentration of volatile compounds in surface waters.

The proposed methods have been used successfully for the determination of volatile hydrocarbons and organohalogen compounds in rivers and sea waters. The results show that the volatile organic compounds are not present at high concentrations in the samples collected, with the exceptions of trichloroethylene in the Kacza River and sea water at Orłowo, and hydrocarbons in the Rozwójka River. Nevertheless all the compounds of this group are considered hazardous to human beings and the environment and the pollution of water by municipal and industrial sewages containing these compounds needs to be controlled.

REFERENCES

1. Dojlido J., (1987) *Chemia Wody*, Arkady, Warsaw.
2. Hellman H., (1987) *Analysis of Surface Waters*, Ellis Horwood, New York.

Table 16.3 Results of the determination of volatile hydrocarbons in natural waters (units µg/l) (Source: Zygmunt)

Sampling site	Sampling date 1994	Hexane	Heptane	Benzene	Toluene	Nonane +Xylene	Decane	Undekane	Chloro-benzene	Iso-octane
Vistula River in Kiezmark	7 March	nd	nd	0.78	2.14	nd	nd	nd	0.89	nd
	14 March	nd	nd	nd	nd	nd	nd	nd	nd	nd
	22 April	nd	nd	nd	nd	nd	nd	nd	nd	nd
	10 June	nd	nd	nd	nd	0.35	nd	0.16	nd	0.23
Radunia River in Straszyn	7 March	nd	nd	1.08	1.23	nd	3.14	nd	0.38	nd
	14 March	nd	nd	nd	nd	nd	nd	nd	nd	nd
	22 April	nd	nd	nd	nd	nd	nd	nd	nd	nd
	9 June	nd	nd	0.6	0.3	nd	nd	nd	nd	nd
Motława River in Gdansk	7 March	nd	nd	0.83	1.22	nd	nd	nd	nd	nd
	14 March	nd	nd	nd	0.07	nd	nd	nd	nd	nd
	22 April	nd	nd	nd	nd	nd	nd	nd	nd	nd
	10 June	nd	nd	nd	nd	0.03	nd	0.18	nd	nd
Kacza River in Orłowo	7 March	nd	nd	0.18	0.93	nd	nd	nd	nd	2.05
	14 March	nd	nd	0.08	0.6	nd	nd	nd	nd	nd
	22 April	nd	nd	nd	nd	nd	nd	nd	nd	nd
Rozwójka River (near Refinery)	7 March	nd	nd	33.7	0.07	nd	48.8	nd	31.3	nd
	14 March	nd	nd	22.3	nd	nd	22.5	6.7	nd	nd
	22 April	nd	nd	25.9	15.2	34.4	9.9	2.9	nd	nd
	9 June	296	nd	165.0	nd	0.1	0.03	nd	0.03	nd
Sea water (Orłowo)	7 March	nd	nd	nd	nd	nd	0.17	nd	nd	nd
	14 March	nd	nd	nd	nd	nd	0.08	nd	nd	nd
	22 April	nd	nd	nd	nd	nd	nd	nd	nd	nd

ᵃ nd = not detected

3. de Kruijf, H.A.M. and Kool, J.H. (eds) (1985) *Organic Micropollutants in Drinking Water and Health*, Elsevier, Amsterdam.
4. Fresenius, W., Quentin, K.E. and Schneider, W. (eds) (1988) *Water Analysis*, Springer Verlag, Berlin.
5. Moore, J.W. and Ramamoorthy, S. (1984) *Organic Chemicals in Natural Waters. Applied Monitoring and Impact Assessment*, Springer Verlag, New York.
6. Nemerow, N.L. (1985) *Stream, Lake, Estuary and Ocean Pollution*, Van Nostrand Reinhold Co., New York.
7. Janicki, W., Zygmunt, B., Wolska, L. *et al.* (1992) *Chem. Anal.*, 37, 599.
8. Janicki, W., Wolska, W., Wardencki, W. *et al.* (1993) *J. Chromatogr.*, 654, 279.
9. Loung, T., Peters, C.J., Young, R.J. and Perry, R. (1980) *Environ. Technol. Letters*, 1, 299.
10. Temmerman, I.F.M.M. and Quaghebeur, D.J.M. (1990) *J. High Res. Chromatogr.*, 13, 379.
11. Biziuk, M. and Polkowska, Ż. (1993) *Pollutants in Environment*, 3, 12.
12. Grob, K. (1984) *J. Chromatogr.* 299, 1.
13. Grob, K. and Habich, A. (1983) *HRC & CC*, 6, 11.

CAUSE ANALYSIS OF MICROBIOLOGICAL POLLUTION AROUND BEACHES ALONG THE GDANSK COAST

by K. Olańczuk-Neyman

17.1 INTRODUCTION

The coastal zone of Gdansk Bay adjacent to the three-city agglomeration of Gdansk, Sopot and Gdynia extends over a length of 35 km and presents an attractive area for tourism and recreation (Figure 17.1). Unfortunately the coastal waters in some places have consistently been polluted, which has resulted in their deterioration and the closure of numerous beaches.

Of the different variables included in water quality assessment for recreation purposes the most important are microbiological indicators (Table 17.1). The selection of variables or pollutants should only include those most appropriate to local conditions.

17.2 SOURCES OF POLLUTION

In the case of the Gdansk coastal waters the principal sources for the poor sanitary condition of these waters are rivers, streams and storm-water drains flowing in the vicinity of the beaches. They carry with them fresh microbiological pollutants to the coastal waters. Of these the most significant are micro-organisms of human and animal faecal origin because they create risk to human health.

Human faeces can contain a variety of intestinal pathogens which cause diseases ranging from mild gastroenteritis to serious, and possibly fatal, dysentery, cholera and typhoid. Depending on the prevalence of certain other diseases in a community, other viruses and parasites may also be present [1, 2].

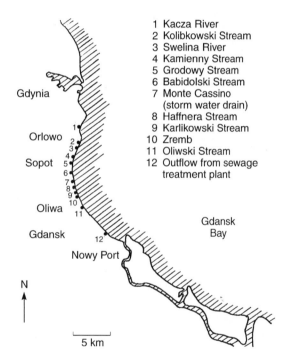

Figure 17.1 Location of sampling points along coast of Gdansk Bay.

1 Kacza River
2 Kolibkowski Stream
3 Swelina River
4 Kamienny Stream
5 Grodowy Stream
6 Babidolski Stream
7 Monte Cassino (storm water drain)
8 Haffnera Stream
9 Karlikowski Stream
10 Zremb
11 Oliwski Stream
12 Outflow from sewage treatment plant

International River Water Quality. Edited by Gerry Best, Teresa Bogacka and Elżbieta Niemirycz. Published in 1997 by E & FN Spon, London, ISBN 0419215409

Table 17.1 Factors to be considered when assessing water quality for recreational purposes

Variables	Pollutants
Physico-chemical	Organic
Temperature	Oil and hydrocarbons
Colour	Surfactants
Odour	
Suspended solids	Microbiological
Turbidity	Total coliforms
Ph	Faecal coliforms
Chlorophyll-*a*	Faecal streptococci
Total organic carbon	Pathogens
Ammonia	
Nitrate/nitrite	
Phosphorus/phosphate	

Table 17.2 Microbiological quality requirements for bathing waters of Poland. Percentages in parentheses are the proportions of samples which must meet the required standard

Parameters	Recommended	Mandatory	Minimum sampling frequency
Faecal coliforms /100 ml	<1000 (100%)	<1000 (70%) <10 000 (30%)	monthly

Source: Polish Sanitary Inspectorate

Municipal raw sewage typically contains 0.3 to 30 million *Escherichia coli* per 100 ml. In addition, up to 1 million enterovirus particles are excreted per g of faeces. It should be stressed that some enteroviruses are highly infectious and under suitable conditions as few as one virus can cause infection [3].

Many water-borne pathogenic micro-organisms can be excreted by an apparently healthy person acting as a carrier and can also be carried by some birds and animals; one to four percent of the human population are carriers. The incidence of *Salmonella* infection is much higher in farm animals than in humans. The daily load of *E. coli* produced by the waste from gulls is equal to human waste, and their droppings are rich in pathogens such as *Salmonella* and *Campylobacter*. Therefore the contamination of water bodies by animal or human excrement introduces the risk of infection to those who use the water for drinking and for recreation [4].

17.3 QUALITY STANDARDS

The microbiological quality requirements for bathing waters of Poland and countries within the EU are given in Tables 17.2 and 17.3.

17.4 EFFECT ON TIDAL WATER QUALITY

The excessive number of faecal micro-organisms in marine recreational water of Gdansk region has resulted in some beaches being closed to the public. The results of investigations carried out have proved that the major sources of bacteriological pollution of the coastal recreational waters are the streams flowing into the sea in the vicinity of the bathing areas. The streams introduce large quantities of bacteriological contamination of faecal origin (Table 17.4).

The bacteriological load discharged through the coastal streams directly to the water of Gdansk Bay depends on the flow of the streams and their degree of pollution. The flow, as well as the bacteriological pollution of the streams, varies in time over a relatively wide range. The variations are mainly due to variable atmospheric conditions and human activity. Table 17.4 illustrates values of the medians of the parameters tested for respective streams measured in 1993 [5]. The results for the Oliwa Stream are for the first half of 1994 [6]. The location of the sampling points are shown in Figure 17.1. The watercourses tested contain a large quantity of bacteria ranging between 2300 and 2.4 million in 100 ml of water. The most contaminated water is in the Monte Cassino main sewer.

To make the interpretation of the results easier the daily bacteriological pollutant loads have been calculated in terms of the number of population equivalents, assuming that the number of *E. coli* excreted in faeces is 1.9×10^9 per person per day [7]. In this way it has been concluded that the largest bacteriological load discharged each day by rivers, streams or storm-water overflows, comes from the Kacza River.

By contrast, Table 17.4 also shows the load disposed of by the Zaspa sewage treatment plant,

Table 17.3 Microbiological quality requirements for bathing waters of countries within the EU. Percentages in parentheses are the proportions of samples which must meet the required standard

Parameters	Recommended	Mandatory	Minimum sampling frequency
Total coliforms/100ml[a]	500 (80%)	10.000 (95%)	Fortnightly
Faecal coliforms/100 ml[a]	100 (80%)	2.000 (95%)	Fortnightly
Faecal streptococci/100ml[b]	100 (90%)	–	–
Salmonella[b] 1/dm^3	–	0 (95%)	–
Enteroviruses[b] PFU 10/dm^3	–	0	–

[a] sampling frequency can be reduced by a factor of 2.
[b] sampling frequency is optional depending on competent authorities.

which gives mechanical and biological treatment to 60 000 m^3 of sewage per day. An average daily load of approximately 7.11 t/day BOD$_5$ of raw sewage flows to the plant. The bacteriological load of coli bacteria in the raw wastewater flowing to the plant reaches 2.04×10^{15} per day but this loading is variable. The greatest reduction in the number of faecal coli bacteria takes place during the biological stage of the sewage treatment when the retention time is about 5.5 hours (approximately three times longer than the equivalent stage in the mechanical process). The sewage treatment efficiency, as shown by the reduction in the number of faecal coli bacteria, is approximately 96%. The number of bacteria at the outflow from the treatment plant is 7.79×10^{13} per day which corresponds to 41 000 population equivalent [8], i.e. nearly four times greater than that of the Kacza River.

Table 17.4 Bacteriological loads from rivers and streams flowing into Gdansk Bay near beaches (Sources: Municipal Offices of Sopot and Gdansk, unpublished results of studies in archival materials; Jones and White, *Water Pollution Control*, **83**, 215–225 (1984))

Sampling point	Name of stream or river	Flow rate (m^3/s)	MPN E.coli in 100 ml of water	Bacteriological load (E.coli/day)	Daily bacteriological load expressed as number of population equivalents
1.	Kacza River	0.105	230 000	2.08×10^{13}	11 000
2.	Kolibkowski Stream	0.0053	2 300	1.2×10^{10}	6
3.	Swelina River	0.024	2 300	4.77×10^{10}	25
4.	Kamienny Stream	0.025	2 300	4.96×10^{10}	26
5.	Grodowy Stream	0.088	2 300	1.75×10^{11}	92
6.	Babidolski Stream	0.0187	59 000	9.53×10^{11}	500
7.	Storm-water drain, Monte Cassino	0.0025	2 400 000	5.20×10^{12}	2 730
8.	Haffnera Stream	0.0029	23 000	5.40×10^{10}	28
9.	Karlikowski Stream	0.0319	23 000	6.34×10^{11}	330
10.	Storm-water drain, Zremb	0.0028	2 300	5.60×10^{9}	3
11.	Oliwski Stream	0.068	2 300	1.35×10^{11}	71
12.	Sewage effluent from Zaspa treatment plant	0.3921	70 000	7.79×10^{13}	41 000

The bacteriological contamination of the streams is caused both by communal sewage, including that from septic tanks, and the inflow of rainwater which washes away the impurities from the streets, including animal faecal matter. The dog population of Gdansk, Sopot and Gdynia is estimated to deposit about 2 t of faeces on the streets each day, much of which is washed away by rainwater into local streams and rivers [9]. This is about 2×10^{13} *E. coli* per day (10 000 population equivalent).

According to the bacteriological results, the effect of the polluted streams on the coastal waters depends on bacteriological loads of the streams and the distance from the shore. The bacteriological quality of the water improves with increasing distance from the shore and increasing distance from the outflow of the streams [10]. The effects are confined to a coastal strip 500 m wide.

The quality of the coastal waters is dependent on the inflow of new bacteriological pollutants. Indicator bacteria in fresh waters survive a matter of days whereas in sea water the salinity reduces the survival time to a few hours.

An additional hazard is the presence of bacteriological pollutants that accumulate in the sand, which is washed by the sea water. It has been shown that the sand from the beach washed out by the sea is more polluted than the water itself. In general, the number of indicator bacteria in the sand washed out by polluted water is up to a thousand times greater than in the water. The bacteriological contamination of sand was measured at three locations: in the area washed out by the sea water; in the middle of the beach; in the dunes – about 50 m from the shore line. The degree of contamination decreases with increasing distance from the source of pollution (i.e. the outlet of the stream). This varies from highly polluted (sand washed by sea water) to non-polluted (the dunes) [10]. Therefore a ban on the use of the beaches should include not only the coastal waters but also that part of the beach which is temporarily washed over by sea water.

Pathogenic and non-pathogenic bacteria and human enteroviruses present in large numbers in the bottom sediments may be released upon resuspension following dredging, boating, storms and other activities. It has been found that a significantly higher number of drug-resistant bacteria carrying R-factors (R^+) occur in sediment than in surface water [11]. R-factors may enhance the infectiveness and virulence of pathogens such as *S. typhi* and *Shigella* [12, 13] and may also carry enteropathogens among *E. coli*, thus making them harmful instead of commensal [14].

The guidelines of the Polish Sanitary Inspectorate, which define the conditions for using the beaches, are based on the results of the bacteriological quality of the coastal waters. According to the guidelines, a ban on using public beaches, including sunbathing, comes into effect if more than 30% of the water samples tested contain over 10 000 faecal coli bacteria in 100 ml and the proportion of water samples containing up to 1000 faecal coli bacteria in 100 ml exceeds 70%. However, the guidelines permit the use of the beach if no more than 10% of the water samples tested exceed 1000 in 100 ml and no samples contain more than 10 000 faecal coli bacteria in 100 ml.

REFERENCES

1. Chapman, D. and Kinstach, V. (1992) The selection of water quality variables, in *Water Quality Assessments*, (ed. D. Chapman) Chapman & Hall, London.
2. Olańczuk-Neyman, K. (1980) Microbiological aspects of marine water pollution, *Inzynieria Morska* No. 11, 393–5.
3. Slyde, J.S. (1981) *Viruses in Wastewater Treatment*. (eds M. Gooddard and M. Butler.) Pergamon Press, Oxford, New York.
4. Gray, N.F. (1989) *Biology of Wastewater Treatment*, Oxford University Press.
5. Municipal Office of Sopot (1994) Unpublished results of studies in archival materials.
6. Municipal Office of Gdansk (1994) Unpublished results of studies in archival materials.
7. Jones, F. and White, W.R. (1984) Health and amenity aspects of surface waters, *Water Pollution Control*, 83, 215–25
8. Sulińska-Bogusz, R. (1993) Bacteriological analysis of the Zaspa wastewater treatment operation. Diploma work, Technical University of Gdansk.
9. Polish Ecological Club (1993) Report on coastal streams.

10. Olańczuk-Neyman, K., Czerwionka, K. and Górska, A. (1992) Bacteriological pollution of coastal water and sand in the region of the Gdansk Bay, in *Proceedings of the International Symposium on Research on Hydraulic Engineering*, Zagreb, Croatia, 7–9 September 1992.
11. Goyal, S.M., Gerba, Ch. P. and Melnick, J.L. (1979) Transferable drug resistance in bacteria of coastal canal water and sediment. *Wat. Res.* **13**, 349–56.
12. Gangarosa, E.J. *et al.* (1972) An epidemic-associated episode? *J. Infect. Dis.* 126, 215–18.
13. Thomas, M.E., Haider, Y. and Datta, N. (1972) An epidemiological study of strains of *Shigella sonnei* from two related outbreaks. *J. Hydrol.* 70, 589–96.
14. Geldreich, E.E. (1972) Water-borne pathogens, in *Water Pollution Microbiology* (ed R. Mitchell) Wiley–Interscience, New York, pp. 207–41.

LEACHING OF NUTRIENTS, HEAVY METALS AND PESTICIDES FROM AGRICULTURAL LAND IN SWEDEN

18

by N. Brink

18.1 INTRODUCTION

In natural soils there are two main streams of substances entering rivers depending on the water flow. In tile-drained soils a third pathway appears, which considerably shortens the distance and time of the substance flux through the soil. The drain systems thus conserve the groundwater but increase the harmful effects on surface water [1].

In the 1960s investigations showed increased amounts of harmful substances in water-courses originating from farmland and settlements. Research was done using whole-field and plot experiments to determine the extent of the contribution from pure arable land. The whole-fields were between 4.5 and 36 ha drained soils, the plots generally 0.3 to 0.4 ha [2]. Some of the fields and plots were tile drained, and the water discharged from each was continuously measured. The objective of this chapter is to discuss the available options in order to diminish leaching losses of nutrients and pesticides.

18.2 LEACHING OF NUTRIENTS

18.2.1 NITROGEN

The nitrogen flux in agriculture is important. The nitrogen balance of an average hectare of arable land is shown in Table 18.1.

Table 18.1 Nitrogen balance of an average hectare of arable land

Input nitrogen (kg/ha per year)		Output nitrogen (kg/ha per year)	
Fertilizer	80	Gases	72
N-fix, legum	25	Leaching agr.	30
Precipitation	10	Sewage water	12
Import fodder	10	Export food	5
		Food waste	5
		Sewage sludge	1
Sum	125	Sum	125

Thus the average total annual input is 125 kg/ha of arable land. Of that 23 kg is used by people and the total loss from soil and stables and animal waste is 102 kg – 72 kg to the air and 30 kg to water. Figures 18.1–18.3 show that the yield components of oats, winter wheat and barley increase considerably at low input and level out at high input, whereas leaching shows the opposite effect. Large cattle stables, dung-yards and silage bunkers are often important polluters of streams [3].

The climate is crucial (Figure 18.4). The long winter in northern Sweden effectively limits the loss, since most of the water, which has a low concentration of nitrogen, runs off on the surface. The loss is by far the greatest in southern Sweden, especially on sandy fields where most of the water per-

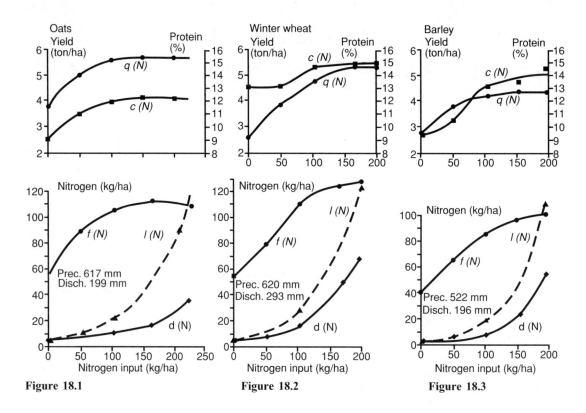

Figure 18.1 Yield components and leaching arising from nitrogenous fertilization of oats (a) yield q(N) and protein content of grain c(N); (b) harvested protein f(N), leaching with drainage water d(N), and total leaching to surface water and groundwater l(N).

Figure 18.2 Yield components and leaching arising from nitrogenous fertilization of winter wheat (a) yield q(N) and protein content of grain c(N); (b) harvested protein f(N), leaching with drainage water d(N), and total leaching to surface water and groundwater l(N).

Figure 18.3 Yield components and leaching arising from nitrogenous fertilization of barley (a) yield q(N) and protein content of grain c(N); (b) harvested protein f(N), leaching with drainage water d(N), and total leaching to surface water and groundwater l(N).

colates through the soil. Also the kind of cropped plant is important. Perennial plants use nitrogen most effectively and annual ones less efficiently. Following ground loses nitrogen rapidly.

18.2.2 PHOSPHORUS

The total annual loss does not differ much between north and south. The variation is mostly between 0.2 and 0.4 kg/ha per year and this depends on the extent of freezing of the soil and the soil type. Thus the erosion from clay is much higher than from sand. Another important factor is the freezing out of phosphorus from plant residues lying on the ground [4, 5].

18.3 LEACHING OF HEAVY METALS

The concentrations and fluxes of heavy metals were measured by Andersson, Gustafson and

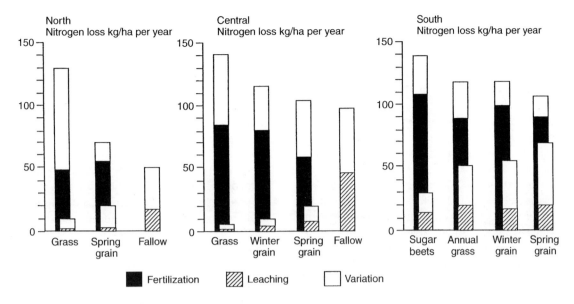

Figure 18.4 Role of the climate and type of crop on leaching losses of nitrogen.

Torstensson [6]. The pH was mostly between 6.6 and 7.9 but at a station with pH equal to 4.7 very high values were found for almost all metals (Table 18.2). This was because of the oxidation of sulphides in the tile-drained soil and the dissolution of the metals by sulphuric acid. The high concentration of manganese measured is close to the toxic level for plants. It was also shown that the leaching of metals is highly dependent on the suspended matter in the water as the metals are mostly adsorbed onto particulate matter.

The high flux of metals from organic soils depends on fluctuating groundwater levels with alternating reducing and oxidizing conditions in the subsoil. The input of metals in rain and snow exceeds that of their output in drainage water. The surplus accumulates in soil and/or is taken up by crops.

18.4 LEACHING OF PESTICIDES

There is a widespread opinion that generally pesticides do not leach through soils to surface and groundwater [7]. Nonetheless, contamination of wells and streams has occurred [8]. This was explained at the time as being the result of incorrect handling of application equipment, careless disposal of residues and spray drift. Such explana-

Table 18.2 Concentrations and fluxes of metals in drainage waters with contrasting levels of acidity (Source: Andersson, Gustafson and Torstensson, Removal of trace elements from arable land by leaching, *Ekohydrologi* 26, 13–22, 1988)

	Chromium	Manganese	Nickel	Copper	Zinc	Cadmium	Lead	pH
Concentration (µg/l)								
Mineral	0.67	41	1.8	2	12.7	0.044	0.55	6.6
Organic	7.6	5410	108	39	311	0.9	0.53	4.7
Flux (g/ha per year)								
Mineral	0.71	39	3.9	4.3	7.5	0.057	0.52	—
Organic	3.09	7300	12.3	12.7	160	0.376	4.15	—

tions might be correct but often it may also be due to the leaching through the soil or run-off on the soil surface.

A monitoring programme was carried out on the leaching of pesticides into rivers [9]. Thirty streams were sampled and 18 compounds identified. The substances most frequently encountered close to the time of spraying were: the herbicides atrazine, the phenoxy acids MCPA and dichlorprop, the fungicide metalaxyl and the insecticide lindane (Table 18.3).

Phenoxy acids disturb the growth of tomatoes, onions and lentils at 2–100 µg/l [10]. Atrazine has similar effects at 0.4–0.6 µg/l [11], inhibits photosynthesis and may affect phytoplankton at 1–5 µg/l [12]. Atrazine is now banned in Sweden. Insecticides such as endosulfan and lindane are toxic to fish. Endosulfan is toxic at 0.003 µg/l [13] which is far below the detection limit of 0.1 µg/l in the method used in this investigation. For several months, lindane exceeded the Swedish Environmental Protection Agency criterion of 0.01 µg/l for aquatic life in fresh water.

18.5 MEASURES TO PREVENT LEACHING

It is an urgent environmental task to minimize the flux of nitrogen to coastal waters and the fluxes of phosphorus and pesticide to rivers and lakes. At the same time, the existing overproduction of cereal crops should be taken into account. There are three possibilities for achieving this in Sweden:

1. Decreasing the intensity of agriculture: The ways to do this include taxes on fertilizers and pesticides. Taxes amount to 10% on the prices of fertilizers (nitrogen and phosphorus) and four Swedish krona per kg active ingredient of pesticides.
2. Changing land use: A payment is made for changing the land use from the production of annual plants to grassland farming and afforestation. The payments range from 700 to 2900 krona per ha annually depending on the agricultural district in which the farm is situated.
3. Special restrictions. Restrictions are placed on handling of manure in catchments of two very sensitive water bodies. In those areas, the stocking rate is to be restricted to 1.6 dairy cows per ha of arable land, and corresponding rates for other animals.

The objectives for the future include a reduction in the use of commercial fertilizers, restricting animal farming and reducing the use of pesticides. As there is a parabolic relationship between intensity and leaching of nitrogen (Figures 18.1–18.3), and a linear one between cultivated area and leaching, a reduction of intensity would be more efficient in minimizing leaching than a reduction of the area fertilized [2]. Among cropping systems analysed, wheat with conifers is the most effective from an environmental point of view, followed by wheat with deciduous trees and wheat with long grass. Wheat with annual grass would not improve leaching losses.

Table 18.3 Maximum pesticide concentrations and numbers of positive samples in Swedish stream waters (Source: Kreuger and Brink, Losses of pesticides from agriculture, 1988)

	Max conc (µg/l)	Number of positive samples						
		Apr	May	Jun	Jul	Aug	Sep	Oct
Pesticide								
Atrazine	6	0	5	9	14	14	13	2
MCPA	8	0	12	45	32	7	3	0
Dichlorprop	16	0	14	44	24	7	5	0
Metalaxyl	1.3	0	0	1	2	1	0	0
Lindane	0.6	2	2	2	2	0	0	
Max.[a]		4	14	25	5	4	10	6

[a] Maximum total concentration in one single stream in (µg/l).

REFERENCES

1. Brink, N. (1986) Factors affecting mass transport from farmland. Int. Sem. Land Drainage, Helsinki Univ. Technol. Finland, 456–73.
2. Brink, N. (1990) Environmental constraints on agricultural production, in *Land use Changes Europe*, (eds F. Brewers, A.M. Thomas and M.J. Chadwick) pp. 235–52.
3. Brink, N. and van der Meulen, K. (1987) Losses of phosphorus and nitrogen to Lake Ringsjön, *Ekohydrologi* 25, 5–15.
4. Uhlen, G. (1985) Nutrient leaching and surface runoff in field lysimeters on a cultivated soil. Reports from Norwegian Agricultural College, Norway, 57–8.
5. Ulén, B. (1985) Erosion of phosphorus from arable land. *Ekohydrologi* 20, 26–35. (Abstract in English.)
6. Andersson, A., Gustafson, A. and Torstensson, G. (1988) Removal of trace elements from arable land by leaching. *Ekohydrologi* 26, 13–22. (Abstract in English.)
7. Hallberg, G.R. 1987. Agricultural chemicals in groundwater. *Alternative Agriculture*, 2, 3–14.
8. Erne, K. (1970) Herbicides in the Swedish environment. Statens Veterinärmed. Anst. Stockholm p. 11 (In Swedish.)
9. Kreuger, J. and Brink, N. (1988) Losses of pesticides from agriculture. Int. Sym. on Changing Perspectives in Agro-chemicals, Neuerberg, 24–7 November 1987, Int. Atomic Energy Agency, Vienna. IAEA-SM-297/2, pp. 101–12.
10. Solyonn, P. (1986) Aspects on irrigation water quality demands. Reports from Agricultural College of Sweden, Uppsala. (In Swedish.)
11. GEFO (1987) Pesticides in surface water and groundwater. Inst. Geiresurs-og Forurensningsforskning, Norway (In Norwegian.)
12. Denoylles, F., Kettle, W.D. and Stinn, D.E. (1982) The response of plankton communities in experimental ponds to atrazine. *Ecol.*, 63, 1285–1293.
13. Frank, R., Braun, H.E., van Hove Holdrinet, M. *et al.* (1982) Agriculture and water quality in Canadian Great Lakes Basin. *J. Environ. Qual.*, **11**(3), 497–505.

NUTRIENT NON-POINT POLLUTION IN EXPERIMENTAL WATERSHEDS IN POLAND

by T. Bogacka and R. Taylor

19.1 INTRODUCTION

Non-point pollution is considered to be a very serious problem. The average riverine contributions of nitrogen and phosphorus from Poland into the Baltic Sea (over 200 000 t of nitrogen and 20 000 t of phosphorus) constitute about 30% and 40% respectively of the total load transported by all rivers entering the Baltic basin. It is estimated that about 60% of the total outflow of nitrogen and 30% of the total outflow of phosphorus from Poland come from non-point sources.

A study was carried out in 1977–90 on small catchments selected from different regions of Poland (Figure 19.1). The catchments represent areas of typical physiographical features. The areas chosen are used for agricultural and non-agricultural purposes and have no significant point sources of pollution. These investigations enabled an estimate of nutrient losses from agriculture into surface waters to be made and also constituted the basis for surface water protection against non-point pollution. The results of experiments carried out in the catchments are presented in Tables 19.1–19.3. Agricultural land use, coefficients of the unit nutrient run-off from catchment, topography and soil permeability were characterized for each watershed. The last two features strongly affect the leaching of nutrients from the catchments into surface waters in the hydrological cycle.

Figure 19.1 Location of experimental watersheds

19.2 EXPERIMENTAL CATCHMENTS AND NUTRIENT RUN-OFFS

The experimental stream watersheds selected in the Raba [1] and Rudawa [2] drainage basins are typical of the piedmont areas and comprise medium-sized, steeply sloping hills, covered with soils of low permeability. The nutrient run-off from

International River Water Quality. Edited by Gerry Best, Teresa Bogacka and Elżbieta Niemirycz. Published in 1997 by E & FN Spon, London, ISBN 0419215409

Table 19.1 Run-off of nitrogen and phosphorus from the experimental watersheds in the Vistula basin. (Source: various)

River, basin, description	Watershed area (km^2)	No. of sub-watershed	Soil permeability	Agricultural land (%)	Year	Water run-off (m^3/ha per year)	Load (kg/ha per year) Nitrogen	Load (kg/ha per year) Phosphorus
Brynica, Czarna Przemsza basin, piedmont plain, dominated by slopes 0–3%	100	I	high (sands, loamy sands)	40.9	1977–8	1835	2.12	0.19
					1978–9	2197	2.98	0.12
Brzezówka Stream, Raba basin, middle mountains, slopes <20%	5.3	I	low (sandstones, slates, clays)	43.5	1979	3217	9.3	0.6
					1980	2586	6.8	0.4
					1981	2807	9.6	0.3
Wolnica Stream, Raba basin middle mountains, slopes 10–20%	5.9	I	low (sandstones, slates, clays)	83.2	1979	3847	9.1	1.1
					1980	4257	10.5	2.2
					1981	3154	11.9	0.5
Rudawa River, Vistula tributary, piedmont, dominated by slopes 5–15%	290	I	low (losses, loans, limestones)	67.5	1987–8	2468	14.6	1
					1989–90	1912	14.5	1.27
	13.9	I	high	78.9	1977		1.06	0.045
					1978		1.2	0.045
	20.2	II	high	83.8	1977		1.44	0.066
					1978		1.55	0.089
	43.6	III	high	75.4	1977		2.09	0.025
					1978		2.37	0.045
	22	IV	high	78.7	1977		1.86	0.05
					1978		1.88	0.078
	31.9	V	medium	78.4	1977		1.26	0.246
					1978		1.54	0.268
Wilga, Vistula tributary, lowland, average slope 2.7%	22.7	I	medium	81	1977–8	2937	10.19	0.9
					1978–9	2988	10.46	0.41
Sucha River, Bzura basin, lowland, average slope 0.5%	13.6	II	low	100	1977–8	3187	16.68	1.69
					1978–9	3000	16.51	0.72
	59.2	III	high	79	1977–8	2364	9.7	1.1
					1978–9	2320	9.56	0.52
Skierniewka River, Bzura basin, lowland, average slope 0.5%	100.8	I	low	84.1	1981–3	1041	2.2	0.1
	14.9	II		83.5	1981–3	657	1.5	0.1
	940	I	very high (gravels, sands)	30	1977–8	1777	0.94	0.171
Wda River, Vistula tributary, outwash area, dominated by slopes 0–3%	1386	II		26	1978–9	1615	0.91	0.162
					1977–8	1825	0.89	0.178
					1978–9	1670	0.82	0.162
	210	I		71	1977–8	3877	2.08	0.26
					1978–9	4092	1.85	0.25
Radunia River, Motława basin, diversified sculpture of the Earth surface, slopes 5–7%	328	II	high	6.47	1977–8	3274	1.93	0.25
					1978–9	3527	1.92	0.27
	31	III		36	1977–8	2205	1.68	0.23
					1978–9	3022	2.04	0.21

Table 19.2 Run-off of nitrogen and phosphorus from the experimental watersheds in the Odra basin (Source: various)

River, basin, description	Watershed area (km²)	No. of sub-watershed	Soil permeability	Agricultural land (%)	Year	Water run-off (m³/ha per year)	Load (kg/ha per year) Nitrogen	Load (kg/ha per year) Phosphorus
Kamienna River, Bóbr basin, mountains (Karkonosze), slopes 7–20%	2.8	I	very low (bedrock, clay)	0	1977–8	11 720	17.44	0.377
					1978–9	12 814	18.57	0.389
	1.2	II		0	1977–8	17 350	26.44	0.557
					1978–9	19 455	27.54	0.552
	5	III		0	1977–8	13 785	20.3	0.43
					1978–9	17 713	25.38	0.528
	3.3	I		64.5	1977–8	660	2.67	0.059
					1978–9	810	3.26	0.071
Piaseczna River, Kaczawa basin, piedmont upland (Sudety) slopes 1.4–7.6%	1.6	II	low (sandstones marls)	91.1	1977–8	3 800	16.65	0.264
					1978–9	4 480	18.4	0.279
	5.3	III		74.5	1977–8	1 700	7.11	0.165
					1978–9	2 000	7.88	0.179
	4.5	IV		82.3	1977–8	1 270	11.22	0.667
					1978–9	1 450	11.91	0.76
	11.8	V		77.7	1977–8	1 380	8.84	0.423
					1978–9	1 540	9.28	0.443
	189	I	medium (loamy sands, clays)	80.4	1977–8	950	1.69	0.15
					1978–9	1 540	2.27	0.13
Noteć Wschodnia, Warta basin, Kujawskie Lake District, slopes 0.5–1%	306	II		81	1977–8	1 000	2.03	0.13
					1978–9	1 460	2.56	0.09
	32.8	I	medium	83.2	1977–8	903	1.98	0.238
					1978–9	1 240	3.89	0.251
Stream from the Kuśmierz, Noteć basin, Kujawskie Lake District, slopes 0.5–1%	42.8	II		84.8	1977–8	910	2.92	0.225
					1978–9	1 250	3.15	0.227
	4	I	medium	98	1977–8	1 168	20.2	0.1
					1978–9	3 554	52.1	0.848
Pomorka River, Noteć basin, Gnieźnieńskie Lake District, slopes 0.5–1%	71.9	II		93.5	1977–8	682	5.4	0.091
					1978–9	1 530	15.2	0.427
Gunica River, Odra tributary, Szczeciński Lowland	11.9	I	high (sands)	14	1975–84	1 104	5.1	0.09
Stobnica River, Odra tributary, Szczecin Lowland	2.7	I	low (clays)	95	1975–84	946	5.5	0.06

Table 19.3 Run-off of nitrogen and phosphorus from the experimental watersheds in the Przymorze region. (Source: various)

River, basin, description	Watershed area (km²)	No. of sub-watershed	Soil permeability	Agricultural land (%)	Year	Water run-off (m³/ha per year)	Load (kg/ha per year) Nitrogen	Load (kg/ha per year) Phosphorus
Reda River, Baltic Sea basin, diversified sculpture of the earth surface, slopes 0.3–5%	6.3	I	medium (sands, loamy sands, clays)	89.5	1981–2	5460	7.29	0.417
					1982–3	2840	6.05	0.193
					1983–4	3720	10.27	0.325
					1986–7	3386	9.22	0.356
					1987–8	4799	13	0.392
					1988–9	2502	7.26	0.325
	35.1	II		65.7	1981–2	4370	5.04	0.377
					1982–3	2430	3.44	0.183
					1983–4	3030	7.02	0.255
					1986–7	2581	5.23	0.266
					1987–8	3278	5.95	0.325
					1988–9	2480	4.89	0.259
	27.3	III		16	1981–2	2580	1.73	0.176
					1982–3	2270	1.33	0.14
					1983–4	2860	2.52	0.188
					1986–7	2492	1.75	0.208
					1987–8	2413	2.23	0.178
					1988–9	2133	1.5	0.155
Gizdepka River, Baltic Sea basin, morainal plain	21.4	I	low (clays)	13.7	1991–2	1245	1.2	0.121
	30.7	II		38.9	1991–2	1442	2.41	0.201
	37.5	III		45.3	1991–2	1385	2.49	0.267
Narusa River, Zalew Wiślany basin, marginal plain	2.9	I	very low (varwe loams)	98	1991–2	2586	12.72	0.897
	3.7	II		39	1991–2	1821	5.97	0.486
	56	III		59.5	1991–2	2206	9.34	0.83
Polder at the Żuławy Wiślane terrain, depressed plain	36.2	I	very low (dusty formations)	84	1981	6192	18.66	0.83
					1982	2708	9	0.25
					1983	2931	9.92	0.33
					1984	2583	7.97	0.33

this area was high and ranged from 6.8 to 14.6 kg/ha per year of nitrogen and 0.3 to 2.2 kg/ha per year of phosphorus. The value was dependent on the amount of arable land in the total area of catchment and on the atmospheric conditions (Table 19.1).

The Brynica River watershed [3] in the Czarna Przemsza drainage basin represents a typical agricultural area of the piedmont plateau and is covered with poor quality, permeable soils. The nutrient run-off was low and ranged from 2.12 to 2.98 kg/ha per year of nitrogen and 0.12 to 0.19 kg/ha per year of phosphorus.

The Piaseczna River watershed situated in the upper Odra drainage basin is characterized by medium slopes covered by rich soils of low permeability (loess); it is used to grow wheat and sugar beet. The run-off for nitrogen and phosphorus varied from 2.67 to 18.4 kg/ha per year and 0.06 to 0.76 kg/ha per year respectively (Table 19.2). The variation of these values resulted from different loading of fertilizers within the catchment.

The Kamienna River catchment (Karkonosze – the upper Odra drainage basin) was selected as being representative of the densely wooded mountainous areas with complete coverage of trees. These areas usually contain reservoirs for municipal supply. The underlying rock is covered with the thin layer of sediment and weathering soils. The catchment is 845–1200 m above sea level and has a low average temperature and very high rainfall. There is a very high run-off of nutrients despite insignificant human influence (17.44–27.54 kg/ha per year of nitrogen and 0.38–0.56 kg/ha per year of phosphorus (Table 19.2).

The Wilga [4], Sucha [5] and Skierniewka River [6] catchments represent the Mazowiecka Lowland District which includes a large part of the upper and lower Vistula drainage basin. These lowland agricultural catchments (75–100% arable land) are characterized by different soil types. The nutrient run-off from these watersheds ranged from 1.06 to 16.68 kg/ha per year of nitrogen and 0.025 to 1.69 kg/ha per year of phosphorus (Table 19.1), and was mainly dependent on the soil permeability.

The Wda and Radunia Rivers – tributaries of the Vistula [7], and the Reda River catchment are typical of the Middle Pomeranian Lake Districts. They are characterized by a diverse surface structure created during the ice age as well as by light, permeable soils of post-glacial gravels, sands and clays. The ratio of the various granulations vary throughout the catchments. Accordingly, the soil areas are characterized by differential infiltration properties. Thus the permeability of the soil of the Wda River catchment is the highest whilst that of the Radunia River is the lowest (Tables 19.1 and 19.3). There is a large groundwater resource and this, together with the water retained in lakes, results in steady flow rates of the rivers and changes in the water levels are smaller than in other parts of Poland. The loads of nitrogen and phosphorus transported by the Wda and Radunia Rivers ranged from 0.82 to 2.08 kg/ha per year and 0.16 to 0.27 kg/ha per year respectively. For the Reda River, the outflow of nutrients was greater and closely related to the applied loading. The values varied from 1.33 to 10.27 kg/ha per year of nitrogen and 0.14 to 0.417 kg/ha per year of phosphorus.

The partial catchments selected in the Great Poland Lake District in Noteć drainage basin have different characteristics [7,8]. The geological structure of the catchments was also created by glaciation: they are largely flat, cut with riverine valleys. Their areas are mostly covered with the low-permeability boulder clays from which fertile brown soils were formed. The area receives the lowest rainfall in Poland, which is both irregular and variable. The groundwater resources are limited and this results in marked changes in water flow because of the lack of buffer capacity. These particular hydrological conditions in the catchments affect the nutrient run-off, which ranged from 1.69 to 52.10 kg/ha per year of nitrogen and 0.09 to 0.85 kg/ha per year of phosphorus (Table 19.2).

The Gizdepka River watershed [9] represents the morainal lowlands of Koszalin and Gdansk coastal area. Here, the soils are fertile but of low permeability. Depending on the land use, the nutrient run-off ranged from 1.2 to 2.49 kg/ha per year of nitrogen and 0.12 to 0.27 kg/ha per year of phosphorus.

The Narusa River flows across a flat landscape which is covered with the very low-permeability and fertile varve silts (brown fen soils) [9]. During the investigation period, the nitrogen and phosphorus run-off ranged from 5.97 to 12.72 kg/ha per year and 0.49 to 0.9 kg/ha per year respectively (Table 19.3).

The plain area of Żulawy Wiślane, with a partial depression, is characterized by rich and low-permeability fen soils [10]. The land is drained and used for agricultural purposes. The nitrogen and phosphorus run-off varied from 7.97 to 18.66 kg/ha per year and 0.25 to 0.83 kg/ha per year respectively (Table 19.3).

Finally, the area of the Szczecin Lowland situated on the west coast is represented by two tributaries of the Odra River [11]. The mean run-off of nutrients varied from 5.1 to 5.5 kg/ha per year of nitrogen and 0.06 to 0.09 kg/ha per year of phosphorus (Table 19.2).

The quality of river water in catchments devoid of point sources of pollution depends usually on agricultural intensity. Of the rivers investigated, those with forested catchments were the cleanest. The mean concentrations of pollutants and nutrient concentrations in these river waters indicated high water quality. Rivers in catchments which were totally or mainly used for agricultural purposes were usually the most polluted. This particularly applied to rivers in areas of fertile soil with low permeability as well as those situated in the piedmont regions with steep slopes covered with low permeability soils. It was found that ammonia nitrogen was the least threatening nutrient from diffuse sources. The concentration of ammonia nitrogen in the run-off water from all the catchments investigated was less than 3 mg/l. The nitrate concentrations in the river waters were the most variable compared to the concentrations of other nitrogen forms. This is closely related to the transport dynamics and the physico-chemical properties of nitrates. During the investigation period, the extreme concentrations of nitrates ranged from 0.02 to 24 mg/l whilst the total nitrogen in the river waters ranged from 0.05 to 26 mg/l. The amplitude of the observed fluctuations depended partly on the nature of the catchment but mainly on the prevailing hydrological conditions (Chapter 27 by Kowalik and Kulbik). Moreover, the maximum concentrations of nitrates and total nitrogen depended on the soil fertility, as well as on the amount of fertilizers used. The lowest nitrate and total nitrogen concentrations were found in those catchments with permeable soils of low fertility (e.g. the rivers of the Middle Pomeranian Lake District), while the highest concentrations always occurred intermittently in some rivers of the Szczecin Lowland, the Pomeranian Lake District, the Warsaw Valley, and in the waters of the Sudetian piedmont catchments. Phosphorus compounds are regarded as potentially more polluting from agricultural sources than nitrogen compounds because phosphorus is usually the limiting nutrient for algal growth in fresh water. Only in the afforested mountain catchments of the Kamienna River were concentrations of phosphorus compounds lower than 0.1 mg/l. In a few cases, mainly in catchments with little arable farming, the average concentrations of phosphorus compounds were as high as 0.4 mg/l. The maximum concentrations were as high as 2.6 mg/l and these high phosphorus concentrations were found mainly in rivers in agricultural catchments covered with loamy soils which have a high content of small sedimented particles. This is particularly so in the piedmont catchments which are characterized by high precipitation and intensive soil erosion. In these catchments, most of the phosphorus is associated with suspended matter.

The proportion of suspended phosphorus to the total phosphorus in river waters varies according to the soil type. It ranges from 26% to 35% for catchments of boulder clays, 35% to 86% for silty catchments and 6.3% to 10.6% for catchments of sands and loamy sands.

The average nutrient concentrations in particular river waters were found to vary seasonally with the changing meteorological conditions. Figure 19.2 exemplifies the annual variation of nitrate and phosphate in the waters of a small stream compared with the rainfall, temperature and flow. It is evident that there is a positive correlation between nitrate levels and flow. Quite a different situation was observed in the case of phosphates, whose concentration fluctuated about an average level and peaked at times of high rainfall and snow melt.

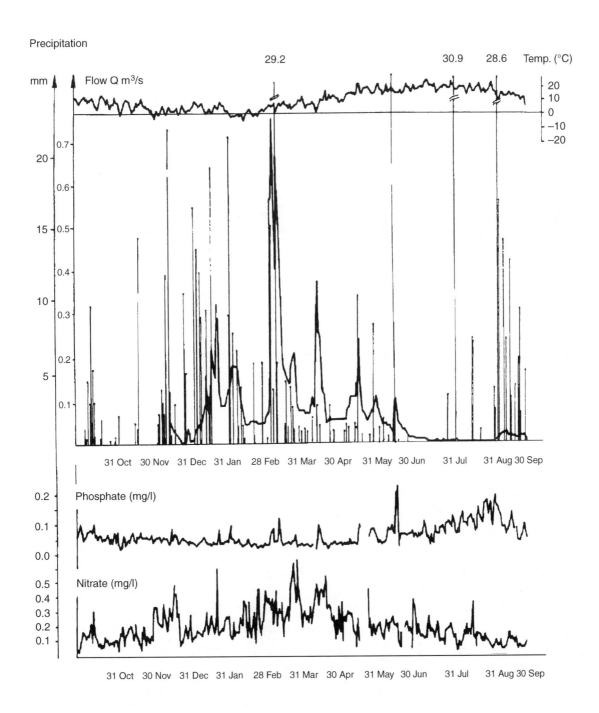

Figure 19.2 Influence of the meteorological and hydrological conditions on the nitrate and phosphate concentration forming in the Krzeszowska Stream 1982–83.

Table 19.4 Mean nitrogen and phosphorus runoff from the experimental watersheds of similar use and characteristics. Vistula basin and Przymorze region terrain

Region	Soil permeability	Agricultural land (%)	Watershed loads (kg/ha per year) Nitrogen	Watershed loads (kg/ha per year) Phosphorus	Run-off (kg/ha per year) Nitrogen	Run-off (kg/ha per year) Phosphorus
Middle mountains:						
Raba basin, slopes 10–20%	low	44	–	–	8.57	0.433
	low	83	–	–	10.5	1.267
Piedmont:						
Rudawa basin, dominated slopes 5–15%	low	68	107	14.6	13.6	1.135
Piedmont plain:						
Przemsza basin, dominated slopes 3%	high	41	124	33.6	2.55	0.155
Lowland:						
Warsaw valley, slopes 0.5–2.7%	high	79	121	30.4	3.27	0.206
	medium	80	117	26	5.89	0.456
	low	92	149	36.5	9.22	0.505
Outwash area:						
Wda basin, dominated slopes 3–4%	very high	28	41	6.9	0.89	0.168
Middle Pomeranian Lake District:						
Radunia basin, slopes 5–7%	high	68	70	23.6	1.95	0.25
Reda basin, slopes 3–7%	medium	16	41	6	1.84	0.174
	medium	66	100	21.1	5.26	0.278
Littoral terrain:						
Gizdepka basin, slopes 3–4%	low	14	12	2	1.2	0.121
	low	42	25	7.7	2.45	0.234
	low	90	164	35.1	8.85	0.335
Marginal lowland:						
Narusa basin	very low	39	56	10.9	5.97	0.486
	very low	59	70	15.1	9.34	0.83
	very low	98	117	26.3	12.72	0.897
Depression plain:						
Vistula marshland	very low	84	110	26.7	11.39	0.435

Table 19.5 Mean nitrogen and phosphorus run-off from the experimental watershed of similar use and characteristics. Odra basin

Region	Soil permeability	Agricultural land (%)	Watershed loads (kg/ha per year)		Outflow (kg/ha per year)	
			Nitrogen	Phophorus	Nitrogen	Phophorus
Mountains:						
Karkonosze, slopes 7.2–18.8%	very low	0	–	–	22.61	0.472
Piedmont:						
Sudety, slopes 1.4–7.6%	low	64	99	22.4	2.97	0.065
	low	76	154	22.4	8.28	0.302
	low	87	288	27.1	14.54	0.493
Wielkopolska Lake District:						
Noteć basin, slopes 0.5–1%	middle	82	131	34.7	2.56	0.18
	middle	96	137	45.7	23.23	0.366
Szczecin Lowland:						
Odra tributaries	high	14	25	3	5.1	0.09
	low	95	162	30	5.5	0.06

The nitrogen and phosphorus run-offs for all the catchments studied ranged from 0.82 to 52.1 kg/ha per year and 0.025 to 2.2 kg/ha per year respectively. Tables 19.4 and 19.5 show the mean nitrogen and phosphorus run-offs, as well as the loadings with nutrients for watersheds of similar management and physiography. From Tables 19.4 and 19.5, the effect of the catchment characteristics on the amount of nutrient washed out from the watershed is clear. It is evident that the nutrient run-off depends on three factors: the soil permeability; the amount of nutrients introduced in fertilizers, manure and precipitation; and the configuration of the catchment. The nutrient run-off from catchments with low-permeability soils and steep slopes (mountain and piedmont rivers) is much greater than from the others. This arises not only from the catchment characteristics but also from additional factors such as the higher nutrient loads introduced with precipitation and the greater run-off.

This is well illustrated by the Karkonosze Mountain catchment which is entirely forested. The nutrient run-off was higher than from catchments with intensive agriculture. This is explained by significant air pollution of the Silesia region caused by nitrogen oxides, high rainfall in the mountains (during the investigation period the mean annual rainfall amounted to 1607 mm) and by water run-off from mountain streams, which was about ten times higher than from the lowland rivers. A good correlation is also evident between coefficients of the unit nutrient loads and nutrient run-off related to the particular physiographical regions.

It was found that the nitrogen compounds, washed out from the fields, are the greatest threat to the waters of lowland watersheds in the Noteć drainage basin. The specific hydrology in the Pomorka River catchment in the Noteć area, as well as fertile soils and intensive agriculture, resulted in nitrogen run-off of 52 kg/ha for one wet year. The agricultural catchments of low- and high permeability soils were characterized by the lowest loads of nutrient runoff (e.g. the Middle Pomerania Lake District and lowland catchments of the Warsaw Valley).

19.3 CONCLUSIONS

River water quality in a watershed with no point sources of pollution depends on the extent of the catchment which is used for agriculture.

The nutrient run-off from non-point sources in the hydrological cycle is determined by various factors. The most important are the permeability and fertility of the soils, topography of the area, land use, water relations hips and meteorological conditions in the watershed.

The greatest amounts of nutrients of non-point origin are discharged from agricultural piedmont catchments with low-permeability soils and from watersheds covered with fertile soils with a high content of easily eroded particles.

REFERENCES

1. Stachowicz, K. (1990) Agricultural non-point pollution in the watersheds of various topography and land uses, in *Proc. of Sem. 26 Non-point Pollution in the Agricultural Watersheds*, Institute of Land Reclamation and Grassland Farming, Falenty.
2. Żeglin, M. (1990) Effect of fertilization and atmospheric conditions on the level and dynamics of nutrient outflow from the agricultural watersheds, IMWM, Kraków, No. 2–3, 115–26.
3. Pudo, J., Erndt, E. and Żeglin, M., (1979) Determination of the agricultural non-point pollutant outflow from the watersheds of different topography and land use (Brynica and Trzonia Rivers), IMWM, Kraków.
4. Guberski, S. and Szperlinski, Z. (1990) Investigation results of the chemical water composition in the upper Vistula basin (in Polish), in *Proc. of Sem. 26 Non-point pollution in the agricultural watersheds*, Institute of Land Reclamation and Grassland Farming, Falenty.
5. Jakubowska, L., (1979) Investigation on non-point pollution influence on the water quality of the Sucha River, IMWM, Warsaw.
6. Dojlido, J., Leszczynski, A. and Włodarczyk, E. (1990) Non-point pollution in the Skierniewka watershed (in Polish), in *Proc. of Sem. 26 Non-point Pollution in the Agricultural Watersheds*, Institute of Land Reclamation and Grassland Farming Falenty.
7. Taylor, R., et al. (1979) Agricultural non-point pollution influence on Pomorka, Wietcisa, Radunia and Wda river water quality, IMWM, Warsaw.

8. Florcyk, H. (1978) Unit runoff coefficients of non-point pollutants from river watersheds of different use characteristics. P-7.01.04.02, IMGW, Warsaw.
9. Niemirycz, E. and Taylor, R., (1992) The influence of watershed use on nutrients runoff. S-1, IMGW, Gdansk.
10. Balcerska, M., and Taylor, R., (1987) Factors shaping the quality of drainage network in Żuławy Wiślane District. Wiadomości IMGW, 10(31).
11. Borowiec, S., Zabłocki, Z., (1988) Agricultural non-point pollutant runoff from agricultural watersheds and draining systems of North-West Poland. Scientific Instalments of Agricultural Academy in Szczecin No. 134, Rolnictwo 14, 27–57.

THE ANALYSIS OF NUTRIENT CONCENTRATION–WATER FLOW RELATIONSHIPS IN A LAKELAND RIVER

by R.J. Bogdanowicz

20.1 INTRODUCTION

Excessive eutrophication is emerging as one of the most significant causes of the deterioration in the water quality in the lakes and reservoirs of the Cashubian Lakeland. The detrimental changes in nutrient levels also create severe problems for the purification of the water for potable supply. A knowledge of both the concentration of the nutrients present and their load (mass transport) is essential for quality considerations. Therefore, the establishment of concentration–discharge relationships undoubtly becomes one of the key issues.

Much of the previous work examining the concentration–flow relationship has concentrated on improving the estimate of the mass load from the available data which typically consists of a set of intermittent concentration values in addition to more frequent measurements of flow [1]. In this chapter the orthophosphate phosphorus and nitrate nitrogen concentrations are analysed in order to establish the nature of the nutrient–discharge curves at selected points in the catchment and to classify them according to influential processes.

20.2 DESCRIPTION OF THE STUDY AREA

The drainage basin of the Radunia River is located in the Cashubian Lakeland, west of Gdansk and has an area of 837 km². Elevations range from 329 m above sea level in the Szymbarskie Hills to 4 m above sea level at the basin outlet to the Motława river (a tributary of the Vistula River). The region is characterized from the hydrological point of view by clear seasonal trends in river flows with the highest flows occurring in the winter half-year (November–April) and the lowest in the summer half-year (May–October). There are 68 natural lakes in the catchment with a total volume of 283.4×10^6 m³, occupying 4.1% of the catchment area [2]. The land use comprises 20% forest and 67% agricultural land.

The watershed contains a few small towns and several small villages of less then 1000 inhabitants which use septic tank systems for sewage disposal. The most important point sources of pollution are the sewage treatment plants located on the Mała Supina (effluent volume of 4840 m³/year) and Strzelenka (1300 m³/year) sub-catchments. The Radunia River is used as a source of potable water for the Gdansk conurbation, therefore the quality of water from the river is one of the most important issues for environmental management in the region (Table 20.1).

20.3 FUNDAMENTALS

The dependence of the ionic concentration on the river flow can be interpreted as a result of several independent factors, the most significant ones being:

International River Water Quality. Edited by Gerry Best, Teresa Bogacka and Elżbieta Niemirycz. Published in 1997 by E & FN Spon, London, ISBN 0419215409

158 *Analysis of nutrients and flows in a lakeland river*

Table 20.1 Annual mean nutrient concentrations at selected sampling sites on the Radumia in 1992

Parameters	Water quality sampling points			
	Ostrzyce	*Borkowo*	*Strzelenka*	*Lniska*
Orthophosphate (mg/l)	0.09	0.19	0.54	0.54
Total phosphorus (mg/l)	0.10	0.16	0.26	0.31
Ammonia nitrogen (mg/l)	0.03	0.07	0.12	0.26
Nitrate nitrogen (mg/l)	0.10	0.40	0.60	0.68
Nitrite nitrogen (mg/l)	0.003	0.009	0.011	0.015
Total nitrogen (mg/l)	2.05	1.86	2.42	2.98

pollution load, rate of self-purification of the river, distance from the source of pollution and the character of the pollution source (point source or diffuse source). Depending on these factors, three main types of curves expressing concentration–discharge relationships can be differentiated [3]. In heavily polluted rivers (type I) where the distance between sampling point and the source of pollution is small, the most important phenomenon is the dilution of a constant load ($C \times Q$). In clean rivers or rivers with diffuse sources of pollution (type II), nutrient transport is influenced by erosion, sedimentation and leaching ($C \times Q^m$, where $m > 0$). In slightly polluted rivers (type III), the relationship is a combination of the first two types. Consequently, in a range of low flows, dilution dominates and the shape of the curve resembles that of type I. However, above a certain threshold flow rate, erosion, resuspension and leaching override the effect of dilution and the concentration increases with an increase of the river flow. Finally, there are also rivers with a comparatively stable nutrient concentration largely independent of the flow rate (C=constant).

20.4 METHOD OF ANALYSIS

In Poland, in order to evaluate the so-called 'reliable concentration', it is assumed that there are statistically significant relationships between the pollutant concentration and the river flow. These relations can be described using the following equations [4]:

$$C = aQ + b \qquad (1)$$

$$C = \frac{a}{Q} + b \qquad (2)$$

$$C = \frac{a}{Q} \qquad (3)$$

$$C = \frac{Q}{aQ + b} \qquad (4)$$

$$C = ae^{bQ} \qquad (5)$$

$$C = aQ^b \qquad (6)$$

$$C = \frac{a}{Q} + bQ + c \qquad (7)$$

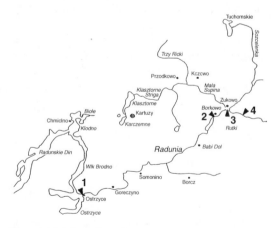

Figure 20.1 Location of water quality sampling points: Ostrzyce (1), Borkowo (2), Strzelenka (3) and Lniska (4).

$$C = \sum_{i=0}^{9} a_i Q^i \qquad (8)$$

$$C = a + bQ + cQ^2 \qquad (9)$$

where,

C = concentration of pollutant
Q = discharge
a, b, c, a_i = parameters established empirically.

In the present study a second-order polynomial regression has been chosen as the most suitable for the analyses (9). This equation is general enough to include some of the equations listed above. Most of the equations above are not applicable to the type III relationship and usually are suitable only to one of the other two relationships. Moreover, the preliminary analysis has proved that the second-order polynomial regression describes the relationship best.

The data for the analyses were for the period 1988–92 and consist of routine monthly measurements at four sites along a 33 km segment of the upper reach of the Radunia River (Figure 20.1). The data include simultaneous measurements of river flow and nutrient concentrations carried out by the National Inspectorate of Environmental Protection at each sampling point.

20.5 RESULTS

The results of the analyses are presented in Figures 20.2 and 20.3. It can be seen that most of the orthophosphate phosphorus–discharge relationships closely resemble a type III relationship, typical for moderately polluted rivers (Figure 20.2). For nitrate nitrogen such a close relationship exist only at Lniska sampling site (Figure 20.3(d)). As no such correlation exists at any other Radunia sampling point it can be assumed that, in respect of nitrate nitrogen, the upper Radunia is characterized by a constant baseflow concentration.

20.6 SEASONAL ASPECT OF NUTRIENT–DISCHARGE RELATIONSHIPS

In rivers where the most important sources of orthophosphate are sewage and effluent discharges clearly the highest concentrations of orthophosphate occur when the proportion of the effluent volume to the river flow is at its highest [5]. This usually occurs in summer, when the lowest flow rates are characteristic. On the other hand, in rivers without significant point sources of nutrients, the highest concentrations of phosphorus are likely to occur during spates, influenced by river bed erosion–sedimentation effects and erosion of surrounding land bringing in particulate phosphorus [6]. However, because river sediments can act as sinks as well as sources of phosphorus [7], phosphate concentrations may both increase and decrease during spates resulting in no clear seasonal pattern.

The occurrence of high concentrations of nitrate nitrogen can be related to the influence of several factors which include the combination of the strong leaching of soluble nitrate ions by water moving through the soil in the winter period and the absence of nitrogen uptake by plants in the dormant season. Moreover, the extent of the drainage network and the volume of saturated soil supplying run-off, which is at a maximum in the winter, allow the tapping of nitrate nitrogen stores which may be unconnected to the main watercourse at other times of the year [8]. Low nitrate nitrogen levels in the summer months are thought to reflect the greatly diminished soil water movement, increased plant uptake of nitrogen and the occurrence of denitrification [9].

In order to elucidate the impact of the seasonal trends on the concentrations of nutrients and the flow regime relationship, the analyses were extended by considering treating the timing of the sample as a second independent variable. In this way a new picture of the relationships at different sampling points was obtained: these are presented in Figures 20.4 and 20.5. These graphs show that in some cases seasonal variations in concentrations correspond with variations in river flow, resulting in clear and stable interrelations. In other cases seasonal patterns are different for different variables

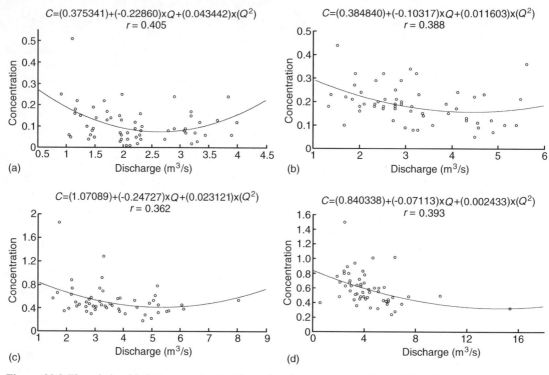

Figure 20.2 The relationship between orthophosphate phosphorus concentration and river flow at (a) Ostrzyce; (b) Borkowo; (c) Strzelenka; (d) Lniska sampling points.

and no relationship can be detected. The result of this analysis proves that the nature of the relationship (the shape of the curve) can be different in different seasons. This means that there are monthly variations in the dominant sources of nutrients and the different processes that influence water quality.

20.7 CONCLUSIONS

No strong correlation was evident between the river flow and the concentration of nutrients in the river selected for examination. However, clearer relationships were established for phosphorus compounds than nitrogen ones. The dynamics of nutrient concentrations in the river were closely related to the physico-chemical properties of nutrients and, consequently, the routes of their transport from the catchment surface to the river. The shape of the curves in most cases approximated to a type III relationship. The results have shown that an analysis of the behaviour of nutrients in a lakeland river cannot be accomplished sufficiently without considering the aspect of seasonality, as regards both the concentration of nutrients and the river flow. Since these variations often have substantial effects on the nature of the phenomena being analysed, long-term and more frequent records should be examined to assess the reliability of the established relationships.

REFERENCES

1. Burn, D.H. (1991) Water quality sampling for nutrient loading estimation, in *Water Pollution: Modelling, Measuring and Prediction*, (eds L.C. Brobel and C.A. Brebbia) Elsevier, London, pp. 505–17.
2. Drwal, J. and Faras-Ostrowska, B. (1987) Hydrological consequences of the development of the Radunia river. *Geography* (16), Gdansk, pp. 105–16.
3. Manczak, H. (1971) A statistical model of a relationship between water flow and pollution concentration in rivers, IMGW, Warsaw.
4. Korol, R. and Kedzia, M. (1994) Monitoring of quality of flowing waters, In: *Proc. of 22nd Seminar on*

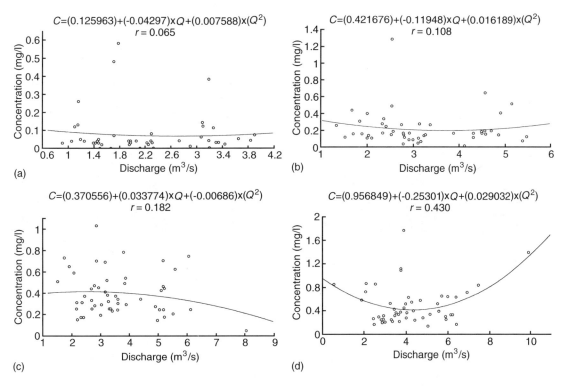

Figure 20.3 The relationship between nitrate nitrogen concentration and river flow at (a) Ostrzyce; (b) Borkowo; (c) Strzelenka; (d) Lniska sampling points.

Contemporary Problems in Hydrology, Madralin, Poland, 1–21.

5. Houston, J.A. and Brooker, M.P. (1981) A Comparison of Nutrient Sources and Behaviour in Two Lowland Subcatchments of the River Wyre. *Wat. Res.* **15**, 49–57.

6. Mohaupt, V. (1986) Nutrient–discharge relationships in a flatland river system and optimization of sampling, in *Monitoring to Detect Changes in Water Quality Series*, IAHS (157), 297–304.

7. Casey, H. and Clarke, R.T. (1986) The seasonal variation of dissolved reactive phosphate concentrations in the River Frome, Dorset, England, in *Monitoring to Detect Changes in Water Quality Series*, IAHS (157), 257–65.

8. Webb, B.W. and Walling, D.E. (1985) Nitrate Behaviour in Stream-flow from a Grassland Catchment in Devon, UK. *Wat. Res.* **19** (8), 1005–16.

9. Betton, C., Webb, B.W. and Walling, D.E (1991) Recent Trends in NO_3-N Concentration and Loads in British Rivers, in *Sediment and Stream Water Quality in a Changing Environment: Trends and Explanation*, IAHS (203).

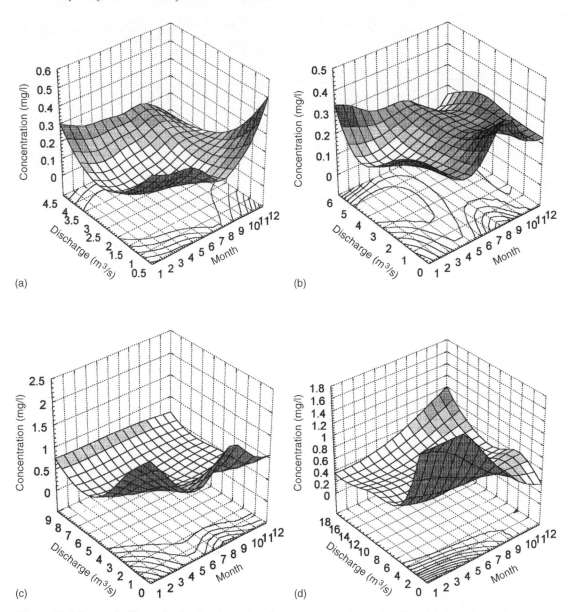

Figure 20.4 Seasonal effects of orthophosphate phosphorus concentration – river flow relationship at (a) Ostrzyce; (b) Borkowo; (c) Strzelenka; (d) Lniska sampling points.

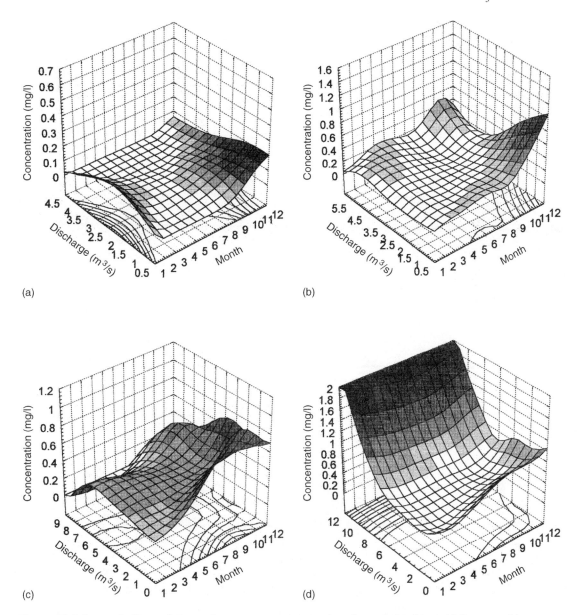

Figure 20.5 Seasonal effects of nitrate nitrogen concentration – river flow relationship at (a) Ostrzyce, (b) Borkowo; (c) Strzelenka; (d) Lniska sampling points.

A PROCESS FOR DEVELOPING WATER QUALITY CRITERIA UNDER THE US FEDERAL CLEAN WATER ACT

by G. McMurray, N.J. Mullane, D.B. Chandler and C. Johnston

21.2 INTRODUCTION

Under US federal legislation, states have the authority to adopt their own ambient water quality standards. In this country, there have been two basic approaches to water pollution control: standards-based and technology-based. When the USA initially enacted the Water Quality Act of 1965, states were directed to develop ambient water quality standards and the programme was largely standards-based. When this approach was not entirely successful, the 1972 amendments (called the Clean Water Act) shifted the programme towards technology-based controls. More recent amendments to the Act have given a new emphasis to standards-based controls.

A water quality standard, as defined by the Act, has two elements. The first element is the type of designated use being protected. Oregon's designated beneficial uses include:

1. public domestic water supply
2. private domestic water supply
3. industrial water supply
4. irrigation
5. livestock watering
6. anadromous fish passage
7. salmonid fish rearing
8. salmonid fish spawning
9. resident fish and aquatic life
10. wildlife and hunting
11. fishing
12. boating
13. water contact recreation
14. aesthetic quality
15. hydropower and
16. commercial navigation and transportation.

The second element in a water quality standard is the specific numeric or narrative criterion or criteria which will provide a sufficient level of water quality to protect that beneficial use. In practice, states are directed to adopt criteria which are protective of the beneficial use most sensitive to that particular water quality parameter. This practice is based on the implicit assumption that, if the most sensitive use is protected, then all uses are protected. For example, salmonid fish spawning or rearing, if present, would be the most sensitive use for the temperature standard.

The overarching policy driving the water quality standards in the USA comes from the interim goals of the Clean Water Act for 'fishable and swimmable' waters. Historically the general policy within the State of Oregon has been for full protection of designated beneficial uses at all times and places. The technical and policy discussions in the triennial review began with this general policy and each existing water quality standard was originally developed within this general policy framework.

The Act requires states to review their water quality standards at least once every three years. This process is called the triennial water quality standards review, or simply the triennial review. The scientific generalizations and data which support water quality and other pollution standards are constantly changing and have become increasingly technical in nature. For example, the US Environmental Protection Agency (EPA) issues technical guidance on an almost continual basis. Additionally, the arena in which the standards are applied is also constantly changing. Thus, the purpose of the triennial review is to keep water quality standards abreast of constantly evolving scientific underpinning and in step with a state's changing needs.

Scrutiny of water quality standards from all sectors has increased greatly in recent years, and has resulted in considerable criticism of Oregon's standards-setting process over the past decade. In 1992, Oregon's Department of Environmental Quality (DEQ) developed a new process to address the difficulties associated with this increased level of attention.

21.2 KEYSTONES OF THE OREGON PROCESS

21.2.1 SEPARATION OF TECHNICAL FROM POLICY ISSUES

Water quality standards and criteria are established presumably using the best available scientific information within a public policy framework. To ensure this, the DEQ established a Technical Advisory Committee (TAC) for the 1992-4 triennial review. The committee was drawn from academe and government, and comprised experts in complementary fields related to water quality criteria. For specific water quality standards, subcommittees with additional expertise were established and included representatives from industry. The role of the TAC and its subcommittees was to conduct a scoped technical review, produce a peer-reviewed technical analysis, and generate criteria alternatives and options for further discussion.

In order to set the process within the appropriate public policy context, the DEQ established an analogous Policy Advisory Committee (PAC). The PAC was drawn from academe, industry and non-governmental orgnizations (NGOs), and provided candid, critical and constructive comments, advice and recommendations on policy issues. This committee reviewed issue papers generated by the technical subcommittees and attempted to reach consensus on the options and alternatives provided by the TAC. The two committees interacted to the extent that the policy committee referred technical questions to and requested technical solutions from the technical committee. Issues with no technical basis for a solution defaulted to the policy committee.

21.2.2 EXTENSIVE PUBLIC EXPOSURE AND INVOLVEMENT INCLUDING NGOS

The new Oregon process is also characterized by an extensive process of public exposure and involvement. All parties involved in past reviews of standards are asked for comment at the inception of the present review. All meetings of the above described committees are announced and open to the public. When draft issue papers have been completed, public workshops are conducted in many Oregon locations. Then, when the proposed criteria are in rule language, the public has a final chance for involvement in public hearings.

Non-governmental organizations played key roles at three specific points in the 1992-4 criteria-setting process. First, when the standards review was announced, the NGOs used this opportunity to bring attention to problems with standards and to urge review of specific standards with the new process. Second, the NGOs had the opportunity to participate in the committee advisory process: they were represented as members of technical subcommittees and were heavily represented on the PAC (taking five of the thirteen seats). Third, the NGOs participated in both public workshops and public hearings as part of Oregon's public involvement process. Thus, the NGOs played an active role in selecting the standards to be reviewed, the actual analysis, and the public review process.

21.2.3 AN ITERATIVE DOCUMENT – THE ISSUE PAPER

The third keystone of the new Oregon water quality standards review process is the resulting document: the issue paper. This document is intended to include all information required to support the proposal and adoption of new criteria. There are chapters for the historical and existing standards and criteria, technical and policy analyses, and recommendations. Records of public response are appended to the document. The conceptual approach is that, when in ten or twenty years, the standard is again reviewed, the issue paper needs only to be updated and is always present to explain the thought and rationale behind the existing standards.

21.3 TECHNICAL ISSUES

A number of fundamental technical issues were identified by the TAC and analysed in each of the parameter-specific issue papers. These fundamental technical issues included:

1. identifying the most sensitive designated uses or resources;
2. assessing natural variability;
3. identifying anthropogenic effects; and
4. providing mechanisms for dealing with interactions with other variables.

Explanations and examples of the fundamental technical issues are discussed below.

21.3.1 MOST SENSITIVE USES OR RESOURCES

Water quality standards generally protect either human health or ecological integrity. With regard to ecological integrity, the most sensitive beneficial use is generally an attempt to simplify the approach to full protection. The implicit assumption is that if the most sensitive use is fully protected, then all uses are fully protected. In some cases the information to identify the most sensitive use with certainty is available; in other cases it is necessary to guess and hope for better information. Take for example the temperature standard. A great number of studies are available that show salmonid fish require relatively cold water, especially during spawning and rearing. Since salmonid stocks in the Pacific North-west region are threatened or endangered, their requirements have taken on added importance. The bull trout is an example of a salmonid requiring extremely cold water. Its distribution in Oregon is limited largely by temperature to headwaters in mountainous areas of the state (Figure 21.1). The temperature requirements of the bull trout are summarized in Figure 21.2. Optimum growth of fry requires temperatures in the range 4–4.5°C (39–40°F). Thus, the TAC subcommittee on temperature has recommended that no net temperature increases take place in stream reaches that remain habitat for bull trout.

21.3.2 NATURAL VARIABILITY

In developing a water quality standard it is necessary to be able to separate natural from anthropogenic effects. An underlying assumption in developing water quality standards is that the natural conditions represent the ideal and will provide full support of beneficial uses. This assumption is primarily supported by the concept that native biological populations are adapted to natural conditions. Thus, an ideal water quality standard takes into account the natural variability inherent in

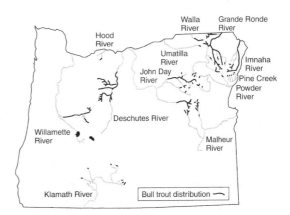

Figure 21.1 Distribution of bull trout (*Salvelinus confluentus*) in Oregon. (Source: Ratliff and Howell, The status of bull trout populations in Oregon, in *Proceedings of the Gearhart Mountain Bull Workshop* (eds Howell and Buchanan); American Fisheries Society, 1992.)

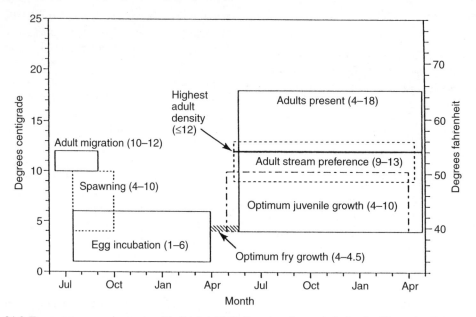

Figure 21.2 Temperature requirements of bull trout (*Salvelinus confluentus*) during its life cycle. (Source: Buchanan and Gregory, Development of water temperature standards to protect and restore habitat for bull trout and other cold water species in Oregon; unpublished manuscript.)

natural conditions, taking place over a wide array of space and time scales.

The State of Oregon covers eight ecoregions [1], each characterized by significant differences in geology and meteorology. For example, rainfall in Oregon ranges from as high as 200 inches (5000 mm) per year in a portion of the Coast Range to as low as 10 inches (250 mm) per year in some portions of the high desert. Spatial variability over basin scales is accommodated in Oregon's water quality standards by developing basin-specific criteria where needed. These basins are shown in Figure 21.3.

Temporal variability is accommodated within the criteria where appropriate. For example, the unit of measure for the temperature standard is the seven-day average maximum, defined as the average daily maximum temperature for the seven warmest consecutive days. The literature reviewed by the TAC subcommittee establishes that if a criterion based on this value protects the most sensitive beneficial use during the low-flow high-temperature period of the year, the use is also fully protected during the rest of the year. Thus, this value takes into account diurnal and seasonal variability. Annual variability is accommodated in the new Oregon temperature standard through a waiver of the standard based on the '7Q10' flow level (i.e. the lowest consecutive seven-day flow in a ten-year period) and/or on the 90%-ile of the seven-day average maximum air temperature.

21.3.3 ANTHROPOGENIC EFFECTS

This issue is closely tied to that of natural variability, in that one must understand enough about the natural variability to begin to identify the anthropogenic effects. Although 'background' is often used as the default, often the question is how a specific variable relates to 'natural'. Substantial databases are often required to support trending analyses. A reasonably straightforward example of the relationship between natural variability and anthropogenic effects may be seen by comparing the thermal regimes of two streams in the Coast Range [2].

Bear Creek, including the upper tributary of Lobster Creek, drains mature secondary growth for-

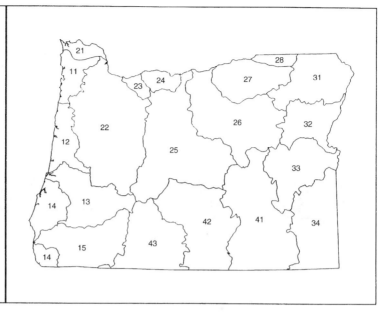

Figure 21.3 Oregon's nineteen river basins for water quality standards' purposes.

est with no streamside degradation. The temperature data in Figure 21.4(a) run from May to November 1992. Three characteristics are apparent: a summer maximum temperature of about 16.5°C; overall summer warming in the range of 5°C; and maximum diurnal temperature excursions in the range of 3°C. These characteristics may be compared to those of the main stem of Lobster Creek, which has been subjected to much anthropogenic riparian disturbance (Figure 21.4(b)). Lobster Creek exhibits: a summer maximum temperature of about 24.5°C; overall summer warming in the range of 10°C; and maximum diurnal temperature excursions in the range of 5°C. Data sets such as these are required to begin to separate anthropogenic effects from natural background variability.

21.3.4 INTERACTIONS WITH OTHER VARIABLES

One of the limitations of parameter-specific water quality standards is that they tend to ignore interactions with other variables. A very simple example is that of decreasing oxygen solubility with increasing water temperature, as shown in Figure 21.5. At sea level (partial pressure of oxygen constant), a temperature increase from 13°C to 18°C results in a decrease in oxygen solubility from roughly 10.5 to 9.5 mg/l. This difference can be significant to some stages of some species. It is arguable that this effect should be accommodated within the temperature standard (see section 21.5).

21.4 POLICY ISSUES

In a manner analogous to that of the technical group, the PAC also found recurrent themes. These were:

1. identifying the acceptable level of protection or risk;
2. providing for equity among users;
3. resolving implementation issues; and
4. providing flexibility in the standard.

21.4.1 LEVEL OF PROTECTION OR RISK

The choice of the level of protection or risk to the resource (or beneficial use) is doubtless the most important single policy issue to be addressed in a water quality standard. Although the goal of all criteria is full protection of the most sensitive

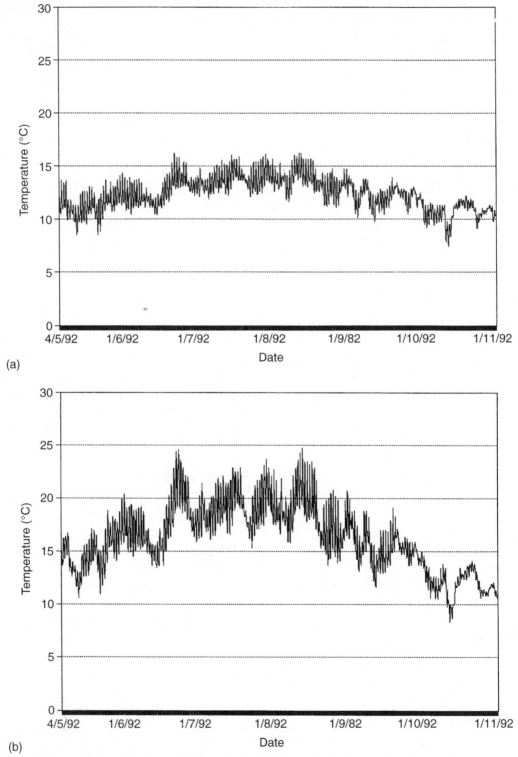

Figure 21.4 Seasonal thermal regime at (a) Bear Creek (b) Lobster Creek, Oregon Coast Range. (Source: Gregory; unpublished data, Oregon State University, 1992.)

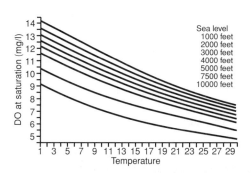

Figure 21.5 The theoretical relationship between temperature and dissolved oxygen at saturation. (Source: Sturdevant *et al.*, Temperature issue paper; Oregon Department of Environmental Quality, 1994.)

beneficial use, full protection can be defined along a continuum of risk. The goal of the criteria, defined by the PAC, is not to protect every individual organism at every place and time, but rather 'to fully protect Oregon's native fauna by providing the habitat quality necessary to support viable, sustainable populations of all native species at levels that fully utilize the habitat potential of a basin or ecoregion'. For example, there is no threshold temperature in the chronic portion of the dose–response relationship with salmonid fishes, but a continuum of increasing risk with increasing temperature and length of exposure. The TAC found that any temperature of 18°C or lower would fully protect the use under the above definition.

21.4.2 EQUITY

Another policy issue recurring in the Oregon standards' review is that of the comparative impact of water quality standards on sources of pollution. Under the Clean Water Act, the states have direct authority to control pollution from point sources, but only very indirect authority over the practices of non-point sources. This has led to a situation in which the pressure to reduce loads from point sources has increased while remaining essentially absent in the case of non-point sources. For temperature, this situation is somewhat exacerbated because the thermal load in headwaters is generally from non-point sources (principally forestry), whereas in downstream waters it is from point sources plus agriculture and urban non-point sources. Thus, often the assimilative capacity of downstream point sources with respect to temperature has already been spent by an upstream user. This issue can only be addressed through basin planning processes and improvement in requiring adequate practices for non-point sources.

21.4.3 IMPLEMENTATION AND FLEXIBILITY

Difficulties with implementation comprise a large group of recurring policy issues. Generally speaking, these issues involve the ramifications of actually putting a standard, which is designed to protect the beneficial uses, into practice. Resolution of these issues requires flexibility in the standard. Most of the implementation issues for Oregon's temperature standard arise from two causes:

1. Most Oregon basins are already out of compliance with the standard; and
2. There are many situations in which Oregon streams could not meet the standard under any conditions.

In the former case, the solution is to specify the manner in which the basins are declared water quality limited under the Clean Water Act. This will, in turn, affect the manner in which point sources and non-point sources are treated under the standard. In the latter case, one is given a number of options with which to demonstrate that sources' thermal loads either do not affect the existing thermal regime or do not impact on the beneficial uses.

21.5 COMPONENTS OF OREGON'S PROPOSED NEW TEMPERATURE STANDARD

Oregon's new temperature standard, developed under the process outlined in this chapter, will be complex. The criteria will be based on the seven-day average of the daily maximum temperatures, which has been demonstrated to be biologically meaningful. Although the criteria have yet to be adopted by the state's Environmental Quality Commission, the standard will include variations of most of these components:

- an overall criterion of 18°C based on the average daily maximum temperature for the seven warmest consecutive days;

- a criterion of 10°C for stream segments in which bull trout are found;
- a criterion of 13°C for stream segments supporting native salmonid spawning, egg incubation and fry emergence;
- a limit of 1°C increase in the temperature of designated cold-water refugia;
- no temperature alterations in stream segments containing federally listed threatened or endangered species which adversely affect the listed species;
- no measurable temperature increases when the dissolved oxygen concentration is within 0.5 mg/l of the dissolved oxygen criterion;
- no increases in temperature due to anthropogenic activity in natural lakes, wetlands, estuaries or marine waters;
- a clause waiving the standard if/when flows are below the 7Q10 level, and/or air temperatures are above the 90%-ile of the seven-day average maximum air temperature;
- a provision for a limited temperature increase for point sources downstream from the point of equilibrium with the air temperature;
- a provision for management plans to meet standards within five years for sources within segments which do not meet the standard;
- a provision for exceptions to the standard to be made by the Environmental Quality Commission; and
- a preamble directing temperature sources to adhere to the state's anti-degradation policy.

The finishing touches to this standard are still being worked on at the time of writing. The temperature standard is expected to be adopted by the Environmental Quality Commission during early 1995.

26.1 CONCLUSIONS

The first iteration of Oregon's new triennial water quality standards' review process has been greeted with great enthusiasm by industry and the environmental community alike. It has provided an opportunity for NGOs and other interested parties to participate fully in all aspects of the process, both technical and policy-oriented. The results appear to be leading the State of Oregon towards standards that are more flexible and that tend to place the burden of increased risk on the sources. The new standards also tend to be somewhat more complex than the earlier, more universal standards. The cost of this process has been high in terms of labour allocation, but appears justified based on the early responses to the process.

REFERENCES

1. Omernik, James M. and Alisa L. Gallant. 1986. Ecoregions of the Pacific Northwest. US Environmental Protection Agency EPA/600/3-86/033.
2. Sturdevant, D., S. Gregory, C. Andrus, R. Beschta, D. Buchanan, C. Frissell, D. McCollough, W. Meyer, C. Schreck, K. Sullivan and D. Wilkinson. 1994. Temperature Issue Paper. Oregon Department of Environmental Quality, Portland.

DIFFERENCES BETWEEN POLISH STANDARDS AND EC DIRECTIVES ON WATER QUALITY

22

by R. Korol, E. Jaśniewicz and M. Kędzia

22.1 INTRODUCTION

In Poland, river pollution has been measured for over 30 years and the results published in yearbooks of river water quality. In the period 1962 to 1988, the number of rivers under investigation increased continually: they now include very small, high-quality streams, which markedly improve the water quality statistics. The length of rivers investigated was greatest from 1975 to 1985, amounting to 17 774 km. Of this, very small streams of local significance, with catchment areas smaller than 1000 km^2, accounted for as much as 43% (Figure 22.1).

Since 1989 a unified programme has been in use. It specifies the principles of, and the methods for, monitoring flowing surface waters. The national monitoring system includes international and national investigations, covering both benchmark and basic networks. Hence, national monitoring is governed by the regulations of the Chief Pollution Control Inspector, whose department specifies the parameters that are to be determined as well as the frequency of their determination.

Thus the pollution level in Poland's rivers is assessed on the basis of:

- rules and regulations defining the permissible values for surface water pollutants, according to quality classes;

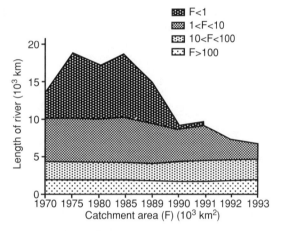

Figure 22.1 Total length of rivers and catchment areas 1970–93.

- results of water-quality determinations, carried out according to relevant standards;
- quantitative measurements concomitant with sample collection.

22.2 PRINCIPLES OF WATER QUALITY CLASSIFICATION

22.2.1 POLISH REGULATIONS

In Poland regulations defining the number of standard pollutants and the values of permissible con-

---

International River Water Quality. Edited by Gerry Best, Teresa Bogacka and Elżbieta Niemirycz. Published in 1997 by E & FN Spon, London, ISBN 0419215409

Table 22.1 Water quality classification system

Class	Purposes for which fit	Favourable conditions created for
I	Drinking Clean industries	*Salmonidae* habitat
II	Farming (stock-breeding) Recreation (swimming pools, aquatics)	Non-*salmonidae* habitat
III	Industrial Agricultural and horticultural (irrigation, glasshouses)	–

centrations have been subject to frequent amendments. Since 1970 however, there has been a classification system in force which distinguishes three classes of water quality (Table 22.1).

The permissible concentration values for each quality class are determined by governmental decrees. The latest decree of 1991 includes 57 parameters of water quality (Table 22.2).

However, the formulations of some of the rules are not sufficiently clear. This lack of clarity applies to the following.

- There is no gradation in the concentrations of 15 of the pollutants across the three quality classes;
- The definition of saprobity does not indicate what it applies to, e.g. seston, periphyton, etc.;
- The number of pathogens required for an adequate assessment of water quality is not specified;
- The method of assessing surface water quality is not described unequivocally.

22.2.2 COMPARISON WITH EUROPEAN LEGISLATION

A comparison of the Polish regulations with EC directives shows a certain analogy in the principles of surface water classification. According to European legislation quality criteria have been devised for:

- water abstracted for drinking purposes [1];
- water for supporting fish life [2, 3]; and
- water used for bathing and swimming [4].

There are differences in the parameters required for water quality assessment between Poland and EU countries. Thus, in Poland the determination of water quality for drinking purposes involves 57 parameters, and in the EU only 31 (Table 22.3). According to EC directives, the assessment of water quality for the support of fish life is based on six parameters only. One of these parameters is dissolved oxygen. For example, the water required to support salmonids and cyprinoids should contain dissolved oxygen concentrations equal to or greater than 9 mg/l and greater than 6 mg/l respectively. Polish regulations allow a limit of 6 mg/l in each case.

22.3 WATER QUALITY INVESTIGATIONS

For monitoring water quality, investigations are carried out both by district Inspectorates of Pollution Control and by scientific and research centres. The responsibility to coordinate these investigations was given to the Institute of Meteorology and Water Management by the National Inspectorate of Environmental Pollution Control. Thus, the analytical methods used need to comply with Polish standards. However, they do not always comply with those of the International Standards Organisation (ISO). The adequacy of determinations is, however, verified by intercalibration and so it can be anticipated that the results of water quality investigations are reliable and comparable with ISO. The data on the quality and quantity of flow are stored in the monitoring database and are verified by mathematical methods. A unified investigations programme yielded a data set characterizing the quantity of flow and the quality of water in the main rivers of Poland. The 1992 data set consisted of a total of 21 030 samples and 302 087 water quality parameter determinations.

22.4 METHODS OF WATER QUALITY ASSESSMENT

22.4.1 POLISH REGULATIONS

Polish regulations do not specify methods of data processing for the purpose of comparison with the

Table 22.2 Parameters of drinking water quality in Poland

No.	Parameter	Unit	Class I	Class II	Class III
1	Temperature	C	≤ 22	≤ 26	≤ 26
2	Odour	–	≤ 3 flora	natural	natural
3	Colour (platinum – cobalt scale)	mg/l	natural	natural	natural
4	Hydrogen ion concentration	pH	6.5–8.5	6.5–9	6–9
5	Conductivity	μS/cm	≤ 800	≤ 900	≤ 1200
6	Dissolved oxygen	mg/l	≥ 6	≥ 5	≥ 4
7	BOD	mg/l	≤ 4	≤ 8	≤ 12
8	Oxidizability (COD–Mn)	mg/l	≤ 10	≤ 20	≤ 30
9	COD–Cr	mg/l	≤ 25	≤ 70	≤ 100
10	Chloride	mg/l	≤ 250	≤ 300	≤ 400
11	Sulphate	mg/l	≤ 150	≤ 200	≤ 250
12	Total dissolved solids	mg/l	≤ 500	≤ 1000	≤ 1200
13	Total suspended solids	mg/l	≤ 20	≤ 30	≤ 50
14	Total hardness (as $CaCo_3$)	mg/l	≤ 350	≤ 550	≤ 700
15	Sodium	mg/l	≤ 100	≤ 120	≤ 150
16	Potassium	mg/l	≤ 10	≤ 12	≤ 15
17	Ammonia as nitrogen	mg/l	≤ 1	≤ 3	≤ 6
18	Nitrite as nitrogen	mg/l	≤ 0.02	≤ 0.03	≤ 0.06
19	Nitrate as nitrogen	mg/l	≤ 5	≤ 7	≤ 15
20	Total nitrogen	mg/l	≤ 5	≤ 10	≤ 15
21	Phosphate	mg/l	≤ 0.2	≤ 0.6	≤ 1
22	Total phosphorus	mg/l	≤ 0.1	≤ 0.25	≤ 0.4
23	Iron	mg/l	≤ 1	≤ 1.5	≤ 2
24	Arsenic	mg/l	≤ 0.05	≤ 0.05	≤ 0.2
25	Boron	mg/l	≤ 1	≤ 1	≤ 1
26	Manganese	mg/l	≤ 0.1	≤ 0.3	≤ 0.8
27	Chromium Cr^{3+}	mg/l	≤ 0.05	≤ 0.1	≤ 0.1
28	Chromium Cr^{6+}	mg/l	≤ 0.05	≤ 0.05	≤ 0.05
29	Zinc	mg/l	≤ 0.2	≤ 0.2	≤ 0.2
30	Cadmium	mg/l	≤ 0.005	≤ 0.03	≤ 0.1
31	Copper	mg/l	≤ 0.05	≤ 0.05	≤ 0.05
32	Nickel	mg/l	≤ 1	≤ 1	≤ 1
33	Lead	mg/l	≤ 0.05	≤ 0.05	≤ 0.05
34	Mercury	mg/l	≤ 0.001	≤ 0.005	≤ 0.01
35	Selenium	mg/l	≤ 0.01	≤ 0.01	≤ 0.01
36	Silver	mg/l	≤ 0.01	≤ 0.01	≤ 0.01
37	Vanadium	mg/l	≤ 1	≤ 1	≤ 1
38	Chlorine (free)	mg/l	undetectable	undetectable	undetectable
39	Cyanides, not bound, complex	mg/l	≤ 0.01	≤ 0.01	≤ 0.01
40	Bounded cyanides/complex	mg/l	≤ 1	≤ 2	≤ 3
41	Fluoride	mg/l	≤ 1.5	≤ 1.5	≤ 2
42	Rhodanate	mg/l	≤ 0.02	≤ 0.5	≤ 1
43	Sulphide	mg/l	none	none	≤ 0.1
44	Formaldehyde	mg/l	≤ 0.05	≤ 0.05	≤ 0.2
45	Acrylonitrile	mg/l	≤ 2	≤ 2	≤ 2
46	Caprolactam	mg/l	≤ 1	≤ 1	≤ 1
47	Volatile phenols	mg/l	≤ 0.005	≤ 0.02	≤ 0.05
48	Detergents (anionic)	mg/l	≤ 0.2	≤ 0.5	≤ 1

49	Detergents non ionic	mg/l	≤ 0.5	≤ 1	≤ 2
50	Subst. after extract by petr. ether	mg/l	≤ 5	≤ 10	≤ 15
51	Faecal coliform index	ml/bacteria	≥ 1	≥ 0.1	≥
52	Chlorophyll-*a*	µg/l	≤ 10	≤ 20	≤ 30
53	Saprobity	–	oligobeta	betamezo	alfamezo
54	Insecticides organophos. and carbam.	µg/l	≤ 1	≤ 1	≤ 1
55	Insecticides organochlor.	µg/l	≤ 0.05	≤ 0.05	≤ 0.05
56	Benzo-*a*-pyrene	µg/l	≤ 0.2	≤ 0.2	≤ 0.2
57	*Salmonella*		none	none	none

Table 22.3 Comparison of parameters used for water quality assessment in Poland compared to those stipulated in different EC Directives (Source: EC Directive 75/440)

| No. | Parameter | Unit | Poland | EU Directives | | |
			Classes 1,2,3	1	2	3
1	Temperature	C	+	+	+	
2	Odour	–	+	+		
3	Colour (platinium – cobalt scale)	mg/l	+	+		+
4	Hydrogen ion concentration	pH	+	+	+	+
5	Conductivity	µS/cm	+	+		
6	Dissolved oxygen	mg/l	+		+	
7	BOD	mg/l	+	+		
8	COD–Mn	mg/l	+	+		
9	COD–Cr	mg/l	+			
10	Chloride	mg/l	+	+		
11	Sulphate	mg/l	+	+		
12	Total dissolved solids	mg/l	+			
13	Total suspended solids	mg/l	+	+		
14	Total hardness (as $CaCO_3$)	mg/l	+			
15	Sodium	mg/l	+			
16	Potassium	mg/l	+			
17	Ammonia as nitrogen	mg/l	+	+	+	
18	Nitrite as nitrogen	mg/l	+			
19	Nitrate as nitrogen	mg/l	+	+		
20	Total nitrogen	mg/l	+			
21	Phosphate	mg/l	+	+		
22	Total phosphorus	mg/l	+			
23	Iron	mg/l	+	+		
24	Arsenic	mg/l	+	+		
25	Boron	mg/l	+	+		
26	Manganese	mg/l	+	+		
27	Chromium Cr^{3+}	mg/l	+			
28	Chromium Cr^{6+}	mg/l	+			
29	Zinc	mg/l	+	+	+	
30	Cadmium	mg/l	+	+		
31	Copper	mg/l	+	+		
32	Nickel	mg/l	+			
33	Lead	mg/l	+	+		
34	Mercury	mg/l	+	+		
35	Selenium	mg/l	+	+		
36	Silver	mg/l	+			
37	Vanadium	mg/l	+			

38	Chlorine (free)	mg/l	+				
39	Cyanides, not bound, complex	mg/l	+	+			
40	Bounded cyanides, complex	mg/l	+				
41	Fluoride	mg/l	+	+			
42	Rhodanate	mg/l	+				
43	Sulphide	mg/l	+				
44	Formaldehyde	mg/l	+				
45	Acrylonitrile	mg/l	+				
46	Caprolactam	mg/l	+				
47	Volatile phenols	mg/l	+	+	+	+	
48	Detergents (anionic)	mg/l	+	+			
49	Detergents non ionic	mg/l	+				
50	Subst. after extract. by petr. ether	mg/l	+	+			
51	Faecal coliform index	ml/bacteria	+	+		+	
52	Chlorophyll-*a*	µg/l	+				
53	Saprobity	–	+				
54	Insecticides organophos. and carbam.	µg/l	+				
55	Insecticides organochlor.	µg/l	+				
56	Benzo-*a*-pyrene	µg/l	+				
57	*Salmonella*		+	+		+	

available standards. It is conventional to assess the pollution level in terms of the so-called 'reliable concentration' which occurs at the mean low flow. For the purpose of monitoring, assessments are additionally made on the basis of 'guaranteed concentrations' (90% probability of occurrence).

To determine a reliable concentration it has been assumed that there exists a mathematical relationship between the mass (or concentration) of the water component and the quantity (rate) of flow. The relationship is described by a number of equations:

$$y(x) = ax + b;$$

$$y(x) = \frac{a}{x} + b;$$

$$y(x) = \frac{a}{x};$$

$$y(x) = \frac{x}{ax+b};$$

$$y(x) = ax^b;$$

$$y(x) = ae^{bx};$$

$$y(x) = \frac{a}{x} + bx + c;$$

where:

$y(x)$ is the concentration of the component or value of the parameter;
x denotes the quantity of actual flow; and
a, b, c, are real numbers determined empirically.

Curves are fitted using the coefficient of determination:

$$r^2 = 1 - \frac{\sum_{i=1}^{n}(y(x_i) - y_e(x_i))^2}{\sum_{i=1}^{n}(y(x_i) - y_m(x))^2}$$

where:

n denotes the number of observations;
$y(x)$ are the measured values;
$y_e(x_i)$ indicates the estimated values; and
$y_m(x)$ is the mean of the measured values.

The ideal value of the coefficient of determination is 1. Its utility as a measure of curve fitting is emphasized in literature.

Calculated reliable concentrations, when plotted on a river profile, provide the hydrochemical profile of the river. This shows the variability of individual parameters along the entire length of the river.

Statistical methods are used to determine the concentrations of a 90% and a 95% probability of occurrence. Guaranteed concentrations are calculated in terms of the simplified Nesmerak method, and thereafter they are assigned empirical probabilities.

Calculations are performed using the software developed by the staff of the Wrocław branch of the IMWM.

22.4.2 EC DIRECTIVES

According to EC directives:

1. Water is fit for drinking purposes if 95% of samples comply with the mandatory values and 90% of samples comply with the more stringent guideline values of the water quality parameters concerned. The frequency of sampling varies from 1 to 12 determinations yearly, depending on the group of parameters and the size of the intake.
2. Water is fit for supporting fish habitat if 95% of samples (taken at least once a month) do not exceed the limit values.
3. Water is fit for bathing and swimming if 95% of samples comply with the mandatory values of a standard set of parameters, and if 90% of samples comply with stricter guideline values, with the exception of total coliform and faecal coliform numbers.

22.4.3 MAKING COMPARISONS

From the foregoing analysis of the water quality assessment methods it follows that only the guaranteed concentration method with 90% or 95% probability will yield a classification which is comparable to the EC directives.

Table 22.4 Assessment of river quality according to reliable concentrations

	Percentage of total length monitored (7005.6 km)					
Class	Organic	Salinity	Suspended solids	Biogens	Faecal coliform numbers	Overall assessment
I	16.4	69.3	49.9	4.4	0	0
II	53.7	13.6	21	21.3	1.5	1.5
III	19.7	5.1	22.9	21.9	10.4	7.1
Unclassified	10.2	12	6.2	52.4	88.1	91.4

Table 22.5 Assessment of river quality according to guaranteed concentrations

	Percentage of total length monitored (7005.6 km)					
Class	Organic	Salinity	Suspended solids	Biogens	Faecal coliform numbers	Overall assessment
I	2.6	54.9	13.9	0.8	0	0
II	36.5	23.9	21.5	4.5	1.4	0
III	35.1	5.5	25.8	19.4	12.6	4.1
Unclassified	25.8	15.7	38.9	75.3	86.1	95.9

22.5 CLASSIFICATION OF POLAND'S RIVERS

The assessment of the pollution level in those rivers which were subject to basic monitoring in 1992 brought about the following classification for 27 rivers. (A length of 7005.6 km was monitored out of a total of 7550.5 km.)

class I – 44.7% (3129.8 km);
class II – 51.7% (3624.7 km);
and class III – 36% (251.1 km).

Water quality estimates were also established for groups of characteristic pollutants: organic substances (BOD_5, permanganate COD, dichromate COD, dissolved oxygen); salinity (chlorides, sulphates, dissolved compounds); suspended solids; biogens (nitrogen compounds, phosphates, total phosphorus); and faecal coliform numbers. These estimates resulted in an overall assessment comprising all the parameters under analysis (Tables 22.4 and 22.5).

The results obtained show that river pollution levels in Poland appear to be lower using EC than using the Polish classification system. This is because the waters of the rivers examined were partly fit for drinking purposes. Thus using EC standards:

- Category A2 water (normal physical treatment, chemical treatment and disinfection) accounted for 1.7% of the total length of river (115 km);
- Category A3 water (intensive physical and chemical treatment, extended treatment and disinfection) accounted for 1.5% of the total length of river (109 km);
- Water suitable for the support of salmonids accounted for 1.6% of the total length of river (111 km);
- Water suitable for the support of cyprinid life accounted for 15% of the total length of river (1048 km).
- Water fit for bathing purposes accounted for 1.5% of the total length of river (104 km).

From these data it is clear that, if the river water quality were to be assessed in terms of EC standards, Poland would have a larger volume of river water fit for drinking purposes and other uses. The basis of this classification system however is the assumption that there is adequate treatment technology in every instance and this is by no means the case.

REFERENCES

1. European Community Directive 75/440/EEC.
2. European Community Directive 78/659/EEC.
3. European Community Directive 90/656/EEC.
4. European Community Directive 79/923/EEC.

WATER QUALITY STUDY DESIGN FOR THE WILLAMETTE BASIN, OREGON, USING A GEOGRAPHIC INFORMATION SYSTEM

23

by M. A. Uhrich and A. Wentz

23.1 INTRODUCTION

The National Water Quality Assessment (NAWQA) programme of the US Geological Survey (USGS) has selected 60 large watersheds and aquifer systems throughout the USA to determine the status and trends of surface and groundwater quality of the country's water resources. These watersheds and aquifer systems, called study units, represent 60–70% of the country's public water supplies [1]. The NAWQA programme provides information on contaminants that occur in rivers and aquifers of the study units by describing where and when the contaminants are found, what concentrations are present and how their distributions are related to natural and anthropogenic factors. Results from multiple study units will be synthesized at national and regional scales in order to compare environmental and hydrological conditions throughout the US. The long-term goal of the programme is to provide information to national, state, and local watershed managers and to increase understanding of the land-related influences that affect water resources.

The Willamette Basin study unit, which includes the Willamette and Sandy River Basins in Oregon, began in 1991 as part of the NAWQA programme (Figure 23.1). Water quality concerns in the Willamette basin include loss of aquatic habitat; impacts of land-use changes and their effects on erosion, sedimentation, and river ecology; interactions between surface and groundwater; eutrophication of reservoirs and streams and increased nutrient levels in groundwater; and occurrence of trace organic compounds and trace elements in surface and groundwater [2]. The primary goals of research in the basin are:

1. to determine the occurrence and distribution of pesticides and nutrients in aquatic systems; and
2. to relate their occurrence and distribution to land-use practices, including pesticide and fertilizer application rates, and to natural factors, such as geology, soils and climate.

This chapter describes the use of a geographic information system (GIS) as a method of selecting sampling sites for evaluating water quality in relation to land-related features in large watersheds such as the Willamette basin. Included is a discussion of the application of a GIS to verify sites selected for sampling stream-bed sediment, fish, and macro-invertebrates in the Willamette basin during 1992–93.

23.2 THE WILLAMETTE BASIN STUDY UNIT

The Willamette River, nearly 440 km in length and with a drainage area of 29 700 km^2, is a prin-

---

International River Water Quality. Edited by Gerry Best, Teresa Bogacka and Elżbieta Niemirycz. Published in 1997 by E & FN Spon, London, ISBN 0419215409

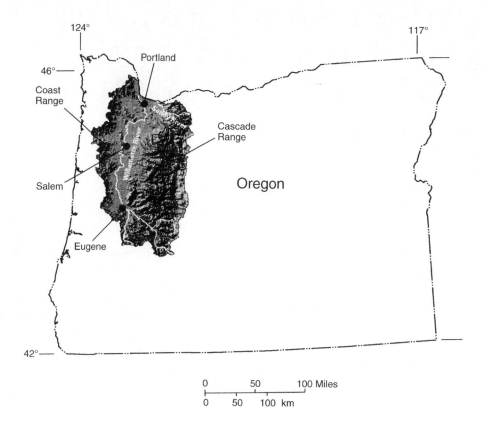

Figure 23.1 Location of the Willamette basin, Oregon.

cipal tributary of the Columbia River and is one of the few major north-flowing rivers in the world. Average annual discharge for 1973–93 at Portland was 903 m³/s, the 13th largest in the United States [3, 4]. The lower Willamette River is a shipping channel and port-of-call for cargo-carrying Pacific Ocean vessels. Imports and exports at ports along the river led the West Coast in shipping tonnage in 1993 at over 18 million tonnes [5]. The Sandy River basin, located adjacent to the north-east portion of the Willamette basin, also is a tributary of the Columbia River; it has a drainage area of 1300 km². The Sandy River basin is included as part of the Willamette basin NAWQA study unit because its three reservoirs, usable storage capacity totalling 55 million cubic metres, provide much of the drinking water for the city of Portland.

The Willamette basin is home to 70% of Oregon's residents (2 million) and includes the three largest metropolitan areas in the state. Nearly 82% of the residents use public water supplies, amounting to 1.1 million cubic metres per day (62% of the public water use in the state), 90% of which is from surface water sources [6].

The Willamette basin has a modified, maritime-temperate climate. Dominant westerly air currents cross the Pacific Ocean (approximately 130 km to the west) in winter, carrying cool, moist air. Precipitation is usually caused by the dominant Aleutian low, a semi-permanent cell of low pressure centred in the Gulf of Alaska. This anticlockwise circulation of air strengthens in October and moves southwards, causing much of the winter rain pattern in the Willamette basin. In April, normally the Aleutian low weakens and retreats northwards leaving warmer, drier air during the summer. Eighty percent of precipitation in the basin falls from October to March, with only 2% occurring during July and August.

Average annual precipitation for the Willamette basin during 1961–90 was 1600 mm, with ranges of 760–1300 mm in the Willamette valley and 2500–4400 mm in the Coast Range and Cascade Range mountains [7]. Maximum stream flow usually occurs in winter from a combination of rain and snow-melt. Mean monthly air temperatures range from 4–20°C on the valley floor to 0–17°C in the foothills and mountains [8].

23.3 APPLICATION OF GIS TO STUDY UNIT

In order to design a water quality sampling strategy from a watershed perspective and to relate the sampling design to land-based factors, a GIS was used. This approach provided a standard method to begin analysis and interrogate resource features, and formed the foundation for selection of sampling sites in NAWQA study units. Application of a GIS to all 60 study units allows the synthesis of data at the national level to occur more efficiently and with improved consistency. All study unit products share a common framework with identical attribute names and tabular databases.

23.3.1 WILLAMETTE BASIN NATURAL RESOURCE LAYERS

Initial assessments of water using a watershed approach must first consider the resource features of the study basin. These resource features can be manipulated by using GIS tools to facilitate geographic overlays and interpretations. In particular, digital resource layers can be stacked and sequentially aligned to allow mathematical and statistical analysis of the layers. The primary natural resource layers used for the Willamette basin are described below.

(a) Basin boundaries

The first GIS process was to digitize basin boundaries for the Willamette River and relevant sub-basins. This was accomplished after basin boundaries were delineated on 1:24 000 or 1:62 500 scale USGS topographic maps. In total, more than 260 1:24 000 scale maps fall within the extent of the Willamette basin; 85 of these maps comprise the basin boundary. Once digitized, the boundary lines were joined to form one continuous polygon. Sub-basin boundaries were assembled using a similar process. The basin and sub-basin boundaries were used as clip polygons on other natural resource layers as studies and sites dictated; that is, the boundaries were applied like a 'cookie cutter' to clip or extract appropriate data for selected basins and sub-basins.

(b) Hydrography

Hydrography includes rivers, streams, and reservoirs in the Willamette basin. Each segment of river between node points (node points usually occur at stream intersections) was assigned attribute data corresponding to its specific location. Attributes included stream name and length; state, county, and agency codes; political designations; and natural resource identifiers. Hydrography was acquired as digital line graphs (DLGs) from the USGS National Mapping Division [9]. A digital line graph is line work scanned from USGS 1:24 000 or 1:62 500 maps. The line work was edited to remove certain features, such as very small lakes and unconnected streams, resulting in a final scale of 1: 100 000. The DLGs were then converted to digital maps or coverages using GIS utility commands [10].

(c) Precipitation/run-off

Data for precipitation were based on mean monthly and annual totals collected from over 200 sites in the Willamette basin during 1961–90. Isohyetal lines were generated by the PRISM model at a grid resolution of 8 km and converted to ARC/INFO format [11]. The model corrects for orographic and terrain effects. Run-off data were represented by digitizing published maps for the period 1928–63 [12].

(d) Ecoregions

Ecoregions are based on homogeneous patterns of a combination of geographic, hydrological, biological, and climatic characteristics. Ecosystem and watershed relationships are considered, as are physical and landscape features. The principal ecoregions in the Willamette basin are the Coast Range (8% of basin area), Willamette valley and foothills (42%), and the Western and High Cascades (50%). A 1:2 500 000 digital map of Oregon ecoregions was acquired, then clipped with the Willamette Basin boundary polygon (Figure 23.2).

(e) Hydrogeology

The two principal aquifers of the Willamette basin are the alluvial aquifer and the Columbia River basalt aquifer, both found primarily in the

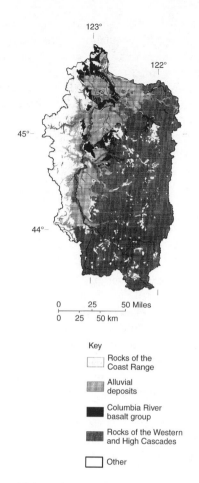

Figure 23.3 Surficial geology of the Willamette basin. (Source: Walker and MacLeod, Geological map of Oregon; US Geological Survey, 1991.)

Willamette valley. A 1:500 000 geological map for western Oregon was digitized, then clipped with the Willamette basin boundary polygon (Figure 23.3).

(f) Land use and land cover

Digital land-use and land-cover data (hereafter referred to as land use) for Oregon were acquired from the USGS National Mapping Division, as part of the Geographic Information Retrieval and Analysis System (GIRAS). Land-use categories for the late 1970s were determined based on the Anderson Level II classification system [13]. The

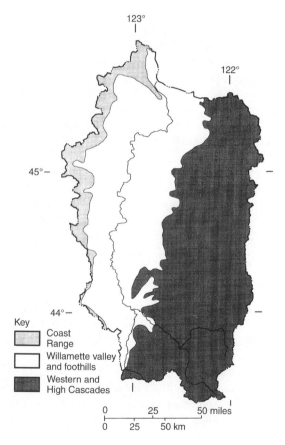

Figure 23.2 Ecoregions of the Willamette basin. (Source: Omernik and Gallant, Ecoregions of the Pacific Northwest, report EPA/600/3-86/033; US EPA, 1986.)

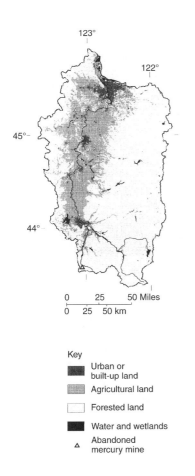

Figure 23.4 Land use and land cover of the Willamette basin. (Source: Fegeas *et al.*, Land use and land cover digital data; US Geological Survey, 1983.)

final map was compiled from a mosaic of NASA high-altitude U2/RB–57 aerial photographs at scales smaller than 1:60 000, and these were registered to 1:250 000 topographic base maps. Land use for the Willamette Basin was clipped from the original Oregon State coverage (Figure 23.4)[13].

23.3.2 SITE SELECTION USING THE GIS

Sites were chosen for monitoring of water quality based on prior knowledge of the Willamette basin, input from local agencies, information on current or historical sampling locations, and NAWQA goals and objectives. Ecoregions, hydrogeology, and land use were the primary GIS resource layers used to verify the selection process. Figure 23.5 shows these resource layers as a block diagram and illustrates the stratification scheme used to categorize sampling sites within the Willamette basin. The precipitation and run-off layers were used to help characterize sub-basins after sites were selected.

To the extent possible, sampling sites for water quality were selected in areas with homogeneous natural resource characteristics and land-related features. To verify this, the natural resource layers described above were aggregated by using GIS commands. As a result, sub-basin areas contained information about each resource layer, and that information was used when selecting sites with specific natural resource characteristics. For example, a site selected to evaluate the effect of pesticides on water quality should have physiographical, geological, and land-use features that best represent conditions where pesticides are applied. In this case, the site would most likely be located in the Willamette valley (ecoregion layer), in either of the two principal aquifer units (hydrogeology layer), and in an agricultural area (land-use layer).

Aggregation of the resource layers (Figure 23.6) was accomplished by using GIS overlay commands. Areas with common ecoregion, hydrogeology, and land-use attributes were grouped and selected, and basin and sub-basin boundaries for the sampling sites were delineated and digitized. Percentages of the above three resource layers within each basin area also were computed using the GIS.

Criteria were then established to describe the dominant resource type represented by each sub-basin. For a basin to be considered representative of a specific ecoregion or hydrogeological unit, the percentage area of that resource type had to be greater than 50%. For example, for a basin to be considered a Willamette valley ecoregion with an alluvial aquifer hydrogeological unit, then at least 50% of its area had to be within the Willamette valley ecoregion and at least 50 percent had to be within the alluvial aquifer. The same was true for other ecoregions and hydrogeological units.

A somewhat more complicated approach was used to categorize land use. For a basin to be con-

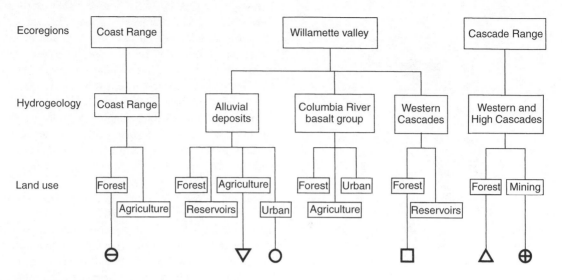

Figure 23.5 Stratification scheme for water quality sampling in the Willamette basin. (Symbols indicate sites sampled for chemical analysis in Figure 23.7.)

sidered representative of agricultural land use, more than 50% of its area had to be agricultural and less than 25% urban. Conversely, for urban basins, more than 50% of the area had to be urban and less than 25% percent agricultural. Forested basins had to be greater than 90% forested land to be considered representative of this resource type. If none of the above criteria were met, then a mixed classification was given.

Figure 23.7 illustrates the classification process for sites sampled for chemical analysis of stream-bed sediment, fish, and macro-invertebrates in the Willamette basin during 1992–3. Different site symbols represent the various categories of the stratification scheme presented in Figure 23.5.

Besides determining basin characteristics and depicting homogeneous natural resource areas for existing sites, a GIS could also be helpful when selecting water quality sites. For example, if sites are desired within the Willamette valley ecoregion, alluvial aquifer hydrogeological unit and agricultural land use area, then a GIS can be used to identify these regions, aggregate them, and plot appropriate sampling points within the aggregated regions.

23.4 OTHER USES OF GIS IN RIVER BASIN MANAGEMENT

Water quality studies conducted with a river basin approach, particularly studies of non-point source contamination, intuitively involve some form of spatial analysis. A GIS is ideally suited as a tool to initiate and drive this type of investigation. Statistical analyses and computations of natural resource percentages on basin-wide scales, along with regional or national spatial analysis, can be accomplished easily using a GIS. Also a GIS could be used to provide input parameters to drive basin-wide hydrological models. Parameters, such as elevation, slope, and aspect, are key components of such models and can be generated using GIS programmes [14]. For example, using these parameters, along with soil and vegetation classifications, a GIS has been used to determine hydrological response units (homogeneous areas that respond similarly to hydrological driving forces), which serve as inputs to a stream-flow model of the Willamette basin [15].

Watershed management relies on the accurate depiction of a basin to support land-use and water-use decisions. The data used to arrive at

Other uses of GIS in river basin management 187

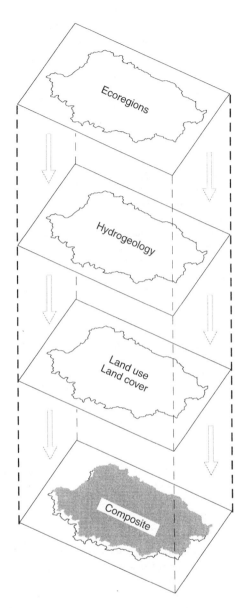

Figure 23.6 Aggregation of Willamette basin resource layers.

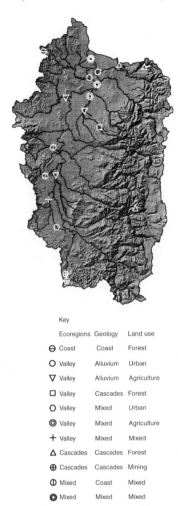

Figure 23.7 Categorization of sites sampled for chemical analysis of streambed sediment, fish and macroinvertebrates in the Willamette basin 1992–3.

these decisions can be displayed and managed using a GIS. Specific areas within a basin, such as riparian zones, can be derived then further categorized based on other geographic and environmental features, such as slope or vegetation species. Finally, biological indices, risk assessment, and suitability rankings for specific physi-

cal or ecological conditions can all be determined using a GIS [16–18].

23.5 KEYS FOR DEVELOPING A GIS DATABASE

GIS database development is an evolving process that needs both a solid framework to build upon and the flexibility to expand. As cartographic data are digitized and a GIS database is developed, it becomes important to focus on certain key elements. First, the digital data assembled for the project should be organized in a systematic directory structure that will facilitate accessibility and maximize storage capacity. Second, the long-term uses of the attribute data must be considered in the project design in order to meet future demands of the project. These demands include revisions and future expansions of the database. Attribute data should be structured in a usable, consistent, and understandable format and be documented, yet also provide a means to update existing information. Some keys for accomplishing this are selecting map projections that minimize error and coordinate drift, providing additional attributes for future updates and documenting processing steps so that intermediate products can be rendered. The final product should provide an assessment of current conditions and provide adaptations to allow for future changes in these conditions as they occur.

Lastly, the final product of the project should be kept in focus. If the eventual goal is to relate water-quality data to land use, as with the NAWQA program, then a plan should be developed to acquire the most current land-use information, perhaps by updating existing digital data. For example, the Willamette NAWQA updated the 1970s GIRAS land-use data by using 1990 census data to revise urban areas and by scanning and digitizing additional maps that categorized agricultural areas as irrigated or non-irrigated.

23.6 CONCLUSIONS

A goal of the NAWQA programme is to relate water quality data to land use. Water quality sampling sites were selected throughout the Willamette basin on the basis of local agency recommendations and the historical sampling record. In order to categorize the basin and help determine sampling locations, natural resource data layers within the basin and sub-basins were aggregated to define areas representative of specific ecoregions, hydrogeological units, and land uses. A GIS was used to perform this aggregation and to verify that sampling locations were within specified limits for percentages of these common natural resource areas. Other ways to use a GIS in watershed characterization and management include statistical analysis of spatial characteristics, driving hydrological models, and displaying data related to land and water use. Systematic directory structures and early consideration of long-term needs are keys for creating a reliable and versatile GIS database.

REFERENCES

1. Leahy, P.P. and Thompson, T.H. (1994) National Water-Quality Assessment Program: Open-File Report 94–70, US Geological Survey.
2. Wentz, D.A., and McKenzie, S.W. (1991) National Water-Quality Assessment program – The Willamette Basin, Oregon: Open-File Report 91–167, US Geological Survey.
3. Kammerer, J.C. (1990) Largest rivers in the United States: Open-File Report 87–242, US Geological Survey.
4. Hubbard, L.E., Herrett, T.A., Kraus R.L. *et al.* (1994) Water Resources Data, Oregon, Water Year 1993: US Geological Survey Water-Data Report OR-93–1.
5. Warren, W. and Ishikawa, T.L., (1993) Oregon Handbook, Moon Publications, Chico, California.
6. Broad, T.M., and Collins, C.A. (1993) unpublished data.
7. Taylor, G.H. (1993) Normal annual precipitation map (1 sheet) for State of Oregon, 1961–90: State Climatologist's Office, Corralis, Oregon. Unpublished.
8. National Oceanic and Atmospheric Administration (1992). Climatography of the United States (81), Monthly normals of temperature, precipitation, and heating and cooling degree days, 1961–90. National Climatic Data Center, Asheville, North Carolina.
9. US GeoData (1989) Digital line graphs from 1:100 000-scale maps – Data user's guide 2: Technical Instructions, US Geological Survey, National Mapping Division.
10. Fisher, B.J. (1996) Methodology used to produce an encoded 1:100 000-scale digital hydrographic data layer for the Pacific Northwest, Water-Resources Investigations Report 94-4043, US Geological Survey.
11. Daly, C., Neilson, R.P. and Phillips, D.L. (1994) A statistical–topographic model for mapping climatological precipitation over mountainous terrain, *Journal of Applied Meteorology*, **33**(2), 140–58.
12. Oster, E.A. (1968) Patterns of runoff in the Willamette Basin, Hydrologic Investigations Atlas HA-274, 1 plate, US Geological Survey.
13. Fegeas, R.G., Claire, R.W., Guptill, S.C., *et al.* (1983) Land use and land cover digital data, Circular 895-E, US Geological Survey.

14. Jeton, A.E. and Smith, J.L. (1993) Development of watershed models for two Sierra Nevada basins using a geographic information system, in (eds J.M. Harlin and K.J. Lanfear), *Proceedings of the Symposium on Geographic Information Systems and Water Resources*, American Water Resources Association, Mobile, Alabama, March 14–17, 1993, pp. 251–8.
15. Laenen, Antonius and Risley, J.C. (1996) Precipitation-runoff and streamflow-routing modeling as a foundation for water-quality simulation in the Willamette River Basin, Oregon, in (ed. Laenen, Antonius) *Proceedings of the Poland–USA International River Quality Symposium*, Portland, Oregon, 21–25 March 1994, Lewis Publishers.
16. Phillips, N.J. and Berbrick, M. (1993) Use of geographic information systems to conduct risk assessments–non-point source applications, in (eds J.M. Harlin and K.J. Lanfear), *Proceedings of the Symposium on Geographic Information Systems and Water Resources*, American Water Resources Association, Mobile, Alabama, 14–17 March 1993, pp. 45–51.
17. Montgomery, D.R. (1994) Use of GIS in watershed analysis (abs), EOS, Transactions American Geophysical Union, **75**, (16), p. 175.
18. Van Steeter, M. and Pitlick J. (1994) Analysis of changes in channel morphology of the Upper Colorado River in relation to endangered fish habitats (abs), EOS, Transactions American Geophysical Union, **75**, (16), p. 176.

THE ROLE OF BENEFIT–COST ANALYSIS IN WATER CONSERVATION PLANNING

by R.G. Sakrison

24.1 INTRODUCTION

Since 1990, managers of public water systems in Washington State have been required to evaluate all recommended conservation measures, implement those that are required and those that meet the public water system needs. Managers must explain decisions not to implement measures they are required to evaluate. The appropriate mix of measures and the degree to which they are implemented depends on the size of the system, type of customer base and the degree to which supply options are independent of the regional sources of supply.

To date there has been little technical assistance offered to public water system managers to help them develop their conservation plans. The purpose of this chapter is to provide technical assistance to managers of public water systems to help them evaluate and select conservation measures. The chapter addresses whether a benefit–cost or cost-effectiveness analysis is appropriate. In addition it provides guidance on identifying factors that should be included in the analysis and suggests that public water system managers can determine optimum levels of conservation given adequate information on their local and regional sources of supply.

24.2 CONSERVATION IN PUBLIC WATER SUPPLY PLANNING

Public water system managers are responsible for providing a safe and reliable water supply. To meet current and future needs, they must evaluate different sources of supply, including demand management options. Water conservation has become an increasingly important means of reducing future demands. In some extreme cases where no additional water is available it is the only means of meeting future needs.

In 1980, water conservation was defined by the US Water Resources Council [1] as 'activities designed to

1. reduce the demand for water;
2. improve the efficiency in use and reduce losses and waste of water; or
3. improve land management practices to conserve water.'

A more recent definition is 'any "beneficial" reduction in water use or water losses' [2]. Conservation measures may be either demand-side measures that reduce customers' demand for water (e.g. replacing turf with low-water-use native plants) or supply-side measures that increase a water system's delivery efficiency and enhance supplies (e.g. leak detection).

The focus of this chapter is long-term conservation plans directed at permanent reductions in water use, which can be accomplished by changing customers' behaviour through public education and technical assistance, or by investing in new water-efficient capital stock (e.g. low-flush toilets). Short-term curtailment plans address shortages during extreme drought events or other periods when the immediate water supply is inadequate. Water shortage response plans may include restrictions on customers' water use (e.g. ban on outdoor lawn watering) which would not be part of a normal long-term conservation plan.

24.3 WASHINGTON STATE CONSERVATION PLANNING REQUIREMENTS

There are over 14 000 public water systems in Washington State. The Washington Department of Health defines a public water system as two or more connections. The majority of these systems are very small, typically less than 15 connections, have relatively limited conservation opportunities and require little analysis for conservation planning purposes. The 200 largest systems, which service over 90% of the state's public water service, have many more conservation options available. The analysis of conservation alternatives for larger systems can be complex.

State conservation planning requirements [3] call for 'an analysis of measures in a conservation programme' This analysis is specific to the requirements and opportunities of individual water systems:

> The specific criteria to be used in the evaluation of the water conservation programme are identified [in the guidelines]. In general, the selection and implementation of conservation measures should be determined by the cost of a measure in relation to the value of the water conserved, i.e. by the relation of benefits and costs.

The recommended conservation measures were chosen by a group of water utility specialists and agency staff from a range of options that have been shown to have potential for successfully reducing water use.

24.4 ECONOMIC EFFICIENCY

The role of economics in water allocation has been recognized for nearly four decades [4, 5]. The federal practice of requiring an orderly measurement and discounting of future benefits and costs of water resources development projects to present values, for the purpose of assuring feasibility, dates back to the first issuance of the *Green Book* [6]. The concepts first enunciated in the Green Book are still found in the federal guidance, *Economic and Environmental Principles and Guidelines for Water and Related Land Resources Implementation Studies*. [7]

24.4.1 SUPPLY AND DEMAND FRAMEWORK

Traditionally, water utilities have looked at new sources of supply in a supply and demand framework, which brings together source options, demand projections and water rates. The theoretical answer to how much a water utility should sell and at what price is 'jointly determined at the intersection of the marginal cost curve and demand curve' [8]. This is the rule of marginal cost pricing. Only those costs and benefits that are incremental to the decision are to be taken into account as relevant to the decision. Sunk costs are ignored.

24.4.2 DEMAND CHARACTERISTICS

Water systems are designed to meet both peak and off-peak (i.e. base) demands. There are two types of peak demand: time-of-day and maximum day (or seasonal). 'Load factor' is the ratio of average to peak demand. The primary contributor to residential peak demand, which causes most water system peak demands, is lawn and garden sprinkling. Water system components are designed to meet different demand characteristics (Table 24.1).

24.4.3 COSTS OF NEW SUPPLY

Water system costs are typically:

1. production, transmission and treatment costs;
2. distribution costs; and
3. customer service.

Investment and operating costs will vary from system to system depending on the size of the system, the particular location and quality of the source of the water, the terrain, customer population density and demand characteristics.

Table 24.1 Water system components and demand characteristics

Water system component	Demand characteristic
Supply source development	Average annual daily
Raw water storage reservoirs	Average annual daily
Transmission	Maximum day
Treatment	Maximum day
Feeder mains	Maximum day
Distribution mains	Maximum hour, or
Pumping stations	maximum day plus
Local storage facilities	Fire protection flow

The costs of water utility service are classified according to customer, capacity (demand), commodity (operating) and fire flow costs:

- Customer costs – metering, billing, collection and customer service;
- Capacity costs – physical plant required to meet peak demands;
- Commodity costs – treatment chemicals, energy (e.g. pumping) and other volume-related;
- Fire flow costs – Flow and storage requirements needed to fight fires.

24.4.4 WATER RATES

Proper cost allocation to customers is fundamental to fair rates and appropriate prices [9]. Rates can be designed a number of ways; some can be used to influence conservation. Costs are allocated to services or customer classes. Residental, commercial, industrial and public/governmental are examples of customer classes.

The usual steps in utility ratemaking are:

1. determining the water systems revenue requirements;
2. allocating costs among different customer classes;
    - customer costs assigned on basis of service connections;
    - commodity costs on the basis of usage;
3. designing rates to cover costs;
    - ideally from those who cause costs;
    - costs identified with a customer class;
    - costs vary with usage associated with customer class;
    - costs assigned depending on whether new capacity is required;
    - costs assigned to the customers that benefit [10].

The rate structure should encourage the efficient use of water. At the margin, the water rate should reflect the full private and social marginal cost of supply [11]. These goals move away from the narrow focus of meeting the utility's revenue requirements to a broader set of goals that include economic efficiency and the promotion of conservation.

(a) Marginal cost pricing

The central argument for marginal cost-based rates is that commodities should be produced and allocated to the point at which their marginal benefit equals their marginal cost. Marginal benefit, the extra benefit from producing and consuming one more unit of the commodity, can equally be defined as the benefit lost and cost saved of producing one less unit. With given productive capacity, the decision to produce one more unit of a commodity implies a decision to produce less of all other commodities. The opportunity cost is the value of the alternative foregone. In this sense marginal cost is synonymous with avoided cost. The distinction between private and social marginal benefits and costs involves the concept of externalities. In the modern theory of pollution, when there is a divergence between private costs of production (which exclude the effects of pollution emissions on other people) and social costs (which include the effects of pollution emissions on other people) prices should follow marginal social rather than private costs [12].

(b) Short-run v. Long-run marginal cost pricing

Short-run marginal cost is always less than long-run marginal cost, owing to the distinction between fixed and variable costs. Fixed costs do not vary with the quantity of service provided, whereas variable costs do change with the quantity of service provided. In the long-run, the capital stock can be replaced and expanded, and the marginal cost includes not only operation and maintenance costs but also capital costs. The distinction matters a great deal in the water supply industry because of its unusually high capital intensity [13]. The ratio of capita assets to annual revenue in the water industry is estimated at $10–12, which is 3 to 4 times the capital intensity of telephone and electric utility industries, and about 5 to 6 times that of the railroad industry.

(c) Rate setting

Setting rates equal to the long-run marginal cost may guarantee that the price of water reflects the opportunity costs associated with the variable inputs and capital assets required for their production. However, this analysis requires detailed cost allocation studies to model future consumption patterns, conservation programmes and supply development. There are a number of generic problems that arise in rate studies that have a bearing on economic efficiency.

Average or embedded cost techniques

Charging a single, average rate for a commodity, which is known as embedded cost, pools many different costs into a single average. Embedded cost emphasizes the estimation and allocation of historical average cost. Average cost may be higher or lower than short-run marginal cost; it will be lower than long-run marginal cost in a dynamic system. Embedded cost technique has one advantage for rate setting: it guarantees that revenues (in the short-run) will equal costs, thus satisfying the requirement that rates return sufficient revenue without collecting too little or too much.

Marginal cost techniques

If rates are set equal to long-run marginal costs, which will normally exceed short-run marginal costs for expanding systems, excessive revenue will be returned to the utility. If rates are set equal to long-run marginal cost there is often a revenue crisis as customers react to higher rates. Ratemaking has attempted to address these problems, mindful of economic efficiency, in a variety of rate designs.

- Two-part tariff. Most water bills consist of a combination of fixed charges and usage charges which vary with consumption. If the fixed charges are set too high the rate essentially becomes a flat-fee rate structure;
- Increasing block rates. Associated with long-run marginal cost; high-consumption customers cause utility to add capacity. Seattle uses increasing block-rate structure with the last block priced at the long-run marginal cost of new supply;
- Seasonal rates. Rates correspond with periods of peak use and reflect system costs required to meet peak demands;
- Goal billing. Incentives are offered to customers who reduce their water use by a certain amount;
- Drought surcharges. During periods of rainfall shortage surcharges are added to rates. The effect is to reduce demand and to recover a revenue requirement shortfall due to conservation. Tacoma and Seattle implemented drought surcharges in 1992;
- Ramsey pricing. This is named after an English economist who died in 1928. In the presence of decreasing average costs only water rates which mark up prices above marginal cost in inverse proportion to the elasticity of demand will maximize welfare and break even. Customers with high elasticity of demand would be charged more.

Other rate designs

Many innovative rate designs have been tried, or are theoretically possible: indoor–outdoor, sliding scale, spatial pricing, scarcity pricing, capital contributions, hook-up fees, and others [9, 10, 14].

24.5 PLANNING FOR ADDITIONAL WATER SUPPLIES

It is efficient for a water utility to 'adopt any project for which the present value of the associated stream of net benefits or new receipts, discounted at the appropriate rate of interest, is greater than zero'. This is known as the present value rule and governs whether projects are economically feasible [15].

Does the application of the present value rule to conservation options meet the requirement for public water systems to 'select and implement conservation programmes based on the relationship of benefits to costs?'.

Whether the routine planning activities of a water utility will identify feasible solutions consistent with the present value rule depends on the particular approach the water system manager takes to planning for future supply options. Water system managers who do not maintain accurate water-use data, do not forecast future water demands based on observed water use or fail to examine demand management and supply-side efficiency improvements on an equal basis are probably not in a position to understand the relationship between benefits and costs of alternative means of meeting future needs. The quality of data collected and analytical methods used to identify feasible options varies greatly. Even with programmes in place to accomplish these ends, many difficult analytical issues in conservation planning have yet to be fully resolved.

24.5.1 TRADITIONAL WATER SUPPLY PLANNING

Utilities have traditionally concentrated on internal issues and emphasized goals of minimizing water rates. The planning processes undertaken by water utilities have been traditionally dominated by supply considerations. The result has been an emphasis on maintaining dependable water supplies and, accordingly, the engineering of facilities for source development, treatment and storage, and transmission and distribution of water [16]. In contrast, using an iterative process can properly account for the interrelationships between rates, water demands, system size, and long-range costs [17].

24.5.2 LEAST-COST PLANNING

Concern over rising costs of capital additions, self-serving demand forecasts, and inattention to environmental externalities was responsible for the emergence of least-cost planning in the 1980s [15].

Least-cost planning is a way of analysing the growth and operation of utilities that considers a wide variety of both supply and demand factors so the optimal way of providing electrical (or water) service to the customers can be determined. A path is chosen that will ensure reliable service for the customers, economic stability and a reasonable return on investment for the utility, environmental protection, equity among ratepayers, and the lowest costs to the utility and the consumer. A least-cost plan balances three interests (reliability, profitability and affordability) while keeping a sharp eye on the risks and uncertainties associated with each component of the plan [18].

Least-cost planning has been an essential tool of economic regulation of electricity utilities. In many states it has become the fundamental regulatory standard and considerable attention has been given to its evolution by the National Association of Regulatory Utility Commissioners. Many useful practices have come out of least-cost planning, including the treatment of externalities [19] and the realization that all demand-side activities that decrease the demand for utility services tend to affect supply, since existing system capacity is released for other customers and other uses.[20]

Water utilities have been reluctant to institute least-cost planning practices to any significant extent. In Washington State there are less than 200 private water companies regulated by the Utilities and Transportation Commission. Without a regulatory agency requiring least-cost plans, managers of public water systems are less likely to adopt least-cost planning practices. Water utilities are more likely to react to the public's perception that least-cost planning means providing a less reliable service, meeting future needs only through reduced water use and raising water rates to meet revenue shortfalls. The expected reaction from utilities under stress is to try to build new sources of supply at whatever cost.

24.5.3 INTEGRATED RESOURCE PLANNING

Integrated resource planning is a smaller development in water utility planning. It is somewhat more encompassing than least-cost planning. However, both emphasize concurrent consideration of demand- and supply-side options and the importance of decision-making based on the least-cost principle [21]. Integrated Resource Planning seeks to create a more open and participatory process and interaction of the many institutions involved with water resources policy and planning. Particularly important is the relationship of water utility planning to the activities of various governmental agencies whose polices may constrain planning choices.

24.6 SUGGESTED ROLE OF BENEFIT–COST ANALYSIS IN EVALUATION OF CONSERVATION MEASURES

How should public water systems approach the evaluation of water conservation alternatives as a means of meeting current and future needs? Is benefit–cost analysis the appropriate technique for picking the right mix of supply-side and demand-side alternatives, or is a cost-effectiveness analysis preferred?

24.6.1 EVALUATING BENEFITS AND COSTS OF CONSERVATION MEASURES

Conservation measures should not be evaluated merely as an abstract exercise; rather, they should be evaluated as they would be implemented in the context of the individual water system or groups of systems as part of a regional programme. Several different manuals and handbooks are available on conservation and conservation planning for guiding a water system manager through an evaluation [2, 22, 23]. For water system managers with computers, there are several public domain software programs available for evaluating conservation programmes [24, 25].

In general, evaluating conservation programmes must include the following steps [2].

(a) Analysis of water system characteristics

Analyse water use and service area data for selecting and evaluating conservation measures. Knowledge of current water use and historical changes is necessary to identify customer groups that show the greatest potential for water use efficiency improvements. The water system characteristics that should be analysed are:

- disaggregation of total water use by major sectors;
- estimation of seasonal and outdoor components of water use in each sector;
- identification of significant end uses in each sector;
- assessment of existing conservation practices.

(b) Forecasts of water use

Develop base line forecasts of future water use for disaggregated selectors without conservation conditions. Developing long-term forecasts of water use may require:

- demographic and economic growth projections for the service area;
- selection and calibration of water-use models;
- preparation of forecasts for alternative growth scenarios.

(c) Initial screening of potential conservation measures

Identify all known conservation practices that are applicable to water used in the service area that have not been implemented or approved for implementation. The initial screening examines:

- technical feasibility – whether measures will result in significant reduction in water use;
- social acceptability – whether measures will be adopted by customers;
- environmental impact – most conservation programmes rank higher (less impact) when compared with traditional supply options.

The potential water savings of conservation measures can be determined from literature sources, pilot programme results, or empirical studies from water agencies. In the initial screening each conservation alternative must be carefully formulated because the implementation conditions will affect the determination of potential benefits and costs [2].

The procedures manual prepared for the California Urban Water Agencies describes the process for estimating potential water savings. The potential water savings resulting from a conservation measure is a function of the fraction reduction in water use, the market penetration (or coverage) and baseline water use. The estimation of total savings is usually disaggregated by particular use sector (single family residential, industrial, etc.) and a specific water dimension (indoor, outdoor, or peak use).

(d) Economic evaluation of the most likely conservation measures

The accepted approach to the evaluation of likely conservation measures is to use benefit–cost analysis to compare the potential economic impacts of alternative conservation programmes [26–28]. A starting point for identifying benefits and costs for water conservation alternatives would be:

Benefits

- Reduced short-run incremental costs
- Reduced long-run incremental costs
- Energy savings
- Other economic benefits effects
- Environmental quality
- External costs.

Costs

- Utility programme costs
- Customer programme costs
- Other economic costs
- Reduced aesthetic value
- Reduced revenues (without rate adjustments, reduced water use leads to reduced revenue).

'Financial feasibility' means that a project generates enough revenues to cover all costs, including interest on funds borrowed to finance the project. Economic feasibility means that the economic valuation of benefits, to whomever they accrue, exceeds the economic valuation of the costs, to whomever they accrue. In economic feasibility the valuation of benefits and costs may be different to the actual monetary receipts and outlays, and the discount rate may diverge from the monetary or market borrowing rates. The distinction usually made between financial and economic feasibility is that certain economic values are not shown in the financial accounts of the agency making the decision [29].

Consumer surplus

In theory, at least, the benefit valuation should include the aggregate value benefits over and above the price set for the last small increment consumed. In theory, a producer's surplus should also exist when costs are below the market price. The producer surplus is given by the difference between the revenue received by the water utility and the cost of supply. Instead of multiplying the quantity of water consumed by the marginal price, the aggregate value is measured by the quantity multiplied by some average value that lies between the value of the first small increment and the last. To do otherwise would take too conservative a view of the benefits. In theory, consumer surplus is made up of two components: willingness to pay (equivalence) and willingness to accept (compensation). Hanneman has shown that the rationed quantity must have some closure substitutes for the various welfare measures to be equal in magnitude [30]. Willing has shown that ordinary or Marshallian consumer surplus is probably adequate for commodities that have small income elasticity of demand, such as water [31].

The benefits of conservation programmes are based on reductions in costs of providing water supplies:

Utility cost savings can be the result of reduced water purchases, reduced operation and maintenance expenses, and deferred, eliminated, or down-sized capital facilities. Customer benefits accrue from reduced water bills, reduced wastewater service bills and reduced energy bills.

The cost of conservation programmes include utility and customer costs:

Utility programme costs result from administrative costs, unit costs (cost per site), field labour costs, publicity costs, evaluation costs, and decreased utility revenue. Customer programme costs result from equipment, materials and installation costs, and operation and maintenance costs.

Primary and secondary benefits and costs

The literature of benefit–cost analysis and conservation planning evaluation is mainly concerned with defining categories and types of benefits and cost. This was the case when benefit–cost analysis was first developed by the federal government [6] and has continued to this day [32]. In order to understand the feasibility of a given conservation measure it is necessary to know whether implementing it will result in a 'beneficial reduction in water use or losses'. However, merely accounting that the total beneficial effects (benefits) are greater than the adverse effects (costs) will result in too conservative a measurement of net benefits.

Changes in consumer surplus

The benefits and costs identified by using conventional techniques that depend on identifying outputs, products and associated costs, will, in theory, underestimate the net benefits. To accurately measure the true economic impact it is necessary to translate the benefit and cost information into price effects. Once the effect on rates is determined, the total net change to consumer surplus can be calculated [31]. Several recent studies have addressed changes to consumer surplus from different types of utility responses to drought conditions: Narayanan et al., in studying the effectiveness of drought policies for municipal water management, found that few water authorities have used pricing to ration water use during a drought [33]; Dandy, in assessing the economic cost of restrictions on outdoor water use, found that reductions in consumer surplus from mandatory restrictions exceed reduction in consumer surplus from price increases [34].

Intangibles

The early discussions in the US Congress on the economic rules for river basin planning debated whether project effects that cannot be easily quantified should be included in routine economic analysis. As long ago as the New Deal in the 1930s, it was recognized that resolution of difficult-to-measure issues was better handled in a political process than in an analytical process led by economists [35]. The orginal hearings on the federal benefit-cost analysis occurred during the authorization of the Flood Control Act of 1936. Intangible effects can include sites of scenic and historical value, national security, regional development patterns, and the enjoyment of recreation and wildlife.

Electrical planning is an area that has provided models for incorporating environmental costs into decisions on future sources of supply. The Northwest Power Act defines system costs as: direct costs, indirect or future costs and such 'quantifiable environmental costs and benefits that are directly attributable to the measure or resource'[36–38].

There has been a very lively debate in the regulated electrical industry over the role of environmental externalities. Some state regulatory agencies are requiring utilities to place monetary environmental extras on top of direct or private costs to reflect uncompensated external costs and result in a full social cost 'decision price'[39–41].

Environmental and social values are less commonly addressed in water supply planning. The practice is beginning to be found in utilities that are following a least-cost or integrated planning process. Seattle and Portland are two water supply utilities that have attempted to incorporate environmental considerations into their resource acquisition strategy [42]. Seattle adopted a 10% bonus for conservation measures to reflect expected environmental benefits of demand-side management options when compared to new sources of supply. This method is parallel to the 10% 'conservation bonus' used by the Power Planning Council and BPA in resources planning.

The pattern in years to come will be to include more intangibles explicitly in the analysis of benefits and costs. The techniques fall into two categories:

1. indirect market methods, such as travel-cost, hedonic valuation techniques, and market-based mechanisms; and
2. direct questioning approaches [43].

The most important direct-question technique commonly used is contingent valuation [30]. The application of these techniques to water conservation planning is still in its infancy and is beyond the scope of this chapter.

Discount rate

Another major issue in benefit–cost analysis is the discussion on whether future benefits and costs should be discounted and, if so, what the proper social discount rate should be. Discounting future benefits and costs reflects the time value of money. Nominal rates include the expected rate of inflation and are considered the same as market interest rates. The real rate of interest is adjusted to eliminate the effect of the expected rate of inflation. A real discount rate of 7% is recommended in federal regulations [44]. It has been common practice for water resources development projects to use lower-than-market social discount rates. Given the long project life of major water projects, this practice will give relatively more weight to future benefits and costs. This chapter does not attempt to address the many public policy issues involving discounting future benefits and costs.

24.6.2 BENEFIT–COST ANALYSIS V. COST-EFFECTIVENESS

A perplexing issue for water utilities is whether to conduct a full benefit–cost analysis or an abbreviated cost-effectiveness study. The state guidelines allow flexibility in the selection of measures and level of implementation. This flexibility is intended to accommodate differences between public water systems, competition for resources, and problems and opportunities of specific water systems [23]. The level of benefit–cost analysis should reflect the size of the system and the level of technical analysis and data available on the system.

Benefit–cost analysis is a systematic quantitative method of assessing the desirability of projects or policies when it is important to take a long view of future effects and a broad view of possible side-effects. **Cost-effectiveness analysis** is a systematic quantitative method for comparing the costs of alternative means of achieving the same stream of benefits or a given objective [45].

(a) Why has benefit–cost analysis been attempted?

Benefit–cost analysis is an appropriate method of providing an information base for the public decision-making process of major water resources development policies and plans for states, regions and river basins. The original intent of the principles, standards and procedures for planning for the use of water and related land resources was to facilitate resolution of conflicting needs and objectives as determined through a co-operative, multi-governmental process [46].

> Plans for use of the Nation's water and land resources will be directed to improvement in the quality of life through contributions to the objectives of national economic development and environmental quality. The beneficial and adverse affects on each of these objectives (of alternative plans) will be displayed in separat with other account for the beneficial and adverse effects on regional development and social well-being.

Descendent from the Green Book [6], the original guidelines for water resources planning have been debated continually and were replaced in 1981 by *Economic and Environmental Principles and Guidelines for Water and Related Land Resources Implementation Studies*. In their present form, the *Principles and Guidelines* are still a comprehensive analytical model for comparing beneficial and adverse effects for economic as well as environmental considerations.

The objectives of national economic development and environmental quality give some idea of the appropriate scope of benefit–cost analysis in water resources planning. National economic development reflects the increase in productive output through the increased outputs of goods and services.

> In addition, national economic development is affected by beneficial and adverse externalities stemming from normal economic production and consumption, imperfect market conditions, and changes in productivity of resource inputs due to investment. National economic development is also affected by the availability of public goods which are not accounted for in the national product and income-accounting framework. Thus, the concept of national economic development is broader than that of national

income and is used to measure the impact of governmental investment on the total national output. In addition to the value of goods and services derived by users of outputs of a plan, there may be external gains to other individuals or groups [47].

The environmental objective is enhanced by the management, conservation, restoration or improvement of the quality of certain natural and cultural resources and ecological systems. This objective reflects society's concern for the natural environment and for its maintenance and enhancement as a source of present enjoyment and heritage for future generations. Given the broad and pervasive nature of the environmental objective, it is not practical to identify all possible components of the environmental quality objective individually. As other components are identified they should be accommodated explicitly in the planning process.

(b) Criticism of benefit–cost analysis

Benefit–cost analysis as performed by the federal water planning agencies has been criticized for making essentially erroneous statements about certain fundamental economic principles and faulty application of estimation techniques. Some early critics of benefit–cost analysis criticized the procedures for having biased water resources management in favour of development to the detriment of preservation of the environment [48].

> To avoid transferring resources through public investment from more productive private activity in one sector of the economy to less productive activity in another, both public and private sector investments must have the same objectives – economic efficiency. This coincidence of objectives can be attained only if national economic development, or economic efficiency, is the criterion used to evaluate economic investments [49].

If the other measures of benefits and costs are not integrated into a single economic system, distortions of benefits and costs are inevitable, especially if the same benefits are measured in more than one way and appear in more than one account. The concept of economic benefits and costs is comprehensive and includes all of the real beneficial and detrimental effects. Conceptually, welfare maximization is measurable in terms of willingness to pay. The difficult challenge is measuring the economic value of environmental effects and incorporating them into the valuation of economic efficiency.

(c) When should benefit–cost analysis be attempted?

Benefit–cost analysis is appropriate when the scope of the planning being undertaken is sufficiently complex or inclusive to involve broad public policy choices that require trade-offs among objectives. Benefit–cost is an analytical technique that requires considerable public debate and discussion to assure a balanced evaluation.

Benefit–cost analysis is suitable for regional water resources planning where competing uses and fundamental resource protection issues are being discussed. River basin planning has the proper scope for weighing benefits and costs at a regional level. However, in the State of Washington, the process for regional planning is moving away from economics as the basis for comparison and decision-making, in favour of consensus-building [50].

(d) When should benefit–cost analysis not be attempted?

Benefit–cost analysis is complex and there are no hard-and-fast rules for its application. Individual public water system administrations may not be capable of conducting the analytical procedures required in benefit–cost analysis. Utilities without economists on their staff will rely on consultants who may lack sufficient impartiality to make broad recommendations. Utilities may also lack public information capabilities and experience to undertake the extensive public participation process required for a benefit–cost analysis. Many utilities have overcome these obstacles by adding conservation planning staff, and in some instances water utilities have made considerable efforts to educate and involve the public in decision-making [42].

A more systemic problem is that probably a single utility or even several utilities working together are not the decision-makers for the major trade-offs between environmental protection and development. The utility is subject to the control of regulatory agencies who will make the major water resource allocation decisions. A water supply utility is often accountable to an elected council which may have very different goals from those of the utility management or the public that is involved most with the water supply planning process. Other external forces also weigh heavily on utilities' options, such as the courts and federal agencies, over which water supply utilities have little power.

(e) What is a cost-effectiveness study?

Cost-effectiveness analysis is a systematic method of assessing the alternative means of achieving the same stream of benefits or a given objective. With this procedure, a project is judged worthwhile if its cost is less than that of an alternative means of achieving the same objective. Cost-effectiveness is no different from the least-cost planning approach in which no measures of actual benefits are possible.

By itself, however, least cost is not a sufficient determinant of project acceptability. It is not sufficient to select the least-cost alternative unless it is first established that society values the output of a project by an amount equal to or greater than the cost of that project [51].

Three approaches were recommended in the *Principles and Guidelines* for calculating economic benefits. Each has significant shortcomings:

1. willingness to pay (benefit estimate may not be the most efficient);
2. change in net income (private gains based on publicly produced outputs may not be efficient); and
3. least-cost alternative (does not measure value of project to society).

The role economic evaluation will play in regional water resources planning is unclear even as Washington State begins the development of a comprehensive water resources plan for the Central Puget Sound basin [52]. Even if regional water allocation decisions are made without an explicit weighing of values in economic terms, there is great significance in deciding how much water is to be retained in streams and aquifers and how much is available for appropriation.

The establishment of instream production levels and out-of-stream allocations will give public water suppliers firm estimates of the yields and cost of their development alternatives. This indirectly establishes a value of the water left instream and sets a price on the water available for allocation. Although there may be a lack of economic rationale in this method, a price set by any other means is a price nonetheless [53].

(f) When is a cost-effectiveness study appropriate?

Cost-effectiveness studies are appropriate whenever objectives have been externally set or the alternatives result in the same stream of benefits. Some situations are discussed below.

Avoided cost

As discussed above, setting legal limitations on the availability of water sets objectives and costs of supply indirectly. To calculate avoided cost a utility should consider lower-cost supply sources, if these are available in the region, along with any apparently cost-effective conservation measures. The package of conservation measures and supply sources required to meet future needs that has the lowest present value (translated into levellized cost) will determine the water utility's avoided cost standard. Resources whose levellized costs are less than the avoided cost would be candidates for the least-cost plan. Environmental costs should be considered in any decision to develop a resource.

Targets

Conservation planning guidance has stressed setting goals, objectives and targets as the initial step in defining programme needs. These targets are usually set as a reduction need to match supply with demand [54]. Some results of conservation activities have been reported and represent a range of effectiveness and costs. Utilities have selected target reduction levels from these literature sources without explicitly evaluating benefits or costs. Beginning in 1989, model conservation programmes were included in several coordinated water supply plans pursuant to Ch. 70.116 RCW. The conservation programmes contained targets, such as, '5% reduction in per capita use by 1995 and 8.5% reduction in 2000'. King County, Washington established by statute a conservation goal of 10% reduction in per capita water use by the year 2000 for all public water utilities requesting approval of water system improvements [55]. This target clearly sets a performance-type standard for water utilities against which they can evaluate the cost-effectiveness of conservation measures.

Wholesale suppliers and wholesale customers

A third situation occurs when a water utility gets its supply either totally, or in part, from a wholesale supplier. In this case, the contracted price of water that the purchasing customer faces provides an avoided cost threshold for evaluating the cost-effectiveness of conservation measures as well as the feasible development of independent sources of supply. The wholesale price should reflect marginal costs of new sources of supply for the master agency, for it is only the marginal social cost that matters in this case [56]. Fixed costs play no role in determination of the best use of water by wholesale customers; their actions do not alter these costs. Variable costs may also be insignificant in regional supply systems where demand is likely to exceed available supplies (e.g. every year districts will purchase their total allotment of available surface water from the master agency). Variable costs that matter are the distribution costs of getting the water to individual districts and intra-district distribution costs.

It is unlikely that an individual district would bear the full social costs for water that is developed independently from the regional supply system. There is a tendency for districts to overpump groundwater aquifers. This can lead to excessive lowering of the water table, interference with other neighbouring water rights, and impacts on surface waters that are in hydraulic continuity.

Within the Seattle Water Department regional supply system there are several examples of groundwater development by independent water districts and suburban cities with wells that are interrelated and essentially in a steady-state situation. The Issaquah Aquifer is a good example: both the City of Issaquah and the Lake Sammammish Plateau Water and Sewer District have numerous wells in the valley. Pumping has reached a level where safe yields (equal to annual recharge) and well interference are in a tenuous equilibrium. The aquifer is effectively closed to new wells because of the effects of overcrowding.

Imposition of a pumping tax, such as suggested by Brown [56], would equate marginal private cost with marginal social cost of the master water supply agency. The pumping tax (in Brown's example for Kern County Water District in California) is derived from costs normally ignored by the individual districts and incurred by the master water agency for:

- Fixed costs – the present discounted value of all incremental fixed costs that would follow from the increase in lift caused by pumping an additional acre–foot today;
- Variable costs – the present discounted value of all increases in direct pumping costs (the variable costs) that follow from the increase in lift caused by pumping an additional acre–foot today [57].

If each individual district does not determine its annual groundwater withdrawals on the basis of the true marginal cost, ultimately pumping lifts and pumping costs will be too high and total social welfare will not be maximized. Imposing the pumping tax will yield the desired social optimum.

Regional conservation programme.

The tax placed on groundwater withdrawals to equate private marginal cost with social marginal cost might just as easily apply to the conservation programme of individual districts. If districts do not use the long-run marginal costs of the master water agency as the avoided cost in designing their conservation programmes they will fail to achieve optimal levels of conservation and will overuse independent sources of groundwater to their and other districts' cost. This can hasten the development of new sources of regional supply and cause fixed and variable costs for the whole water supply system to increase sooner than is socially desirable. The increased regional supply cost is analogous to the increased pumping cost, and imposition of a conservation tax equal to the ignored costs would result in a socially optimal regional conservation programme.

REFERENCES

1. US Water Resources Council (1980) Guidelines for state management planning, Fed. Reg. 21 July, in *Water Conservation* (1987) (W.O. Maddaus), American Water Works Association, p. 5.
2. Planning and Management Consultants (1992) *Evaluating Urban Water Conservation Programs: A Procedures Manual*, Carbondale, Illinois, p. 85.
3. Washington Departments of Health and Ecology, Washington Water Utilities Council (1994) Guidelines and Requirements for Public Water Systems Regarding Water Use Reporting, Demand Forecasting Methodology and Conservation Programs.
4. Eckstein, O. (1958) *Water Resources Development*, Harvard University Press, Cambridge, Massachusetts.
5. Krutilla, J.V. and Eckstein, O. (1958) *Multiple Purpose River Development* Johns Hopkins Press, Baltimore.
6. Subcommittee on Benefits and Costs of the Federal Inter-Agency River Basin Committee (1950) Proposed practices for economic analysis of river basin projects: report to the Federal Inter-Agency River Basin Committee, Government Printing Office, Washington DC.
7. Economic and Environmental Principles and Guidelines for Water and Related Land Resources Implementation Studies.
8. Hirshleifer, J., De Haven, J. and Milliman, J. (1960) *Water Supply: Economics, Technology and Policy*, University of Chicago Press, Chicago, Illinois.
9. Showman, J. (1992) Water conservation rate design. Research paper Washington Utilities and Transportation Commission, Olympia, Washington.
10. Beecher, J., Landers, J.R. and Mann, P.C. (1991) Integrated resources planning for water utilities. National Regulatory Research Institute, Columbus, Ohio, pp. 48–56.
11. Hanneman, W.M. (1993) Designing new water rates for Los Angeles. Department of Agricultural and Resource Economics, University of California, Berkeley.
12. Baumol, William, J. and Oates, W.E. (1988) *The Theory of Environmental Policy*, 2nd edn, Cambridge University Press. New York, ch. 3–4.
13. Hanneman, W.M. (1994) Discussion of criteria for setting water rates, in Manual on Water Rates, California Urban Water Conservation Council, Barkeley, California (in press).
14. Gibson, J. (1989) Review of potential for improving water use efficiency in Washington: municipal and industrial use. Report for the Water Use Efficiency Committee, Olympia, Washington.
15. Hirshleifer, J., De Haven, J. and Milliman, J. (1960) *Water Supply: Economics, Technology and Policy*, University of Chicago Press, Chicago, Illinois, p. 152.
16. Beecher, and Stanford, J.D. (1993) Integrated water resources planning. Discussion paper. National Regulatory Research Institute, Ohio State University, Columbus, Ohio, p. 1–5.
17. Hanke, S.H. (1978) A method for integrating engineering and economic planning, *American Waterworks Association Journal*, Sep., p. 491.
18. National Association of Regulatory Utility Commissioners (1988) *Least-Cost Utility Planning Handbook, Volume 1*, Washington DC, p. 1.
19. Chernick, P. and Caverhill, E. (1991) Methods of valuing environmental externalities, *Electricity Journal*, March pp. 46–53.
20. Baldwin, L.G. (1983) Evaluating utility options: integrating supply-side and demand-side resource planning, in *Adjusting to Regulatory, Pricing, and Marketing Realities* (ed. H.M. Trebing), Institute of Public Utilities, Michigan State University, pp. 250–86.
21. Beecher, J.A. and Stanford J.D. (1993) Integrated water resources planning. Discussion paper. National Regulatory Research Institute, Ohio State University, Columbus, Ohio, p. 1–9.
22. Maddaus, W.O. (1981) *Water Conservation*, American Water Works Association. Denver, Colorado.
23. Parker, J. and Yelton, T. (1991) *Water Conservation Planning Handbook for Small-to-Medium Sized Public Water Systems*, Dept. of Ecology, Olympia, Washington.
24. California Urban Water Agencies (1992) WATERMAN, one of the first software programs available. Department of Water Resources, Sacramento, California.

25. Planning and Management Consultants (1994) IWR-MAIN, combined demand forecasting and conservation program evaluation, Carbondale, Illinois.
26. Planning and Management Consultants (1992) *Evaluating Urban Water Conservation Programs: A Procedures Manual*, Carbondale, Illinois, pp. 85–118.
27. Washington Department of Ecology (1991).
28. American Water Works Association (1985).
29. Hirshliefer, J., De Haven, J. and Milliman, J. (1960) *Water Supply: Economics, Technology and Policy*, University of Chicago Press, Chicago, Illinois, p. 124.
30. Hanneman, W.M. (1991) Contingent Valuation and Economics,
31. Willing, R. (1976) Consumer's surplus without apology. *Am. Econ. Rev.*, **66**, 589–97.
32. Planning and Management Consultants (1992) *Evaluating Urban Water Conservation Programs: A Procedures Manual*, Carbondale, Illinois, pp. 86–125
33. Narayanan, R., Lawson, D.T. and Hughes, T.C. (1985), *Wat. Resour. Bull.*, **21**(3), 407–16.
34. Dandy, G. (1992) *Wat. Resour. Bull.*, **28**(7), 1759–66.
35. Eckstein, O. (1961) *Water Resources Development: The Economics of Project Development*, Harvard University Press, Cambridge, Massachusetts.
36. Northwest Conservation and Electric Power Act (Public Law 80–299), Section 3(4)(B).
37. Northwest Power Planning Council (1988) The role for conservation in least-cost planning. Staff Issue Paper, Publication 88–17, Portland, Oregon.
38. Northwest Power Planning Council (1989) Accounting for the environmental consequences of electricity resources during the power planning process. Staff Issue Paper, Publication 89–7, Portland, Oregon.
39. Joskow, P. (1992) Weighing environmental externalities: let's do it right!, *Electricity Journal*, May, p. 55.
40. Ferman, A.M. and Krupnick, A.J. (1992) Externality address: a response to Joskow, *Electricity Journal*, Aug./Sep. pp. 61–3.
41. Ferman, A.M., Burtraw, D. *et al.* (1992) Weighing environmental externalities: *how* to do it right, *Electricity Journal*, Aug./Sep. pp. 18–25.
42. Seattle Water Department (1994) Water supply plan, Seattle, Washington.
43. Cropper, M. and Oates, W.E. Environmental economics: a survey, *Journal of Economic Literature*, **30**, 700–39.
44. Office of Management and Budget (1992) Circular No. A-94, Washington DC.
45. Ibid., Appendix A.
46. Water Resources Council (1973) Water and related land resources: establishment of principles and standards for planning. Washington DC.
47. Ibid., p. 32.
48. Cicchetti, C.J., Davis, R.K., Hanke, S.H. and Havemen, R.H. (1993) Evaluating federal water projects: a critique of proposed standards, *Science*, 181, August, 723–8.
49. Ibid., p.723.
50. Washington Department of Ecology (1992) Guidelines for regional planning, Olympia, Washington.
51. Cicchetti, C.J., Davis, R.K., Hanke, S.H. and Havemen, R.H. (1993) Evaluating federal water projects: a critique of proposed standards, *Science*, 181, August, 725.
52. Sakrison, R.G. (1993) Regional water resources planning, for Central Puget Sound, perspectives, College of Architecture and Urban Planning, University of Washington, Seattle.
53. Sakrison, R.G. (1991) Water supply source selection criteria: determination of cost-effectiveness thresholds, in American Water Resources Association Summer Symposium: *Water Supply and Water Reuse*, June 2–6, 1991, San Diego, California. Bethesda, Maryland.
54. Maddaus, W.O. (1987) *Water Conservation*, American Water Works Association. p. 18.
55. King County (1990) King County Code, Washington.
56. Brown, G.M. (1967) A socially optimum pricing policy a public water agency, *Water Resources Research*, **3** (1), pp. 297–307.
57. Ibid., p. 301.

EXPLORING POSSIBILITIES FOR AN INTERNATIONAL WATER QUALITY INDEX APPLIED TO RIVER SYSTEMS

by C. Cude, D. Dunnette, C. Avent, A. Franklin, G. Gross, J. Hartmann, D. Hayteas, T. Jenkins, K. Leben, J. Lyngdal, D. Marks, C. Morganti and T. Quin

25.1 INTRODUCTION

Water quality indices were first seriously proposed and demonstrated in the early 1970s but not widely used or accepted by agencies that monitor water quality. The uses and limitations of a water quality index are often misunderstood and its potential for communicating the current status and trends of water quality overlooked.

Requests for water quality information generally result in the generation of raw analytical data. Without expertise, evaluation of water quality in terms of raw data alone can be misleading and confusing. In most cases, the person requesting the data lacks the knowledge necessary to integrate water quality data or to analyse one parameter with respect to overall water quality. As a result, it is difficult for the general public, environmental advocates, and policy-makers to interpret raw data generated from water quality monitoring. This may result in erroneous conclusions regarding water quality status and management practices. Thus it is difficult to gain public support for water quality improvement programmes when the evidence of progress (or lack of progress) is confusing. As a solution, a water quality index integrates complex analytical data and generates a single number expressing the degree of impairment of a given water body. This improves communication with the public and reinforces public trust.

A review of water quality indices will precede a discussion of the development and design of an international water quality index. Application of a prototype international water quality index to the Willamette (Oregon, USA) and Vistula (Poland) river basins is illustrated and limitations of interpretation of the data are identified. Finally a proposal is made to institute the use of an index among active non-governmental organizations using existing technology.

25.2 WATER QUALITY INDICES EXPLAINED

A water quality index expresses water quality via a single number by combining measurements of selected parameters. The Council on Environmental Quality [1] outlined the characteristics of an ideal water quality index to provide a consensus on index design. The consensus eased apprehension that indices would be misused and that technical information would be lost in the aggregated data. An ideal water quality index improves communication of water quality information to the public and derives from available monitoring data. The index strikes a balance between oversimplification and complex technical characterizations. It provides an

---

International River Water Quality. Edited by Gerry Best, Teresa Bogacka and Elżbieta Niemirycz. Published in 1997 by E & FN Spon, London, ISBN 0419215409

understanding of the significance of the data it represents. Finally, an ideal water quality index is objectively designed, yet can be validated by comparison with expert judgement.

A water quality index has certain limitations because of resource availability and the need to maintain a manageable, yet representative, system. A water quality index cannot determine the quality of water for all uses. Some uses conflict with others. For instance, water quality considerations for agricultural uses are different from considerations for recreational uses. A water quality index cannot provide complete information on water quality. An index provides only a summary of the data. Also, a water quality index cannot evaluate **all** health hazards. A water quality index expresses variation both spatially and temporally. An index allows users to interpret data easily and relate overall water quality variations to variations in specific categories of impairment. A water quality index identifies water quality trends and problem areas. These can be screened out and evaluated in greater detail by direct observation of pertinent data, thus increasing efficiency. Used in this manner, water quality indices provide a basis to evaluate effectiveness of water quality improvement programmes and assist in establishing priorities for management purposes [2].

As an example of how a water quality index communicates information, Figure 25.1 presents the original Oregon water quality indices by stream [3].

It is possible to describe Oregon's water quality index qualitatively as poor to excellent (Figure 25.1). Raw data measurements taken at different locations throughout Oregon produce a range of variation of the water quality index. The mean water quality index of each stream is indicated by the vertical bar in each range.

25.3 INTERNATIONAL WATER QUALITY INDEX DEVELOPMENTAL CONCERNS

25.3.1 POLICY IMPLICATIONS

Policy implications of implementing an international water quality index and variation of water quality standards were considered. While many potential users are cautious of water quality indices because of their simplification, this is generally a result of overexpectation. An international water quality index should act as an indicator of change, not as a catastrophe preventer. It is an assessment of the water quality of the stream, not a regulatory lever. An index that is too technical is likely to create controversy about its broad applicability, accuracy, and unnecessary expense. The challenge lies in striking a balance between complex technical characterization and oversimplification. Other considerations for an international water quality index's acceptance include a reasonably short elapsed time between data collection and genera-

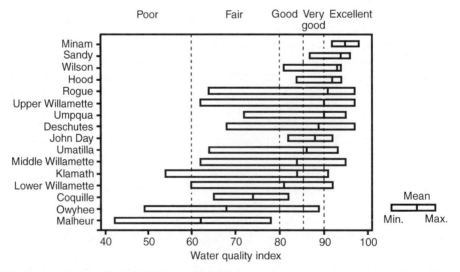

Figure 25.1 Oregon water quality indices by stream 1976–9.

tion of an index value. This depends upon available resources and level of quality assurance of the data. Lastly, an international water quality index should be transportable throughout global river systems. This means that differing financial resources, environmental impacts, and geological and aquatic systems must serve as guidelines for an international water quality index's design. Governmental agencies, with different resources and water quality standards, place different priorities on projects such as index calculation. Non-governmental organizations concerned with monitoring changes in their local, regional, and global environments would be more likely to use an international water quality index.

25.3.2 ADVANCEMENT OF ANALYTICAL TECHNIQUES

The advancement of analytical techniques over the past two decades has pushed detection limits of environmental analytes to background levels. These techniques are a tremendous step forward from the colorimetic tests employed in the 1960s, offering higher sensitivities and greater selectivity. The major drawback of these techniques is their high cost and need for highly trained operators. To make an international water quality index available to all potential users the index should minimize cost while maximizing precision and accuracy. One option may be to develop quality assurance/quality control (QA/QC) codes attached to analytical results. For instance, analyses performed using a colorimetric kit with no QA/QC protocol would receive a lower rating than analyses performed by a sanctioned laboratory with approved QA/QC protocol. Both sets of water quality index results would be usable and information about the quality of the results would be available to the user.

25.4 THE DESIGN OF A PROTOTYPE INTERNATIONAL WATER QUALITY INDEX

In the design of water quality indices several criteria are applied for parameter selection:

- The parameter must have significance for measuring water quality impairment;
- Established water quality criteria and standard methods for analytical measurement must be available;
- The parameter must be widely used and accepted as a measurement of water quality;
- Measurement should be affordable;
- It must be possible to maintain precision and accuracy of analysis to ensure validity of results.

For the prototype international water quality index, types of water quality impairment were considered and four main categories identified: oxygen depletion, eutrophication, physical characteristics and health hazards. The most important water quality parameters representing each category were identified. The Delphi method [4] was employed to select parameters and their relative weights. A survey was conducted, during which Research participants were asked to select parameters that reflect overall water quality best and to assign a relative significance value to each. Results of the survey were returned to the respondents for review. This process was repeated until significant differences in responses were no longer reported.

25.4.1 OXYGEN DEPLETION: DISSOLVED OXYGEN (DO) BOD, AND NITRATE AND AMMONIA NITROGEN (N)

Dissolved oxygen (percent saturation) integrates temperature and takes into account oxygen supersaturation. Biochemical Oxygen Demand (five-day) measures the oxygen demanding capacity of discharged wastes. Wasteload management is largely based on BOD. Nitrification represents a significant oxygen demand in some streams.

25.4.2 EUTROPHICATION: NITRATE AND AMMONIA NITROGEN (N), AND TOTAL PHOSPHATES (PO_4)

Both are indicators of nutrient loading and may be naturally occurring or introduced. Nitrate and ammonia nitrogen represents two impairment categories in one, thus economizing the index.

25.4.3 PHYSICAL CHARACTERISTICS: TOTAL SOLIDS (TS) AND pH

Total solids analysis includes dissolved and suspended solids. Suspended solids analysis correlates well with turbidity, which is a good indicator of erosion processes. Dissolved solids indicate high ionic activity. The pH represents the balance of acids and bases in the system.

25.4.4 HEALTH HAZARDS: FECAL COLIFORM (FC)

Direct search for specific pathogens is slow, costly and impracticable for international applications. Fecal coliform is a good indicator of potentially dangerous microbiological contamination of water. Special health hazards, such as toxicity and radioactivity, are usually localized in duration, extent and intensity in situations with applied control measures.

The proposed international water quality index values range from 0 (extremely impaired) to 100 (ideal conditions). Transform functions convert different units of measurement, e.g. mg/l and percentage saturation, to common sub-index values (Figure 25.2). For illustrative purposes, transform functions for the proposed international water quality index were taken from the original Oregon water quality index [5] for all parameters except total phosphates (which was taken from US Environmental Protection Agency Region VII (6)).

An aggregation function combines sub-index values into a single water quality index value. After comparing existing models, the weighted geometric mean function (IWQT) was selected for its reliable indication of current water quality impairment and of trends in water quality (1).

$$IWQI = (SI_{DO})^{0.3} \times (SI_{FC})^{0.2} \times (SI_{pH})^{0.1} \times (SI_{BOD})^{0.1} \times (SI_{N})^{0.1} \times (SI_{PO_4})^{0.1} \times (SI_{TS})^{0.1}, \quad (1)$$

where $(SI_x)^n$ is Sub-index X with relative weight n.

25.5 APPLICATIONS

A water quality index portrays current water quality status. Its greater value lies in illustrating trends and comparing water quality conditions. Trend analyses can be performed temporally (at a given sample location over a period of time), or spatially (at a given time throughout a basin or other geographic area). Figure 25.3 is a temporal plot of the prototype international water quality index of the Willamette River in Portland, Oregon, over a five-year period.

A significant decrease and subsequent increase in water quality occurred in 1991. A more detailed analysis of data for that period may point to possible causes. The change in water quality may be a result of a normal drought cycle, anthropogenic factors or a combination of both. To resolve this

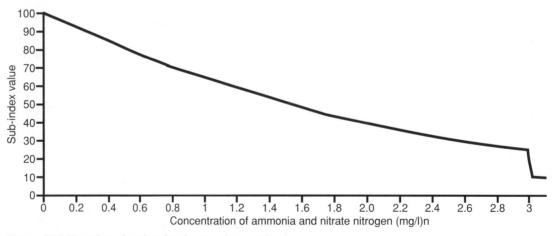

Figure 25.2 Transform function for nitrate and ammonia nitrogen.

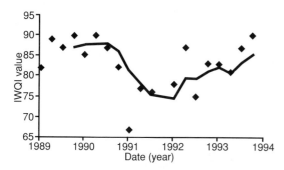

Figure 25.3 Temporal plot of prototype international water quality index. (Source: Quarterly data for Willamette River at Hawthorne Bridge Portland; 1989–93.)

question, the user could incorporate more data as available, and apply a Seasonal–Kendall statistical test to account for seasonal influence on water quality. The user could examine water quality data for the specified time period and data for other sampling sites nearby.

To demonstrate the value of comparing international water quality index information between different basins, Table 25.1 provides data from the Willamette and Vistula rivers. By examining international water quality index values, the Willamette River averages 85 (good) while the Vistula River averages 42 (poor). Closer examination of the water quality parameters contributing to the indices focuses attention on the sources of impairment. The data on faecal coliforms and BOD show that, while agricultural non-point sources of coliform and BOD may be similar in both basins, municipal and industrial sources still have a significant impact on water quality in the Vistula basin. Nutrient values suggest a greater use of fertilizers causing more non-point source pollution of the Vistula. The greater range of dissolved oxygen in the Vistula River supports both of the above hypotheses. Higher pH and total solids in the Vistula basin may reflect different natural conditions, but may also reflect the results of erosion, eutrophication and point source pollution.

An international water quality index can detect trends in water quality and find 'hot spots' or areas of degraded water quality. By reducing the volume of data to a single number, a researcher can identi-

fy these problems more easily and return to the analytical data for further analysis.

25.6 PROPOSALS

Correlation studies were performed during development of the original Oregon water quality index [5]. Six water quality indexes were compared with five versions of the Oregon water quality index to determine effects of parameter substitution and weighting variation on index values. It was concluded: as long as major impairment categories are properly represented, parameter selection and methods of weighting, transformation and aggregation are not critical.

While the benefit of scrutinizing the smallest details of an index may not justify the time spent, significant improvements can be accomplished on a larger scale. The original Oregon water quality index [2] provided the basis for the prototype international water quality index, originally proposed in 1994. In 1995, the Oregon Department of Environmental Quality laboratories revised the Oregon water quality index, incorporating current understanding of water quality impairment (G.A. Pettit, Oregon Department of Environmental Quality, 1995, personal communication). Parameter transform equations were refined to reflect impairment to aquatic life and recreational uses. Dissolved oxygen concentration, temperature and total phosphates were added to the index. The unweighted harmonic square mean formula [7] for combining parameters into a single index value is now in use (2).

$$IWQI = \sqrt{\frac{n}{\sum_{i=1}^{n} \frac{1}{SI_i^2}}}, \qquad (2)$$

where n = number of subindices
SI = Subindex value

In this aggregation function, sub-indices do not carry constant weight. Due to the form of the equation, the sub-index with the lowest score has the greatest influence on the water quality index. This method acknowledges that different water quality parameters affect overall water quality differently at different times and locations. The formula is sensi-

Table 25.1 Willamette and Vistula river water quality comparison 1989–92 (Source: Oregon Department of Environmental Quality, Portland; Institute of Meteorology and Water Management, Warsaw)

	Willamette River Portland, Oregon			Vistula River Warsaw, Poland		
International WQ Index (Mean, 1989–92 data)	85			42		
	Min.	Max.	Mean	Min.	Max.	Mean
Dissolved oxygen (% Sat)	84	118	101	65	124	98
Faecal coliform (per 100 ml)	11	1100	350	30	10^6	1200
BOD_5 (mg/l)	0.7	8.3	2.3	0.6	19.2	7.6
pH	6.9	8.1	7.5	6.9	9.7	8.4
Inorganic nitrogen (mg/l)	0.3	1.1	0.6	nd[a]	16.7	3
Total phosphorus (mg/l)	0.06	0.19	0.13	0.1	1.9	0.6
Total solids (mg/l)	40	110	75	231	1390	652

[a] nd = not detectable

tive to changing conditions and to significant impacts on water quality. Oregon's Department of Environmental Quality laboratories are also currently developing computer software for automatic calculation and report generation of water quality index results. These revisions and continuing research will improve the ability of the original international water quality index to reflect water quality status and communicate water quality information.

Building upon these efforts, further research is necessary to ensure that a water quality index becames applicable globally. More research is needed to produce analytical packages with associated QA/QC protocol, thus giving more people access to a high-quality international index. Research into affordable software for water quality index calculation and analysis is necessary.

Using existing technology, non-governmental environmental organizations could use an international water quality index to examine water quality trends and effectiveness of management practices, to compare water quality across different basins and communicate water quality status to users and consumers.

25.7 CONCLUSIONS

An international water quality index would develop a common language with which to communicate water quality concerns. This prototype international water quality index integrates various impairment categories common to most river basins enabling a standard comparison of water quality to be made. Current knowledge and technology are sufficient for index development, but further research is necessary to maximize the quality of the results and to minimize the cost of applying the index.

REFERENCES

1. Council on Environmental Quality (1974) *5th Annual Report*.
2. Dunnette, D.A. (1980) *Oregon water quality index staff manual*, Oregon Department of Environmental Quality.
3. Oregon Department of Environmental Quality (1980) *Water quality in Oregon 1980*.
4. Dalkey, N.C. (1968) *DELPHI*, The Rand Corp.
5. Dunnette, D.A. (1979) A geographically variable water quality index used in Oregon, *J. Water Pollution Control Federation*, **51**(1), 53–61.
6. McClelland, N.I. (1974). Water quality index application in the Kansas River basin, US Environmental Protection Agency, Region VII, Kansas City, Missouri.
7. Dojlido, J. et al. (1994) Water quality index applied to rivers in the Vistula River basin in Poland, *Environ. Monit. Assessment*, **33**, 33–42.

RIVER WATER QUALITY MODELLING IN POLAND

by M.J. Gromiec

26.1 BACKGROUND DATA ON WATER QUANTITY AND WATER QUALITY

26.1.1 WATER RESOURCES AND DEMANDS

Poland is divided up into 49 provinces (voivodeships). The country is 312 520 km^2 in area; the two most important river basins are the Vistula, with an area of 194 000 km^2, and the Odra, with an area of 110 000 km^2. There are about 9300 lakes, covering an area of 3200 km^2; the Masurian Lake District contains 1063 lakes, the largest of which are about 110 km^2 in size. Lakes and artificial reservoirs have a total storage capacity of 33×10^9 m^3 with smaller lakes and ponds contributing an additional 1×10^9 m^3.

The average annual rainfall in Poland is 597 mm, equivalent to 186.6×10^9 m^3 of water per year over the whole country. Since tributaries from outside Poland yield an additional 5.2×10^9 m^3 of water annually, the total input of water is 191.8×10^9 m^3. Groundwater resources have been estimated at 33×10^9 m^3 per year over an area of 272 520 km^2; the remaining 13% of the total area is dry. The annual dynamic groundwater resources have been evaluated at 9.2×10^9 m^3. However, rivers and streams discharge only about 58.6×10^9 m^3 of water into the Baltic Sea during a mean low-flow year, and about 34×10^9 m^3 in a dry-weather year. Obviously, only a portion of this volume is available for use as about 10×10^9 m^3 per year is necessary as a minimum flow to maintain biological life and for sanitary reasons; therefore the minimum available flow for abstraction is only 24×10^9 m^3 per year.

Poland belongs to the group of European countries most deficient in water resources, ranking 22nd overall. Average annual water resources in Poland (estimated on the basis of atmospheric inputs and the size of the population) amount to 1600 m^3 per capita, compared with 2800m^3 per capita for Europe. In 1992 the total water consumption in Poland was 12.5×10^9 m^3, of which 22.5%, 66.5% and 11% were for municipal, industrial and agricultural purposes respectively. Most of the water for agriculture is taken during the summer months. However, the volume of water available compares unfavourably with the anticipated water demand in the future, particularly during dry seasons.

21.1.2 LEGISLATIVE AND ADMINISTRATIVE ASPECTS

The present basis for legal action in the field of water pollution protection is the Water Act issued by the Polish Parliament in 1974. In 1975, using the new Act, the Council of Ministers announced regulations concerning classification of waters and determination of effluent standards, as well as financial penalties for effluent discharges that do not meet the requirements specified in the regulations. The following classes of surface water quality were established.

- Class I waters are those used for municipal and food processing supply purposes, and for salmon fish growth;

International River Water Quality. Edited by Gerry Best, Teresa Bogacka and Elżbieta Niemirycz. Published in 1997 by E & FN Spon, London, ISBN 0419215409

- Class II waters are intended for use as recreational waters, including water sports and swimming, and for growth of fish other than salmonids;
- Class III waters (the lowest class) are used only as industrial water supplies and for irrigation purposes.

Water quality standards are tailored to meet appropriate use of surface waters. In addition, the following provisions are laid down by the Water Act.

- Industrial plants and other operations which discharge wastewaters to water or to land are obliged to construct, maintain, and use wastewater treatment facilities;
- No industrial plant or any other plant from which wastewater is discharged can be started up unless there is a suitable wastewater treatment system;
- A permit is required to maintain wastewater discharges.

In addition, the two most important decrees include:

- Decree of Minister of Environmental Protection, Natural Resources and Forestry on the classification of waters and on the conditions that must be fulfilled when wastewater is discharged into waters or the ground;
- Decree of Minister of Health and Social Welfare concerning conditions that must be met by drinking water and by water for industrial purposes.

In 1991 the Polish Parliament passed a Bill on environmental policy which determines the general rules, aims and directions of future actions in Poland. Currently a new version of the Water Act is under preparation. The new system of water resources management introduces subdivision of the country's water system into river catchment based areas under the responsibility of River Basin Water Authorities.

26.1.3 WATER QUALITY PROBLEMS

Processes of increasing urbanization, population growth, intensification of agriculture and the growth of industry have resulted in a deterioration of surface water resources, despite the introduction of certain measures to control water pollution. Therefore, the majority of major rivers have experienced serious degradation of water quality as shown in Figure 26.1 and described by Dojlido in Chapter 3. In addition, lakes, especially in the northern region, are threatened by eutrophication with excess phosphorus as the primary cause.

Currently, about 70% of all wastewaters which require purification are treated to varying degrees (Figure 26.2). However, a substantial number of the existing municipal wastewater treatment plants are overloaded.

The problem of water pollution control has become one of the most important environmental problems in Poland, since most industry is situated in the south near the sources of the country's river systems. In addition, the main rivers – the Vistula and Odra – are heavily used for municipal and industrial water supply, agricultural irrigation, cooling purposes for power plants, and navigation; however these rivers also receive discharges of wastewater with varying degrees of treatment, and run-off. These multiple uses impose competing demands on waters, and water resources management must protect many desirable uses.

26.1.4 MONITORING SYSTEM AND DATA ANALYSIS

The rivers, lakes and marine waters in Poland are all subject to regular monitoring as part of an overall surveillance scheme. The country is covered by a network of established measuring stations. The sampling frequency depends on the purpose for which data are recorded, ranging from a minimum of bi-monthly sampling up to daily sampling at some points. The sampling of water is performed simultaneously with rate of flow measurements.

The river monitoring network provides a large amount of observed data. These data are analysed by a statistical method based on the assumption that, at a given cross-section, some correlation exists between the pollutant concentration and the rate of flow. The shape of the curve depends on many factors, such as the degree of water pollu-

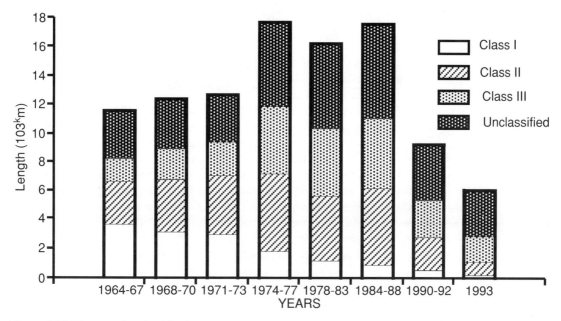

Figure 26.1 Water quality classification: physical and chemical criteria.

tion, the type of pollutant, hydrological characteristics of the river, its self-purification capacity, the distance between monitoring stations etc. From the relationships between the stream flow and the concentrations of water quality constituents, so-called indicative concentrations (IC) for a design

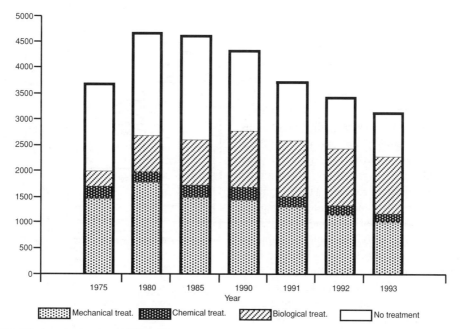

Figure 26.2 Wastewater treatment 1975–93.

flow are established. The mean low stream flow (MLQ) has been selected as the design stream flow at each site, based on the assumption that higher stream flows will result in higher dissolved oxygen (DO) concentrations and better water quality. In other countries, a similar approach has been taken. For example, in the US the design flow is the minimum average seven-day consecutive flow expected once every ten years. However, this is an extremely low stream flow which is exceeded more than 99% of the time. The IC values are plotted along the river for various water quality constituents. The final interpretation of water quality is based on these hydrochemical profiles, and the overall river classification is done after all measured water quality constituents are compared with standards. A compendium of hydrochemical profiles for major rivers and streams is prepared each year by the Institute of Meteorology and Water Management (IMWM), which serves as an overall river classification system.

26.2 DESCRIPTIONS OF WATER QUALITY MODELS

26.2.1 EARLY BOD–DO MODELS

Mathematical modelling of river systems in Poland has become an integral part of water resources planning and water quality management. These models can be used to aid water quality surveillance and to predict future water quantity and water quality conditions [1]. Various computerized models have been applied for water quality simulations in the Odra and Vistula rivers. As an example, a Streeter-Phelps model and QUAL 1 model were used by the IMWM to evaluate concentrations of BOD in the Vistula River reaches. The first model was designed to simulate the spatial and temporal variations in BOD under various conditions of flow and temperature. The second model is capable of routing BOD, DO and temperature through a one-dimensional, completely mixed, branching river system. These early BOD–DO models are representative of non-conservative coupled models. It should be stressed that the predictions obtained from these models are only as reliable as the input data, proper measurement and estimation of the various model parameters.

26.2.2 AN OVERVIEW OF THE QUAL 2E AND QUAL 2E–UNCAS MODELS

The Stream Water Quality Model QUAL 2E [2] is a steady-state model for conventional pollutants in one-dimensional streams and well-mixed ecosystems. The conventional pollutants include conservative substances, temperature, bacteria, BOD, DO, nitrogen, phosphorus and algae (Figure 26.3). The model is widely used for simulation of water quality and for waste load allocations and discharge permit determinations in the USA, and it is a proven, effective analytical tool [3].

A major problem faced by the user when working with a complex model such as QUAL 2E is model calibration and determination of the most efficient plan for collection and calibration of data. This problem can be addressed by application of principles of uncertainty analysis. QUAL 2E–UNCAS is a recent enhancement of QUAL 2E which allows the user to perform uncertainty

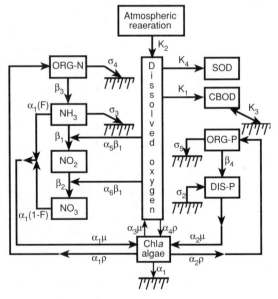

SOD—Soluble oxygen demand
CBOD—Carbonaceous BOD

Figure 26.3 Major constituent interaction in QUAL 2E.

analysis on the steady-state water quality simulations. The above models are available from the Athens Environmental Research Laboratory (AERL) of the US Environmental Protection Agency (EPA), in Athens, Georgia.

26.2.3 THE US – POLISH JOINT PROJECT

A variety of micro-computer models for water quality simulation is supported by the Center for Exposure Assessment Modelling (CEAM) at EPA's Athens laboratory. The Athens Environmental Research Laboratory and IMWM have jointly selected a set of CEAM-supported computer models that address surface water quality problems of mutual interest.

The QUAL 2E and QUAL 2E–UNCAS models have been applied to tributaries of the Vistula River in Poland [4]. About 30 major rivers from the upper, middle and lower part of the Vistula River basin have been chosen for this study. A summary of river characteristics are displayed in Table 26.1.

For the water quality simulation of Polish rivers, the conditions chosen were: steady-state model; water quality constituents: temperature, DO, BOD, nitrogen compounds (organic nitrogen, ammonia, nitrite, nitrate), phosphorus compounds (organic and dissolved), coliforms, chlorophyll-a, conserva-

Table 26.1 Summary of hydraulic characteristics of simulated rivers

Name	Total length (km)	Studied length (start–finish) (km)	Range of flow (m^3/s)	Number of reaches
1. Biebrza	155.3	102–0	1.37–5.7	3
2. Brda	238	161–74	4.93–22.5	7
3. Bug	587.2[a]	580–36	13–81.7	14
4. Dlubnia	38	34–0	0.18–0.49	4
5. Drweca	207.2	185–1	1.7–28.94	12
6. Dunajec	247.1	200–0	17.1–50.5	14[b]
7. Ilownica	26.5	15–0	0.31–1.28	5[b]
8. Kamienna	138.3	128–6	0.1–5.99	9
9. Mała wisła	97.1	96.5–3.5	0.54–9.64	20[b]
10. Narew	448[a]	384–57	4.75–91.3	29[b]
11. Nida	151.2	116–22	0.49–13.4	6
12. Pilica	319	281–1	1.16–27.8	15
13. Poprad	62.6[a]	62–0	13.3–17.6	4
14. Przemsza	87.6	86–6	0.14–17.8	11
15. Raba	131.9	116–2	0.49–15.3	8
16. Radomka	107	92–2	0.14–4.97	7
17. Radunia	104.6	75–18	2.25–4.97	5
18. Rudawa	27.7	16–0	0.72–1.5	2
19. San	442.1	304–4	28.6–81.9	34[b]
20. Skawa	96.4	82–4	0.16–6.72	7
21. Sola	88.9	81–2	0.32–11.7	8
22. Suprasl	93.8	84–0	0.12–5.5	6
23. Tanew	113	70–0	2.74–5.67	4
24. Wapienica	21.1	11–0	0.14–0.37	2
25. Wda	198	62–0	6.61–11.58	5
26. Wieprz	303.2	269–62	1.13–12.9	11
27. Wislok	204.9	168–6	0.95–9.34	11
28. Wisloka	163.6	145–3	0.88–16.2	14
29. Wkra	249.1	210–18	0.84–9.88	13

[a] on Poland territory
[b] number of reaches involves reaches of tributaries

tive constituents (chlorides and sulphates) and arbitrary non-conservative consituents (COD).

The water quality database, which has been used for preparing input files, contains results from field measurements. Sampling points were located at significant points along the river (upstream and downstream of tributaries, point loads and abstraction points) and on tributaries just upstream of the main river. Municipal and industrial wastewater data were obtained from the Central Statistical Office [5]. Hydraulic data (flow, velocity, depth etc.) and meteorological data (dry and wet bulb temperatures cloudiness, wind speed, atmospheric pressure and elevation) were obtained from the hydrological annual report [6] and climatological annual report [7] respectively. A large number of simulations with the QUAL 2E model were performed.

In general, the majority of simulation results for water quality constituents, transformable by biochemical, chemical or biochemical processes, were similar to measured values. However, the temperature results from the simulation were lower than the actual temperatures. There was a problem also with mineral constituents: there is no provision for mineral concentrations to decrease via processes such as settling and adsorption in the model, therefore only dilution is considered. An example of the simulation of various water quality constituents in the Dunajec River is shown in Figure 26.4.

There are three uncertainty analysis options available in the QUAL 2E–UNCAS model: sensitivity analysis, first order error analysis and Monte Carlo simulation. Sensitivity analysis gives the relative change in the value of each output variable resulting from the changes in the value of the input variable. First order error analysis represents the percentage changes in the output variable weighted by the variance of each input. The Monte Carlo simulation gives summary statistics and frequency distributions for the variables at specific locations in the system.

In these studies the first order error analysis with 5% input perturbation was chosen for all input variables (hydraulic, water quality constituents, algae content). For each river in the Vistula River basin an uncertainty analysis was completed. To perform the uncertainty analysis, at least five locations were selected in each river catchment. The criteria for locations included water quality at the beginning of each system, major tributaries (treated as river or point loads), wastewater discharges and reservoirs. For incremental characteristics and headwaters, the variance coefficients were 10%, whereas for point loads they were from 2% to 5% greater. Where tributaries were considered, the variance coefficients for incremental, headwater and point loads had the same values.

In general, first order error analysis has been performed for all of the input variables (hydraulic, reaction coefficients and point load forcing functions) for input perturbation equal to 5%. The calculated relative standard deviations of the water quality constituents are summarized in Table 26.2. The values of relative standard deviation for all of the constituents are in the range of low and typical values. At some sites where there are point discharges, or for upstream locations, the values are higher.

26.2.4 WATER QUALITY MANAGEMENT MODELS

In addition to the prediction of water quality by simulation, mathematical models can serve as the basis for determining the investment policies for the construction of wastewater treatment plants. For example, a water quality management model was applied to the Klodnica River. The main goal was to determine, for the given system of wastewater treatment, the level of efficiency necessary to achieve the required standard of water quality at the least cost. The quality of the water in the Klodnica River catchment has been determined for various levels of improvement to wastewater treatment plants.

It should be stressed, however, that these management models are only tools in assisting management decision-making processes. Final decisions are not usually made solely on the basis of their predictions. Additional data, including socio-political factors, are taken into account. However, in spite of their limitations, these models are the only reasonable means currently available for the prediction of water quality.

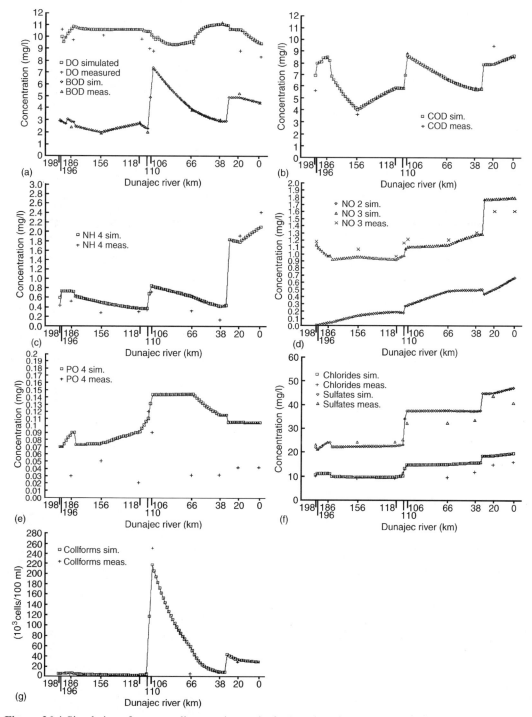

Figure 26.4 Simulation of water quality constituents in the Dunajec River (a) dissolved oxygen and biochemical oxygen demand; (b) chemical oxygen demand; (c) ammonia; (d) nitrite and nitrate; (e) dissolved phosphorus; (f) chlorides and sulphates; (g) coliforms.

Table 26.2 Summary of relative standard deviation for analysed rivers

Name	DO	BOD	Relative standard deviation (%) Organic nitrogen	Ammonia	Nitrite	Nitrate
1. Biebrza	2–6	7–8	8–11	8–12	12–15	7–11
2. Brda	1–3	7–8	9–11	7–11	10–17	8–10
3. Bug	2–11	8–10		10–21	11–20	7–12
4. Dlubnia	2–3	8–10	11–12	12–15	0–22	11–15
5. Drweca	2–5	7–12	10–12	9–18	8–25	7–8
6. Dunajec	2–4	7–21		7–14	12–35	6–9
7. Ilownica	2–3	7–10		9–11	12–16	7
8. Kamienna	2–3	5–8	7–9	8–9	13–15	8
9. Mała Wisla	3–4	7–9		10–14	14–17	8–9
10. Narew	2–45	6–12		9–32	9–15	7–10
11. Nida	2–4	6–7	7–11	8–13	14–20	6–8
12. Pilica	2–3	8–9	9–20	9–13	11–13	8–9
13. Przemsza	0–32	8–11		7–13	13–95	9–11
14. Raba	2–3	8–11	10–12	8–12	12–13	7–8
15. Radomka	2–9	8–10		8–14	12–17	8–12
16. Radunia	3	7–10	7–10	10–15	17	7–11
17. Rudawa	2–3	7–10	7–10	7–10	6–10	7–9
18. San	2–4	6–10	10–20	10–14	10–20	7–10
19. Skawa	2–6	7–8		7–12	10–15	7–10
20. Sola	2–3	8–10				9–10
21. Suprasl	2–5	7–9		9–11	12–26	7–8
22. Tanew	2	9–12		8–12	10–20	7–10
23. Wda	2–4	7–10	11–15	13–14	17–20	8–15
24. Wieprz	3–30	8–11		8–12	12–18	8–10
25. Wislok	2–4	6–8		7–16	8–11	6
26. Wisloka	2–3	7–24		10–41	8–15	7–9
27. Wkra	2–3	8–10	7–10	8–23	10–20	8–13

Name	Organic phosphorus	Phosphate	Relative standard deviation (%) CM-1	CM-2	Coliforms	COD
1. Biebrza		7–12	2–3	2–4	10–14	6–8
2. Brda	10–14	8–10	2	2	14–39	7–8
3. Bug		6–10		2–3	12–28	7–10
4. Dlubnia	12–15	10–14	2–5	3–4	15–21	7–10
5. Drweca		8–11	2–3	3–9	11–35	6–8
6. Dunajec		7–9	2	2–3	12–39	7–12
7. Ilownica		8–9	2–3	2		7–10
8. Kamienna		5–8	2–3	2–3	12–21	7–10
9. Mala Wisla		8–13	3–6	3–4		8–9
10. Narew		8–9	2–3	2–3	8–33	7–12
11. Nida		8–12	2–4	2–4	10–38	6–8
12. Pilica	9–16	8–9	2–3		12–36	7–8
13. Przemsza	7–14	8–9	3–5	2–3	14–19	7–9
14. Raba	8–20	7–8	2	2–3	11–17	7–8
15. Radomka		8–12	3–5	2–4	16–20	8–11
16. Radunia	7–10	10–14	2–3	2–3	11–19	6–9
17. Rudawa	8–10	7–9	2–3	2–3	11–15	7–9

18. San	10–25	7–10	2–4	2–5	12–48	6–10
19. Skawa		7–10	2–3	2–4		7–9
20. Sola		8–10	2–3	2–3		8–15
21. Suprasl			2–3	2–3	12–17	8–10
22. Tanew		7–9	2–3	2–3	12–16	6–10
23. Wda	10–14	9–15	3–5	3–5	13–20	7–10
24. Wieprz		7–10	2–11	2–3	14–15	7–10
25. Wislok	13	7–10	2–3	2–3	12–22	6–8
26. Wisloka		8–10	3	2–3	15–142	7–21
27. Wkra	12–20	7–8	2–3	19–67	8–10	

The stream water quality models QUAL 2E and QUAL 2E–UNCAS are now available in Poland for waste load allocation studies. A number of such applications are currently being performed. These computer programs can be used as an aid for solving conventional and non-conventional pollution problems in Poland.

26.3 CONCLUSIONS

On the whole, Poland is lacking in water resources and there are various conflicts over the use of the resources as well as an increase in development. Water resources are unevenly distributed among different parts of the country, water supplies now appear inadequate in quality, and demand will grow in many regions. The environmental policies of the past in Poland have not protected water quality adequately, with the result that there has been a deterioration in water quality. The negative consequences of this are borne by all water users.

A number of constraints make traditional environmental strategies less effective than in the past. In addition, many traditional strategies for solving water quality problems are extremely costly. In the current economic situation there are not enough financial resources to upgrade all of Poland's treatment capacity at once. In 1995, the investment programme for water pollution control alone requires about 2.1 billion Polish zlotys. Therefore, a new strategy is required for investing the available resources to optimize the environmental benefit.

A water watershed protection approach, with water conservation, pollution prevention measures and the restructuring of industry, is clearly needed to ensure the most cost-effective solutions to the water resource problems in Poland. Water quality improvement programmes should include the improvement of treatment methods, implementation of advanced treatment processes, recovery and water reuse in industry, encouragement of the use of biodegradable detergents and pesticides, and the use of dry technologies.

The catchment protection approach is an integrated strategy for restoring aquatic ecosystems and protecting health more effectively. This approach focuses on river catchment areas rather than on areas arbitrarily defined by administrative boundaries. The above approach places greater emphasis on all aspects of water quality, including physical and chemical quality, habitat quality and biodiversity. The catchment approach provides comprehensive methods for implementing the solutions to water quality problems in Poland.

The most urgent water quality problems will be solved only if clear goals are established and efficient ways to achieve the goals are identified. Setting priorities often involves difficult choices. In many cases, priorities are pretreatment of industrial wastewater, where heavy metals or toxic chemicals threaten the quality of surface waters or the micro-organisms in municipal treatment works. Appropriate investments are required to reduce and treat toxic discharges from industrial plants, and to treat saline waters and other discharges from mines. From a domestic perspective, wastewater investments should focus on upstream water. However, in regional terms, strategies should focus on the overlap between local and trans-boundary impacts of pollution reduction measures. This requires financial resources to be directed towards reducing the flows of nutrients and emissions of harmful substances from municipal, industrial and agricultural sources, in order to achieve an overall reduction of pollutants discharged into the Baltic Sea.

The priority-setting requires basic data about the extent and type of water pollution. Monitoring water quality parameters in drainage basins and watersheds is essential for assessing the status, trends and causes of water quality conditions. However, the best method of conducting such an assessment is not easily determined. This is because of the multiplicity of water quality constituents, the variations in natural water quality with time and space, and the high cost of collecting and analysing samples.

Mathematical modelling of river systems has become an integral part of water quality management. These models can be used to aid water quality surveillance and to predict future water quantity and water quality conditions. They are easily applicable and usable for planning and management by water authorities. There seem to be two general types of water quality models. One has arisen from the desire to achieve a more comprehensive understanding of the physical, biochemical and ecological processes that take place in water ecosystems receiving potential pollutants. The second type is directed towards planning and management and/or real-time control. The preferred model for planning and management will depend on the information needed, which will differ for various water ecosystems, on management alternatives, and on possible institutional objectives and constraints. Only through case studies will it be possible to learn more about how to select appropriate models and how to improve the quality of information derived from models for the planning process.

REFERENCES

1. Gromiec, M.J., Loucks, D.P. and Orlob, G.T. (1983) Stream quality modelling, in *Mathematical Modelling of Water Quality: Streams, Lakes, and Reservoirs* (ed. G.T. Orlob), John Wiley and Sons Ltd, Chichester.
2. Brown, L.C. and Barnwell, T.O. (1987) The enhanced stream water quality models QUAL2E and QUAL2E-UNCAS: documentation and user manual, EPA/600/3–87/007, US Environmental Protection Agency, Environmental Research Laboratory, Athens, Georgia.
3. Barnwell, T.O., Brown, L.C. and Whittemore R.C. (1987) QUAL2E – a case study in water quality modelling software, in *Systems Analysis in Water Quality Management*, Pergamon Press, Oxford.
4. Gromiec, M.J., Russo, R.C., Barnwell, T.O., *et al.* (1994) Final report on innovative microcomputer applications of water quality models, Project MOS/EPA-89-17, Institute of Meteorology and Water Management, Warsaw and US Environmental Protection Agency, Environmental Research Laboratory, Athens, Georgia.
5. Central Statistical Office (1992) Environmental protection and water management data, Warsaw (in Polish).
6. IMWM (1992) Hydrological annual report, Warsaw (in Polish).
7. IMWM (1992) Climatological annual report, Warsaw (in Polish).

MATHEMATICAL MODELLING OF SOIL NITRATE AND PHOSPHATE LEACHING FROM SMALL AGRICULTURAL CATCHMENTS IN NORTHERN POLAND

by P.J. Kowalik and M. Kulbik

27.1 INTRODUCTION

Soil influences the quality of river water such that the latter is an important indicator of the impact of human activities on a catchment environment. Rainfall in a river drainage basin absorbs gaseous and particulate material from the atmosphere and is supplemented by agricultural run-off, as well as discharges from household activity. Rainwater flows down land slopes and river beds, dissolving and transporting various chemical species from the soil environment. 'Rivers are integrative and reflect conditions within their boundaries' [1].

The aim of this chapter is to identify, describe and explain the main factors that affect the observed water quality conditions associated with non-point sources of pollution originating from soils. The first step in the river quality assessment process is an analysis of river hydrology. Water data are useful for assessment of impact only when frequent sampling is conducted to determine the extent of variability. Data analysis can provide a measure of background variability in water quality caused by the variability of natural hydrological processes. This information can then be used to estimate the impacts on river water quality resulting from human activities in the river drainage basin more accurately [1].

Surface water is often a source of water supply for municipal and industrial use. In the Gdansk region (northern Poland) about 1 million people live in cities. The water for the city of Gdansk is supplied from the Radunia River (130 000 m^3/day) and the city of Gdynia from the Reda River (86 000 m^3/day). A knowledge of the variability of the nitrate and phosphate concentrations in the rivers is required. It is important not only to know the loads of pollutants but also to know the extreme values of concentrations and when they occur. It is then necessary to carry out corrective actions to protect the water intakes against concentrations above the critical limit. The main form of dissolved inorganic nitrogen in rivers is nitrate. Polish standards require the concentration of nitrate in water for municipal use to be no greater than 10 mg/l.

How does the leaching of soil nitrate and phosphate to rivers relate to known agricultural catchment characteristics and meteorological variables? Some of the factors which influence nitrate leaching losses from agricultural catchments have been examined. A working hypothesis is that the concentrations of nitrate and phosphate in a river are functions of not only the amount of fertilizer applied but also, and mainly, the flows of water in the soil profile, on fields and in the river. These

---

International River Water Quality. Edited by Gerry Best, Teresa Bogacka and Elżbieta Niemirycz. Published in 1997 by E & FN Spon, London, ISBN 0419215409

flows, in turn, are functions of rainfall and water retention in the catchment. The main factors considered are land use and the effects of climatic conditions, particularly extremes. As a result, a conceptual model for the simultaneous computation of run-off and nitrate and phosphate leaching is described.

27.2 MATERIALS AND METHODS

In Chapter 19, Bogacka and Taylor described the leaching of nutrients into a variety of streams at a number of locations in Poland. In this study, two small catchments were selected: the Pomorka River, a tributary of the Noteć River, and the Wietcisa River, a tributary of the Wierzyca River. The empirical data were collected by the Institute of Meteorology and Water Management, Maritime Branch, regional office in Gdansk [2, 3].

The Pomorka River catchment in northern Poland is of post-glacial loam. The catchment has an area of 72 km^2 and is situated above the village of Brzyskorzystew. Hydrological data on the runoff were collected in the period 1977–82 (five years) as daily values. In addition a subcatchment of an area of 4 km^2 was selected above the village of Paryz and daily values were collected for two years. Water quality was measured every day for two years in both catchments. Only soluble compounds were measured, mainly nitrate, and standard methods were applied.

The catchment of the Wietcisa River in northern Poland is characterized by post-glacial deposits on sandy soils. It has an area of 236 km^2 and is situated above the town of Skarszewy. Daily values of hydrological and hydrochemical data were collected for two years.

Both catchments are used for agriculture; there are no villages or towns. The annual fertilizer application is about 100 kg/ha for the Wietcisa River catchment and about 280 kg/ha for the Pomorka River catchment; the ratio nitrogen/phosphorus/potassium is 3:1:1.

Standard meteorological data of air temperature and relative humidity, wind speed, solar radiation and precipitation were collected as mean daily values from synoptic stations close to the catchments. The catchments have an annual precipitation of 600 mm.

At present there are relatively few data that provide an inventory of atmospheric inputs to catchments in Poland. The nitrate deposition was estimated to be 15 kg/ha per year, similar to that reported for the southern Baltic Sea [4, 5].

27.3 EXPERIMENTAL CATCHMENT DATA

For detailed examination of the cause–effect relationships, a small catchment of about 4 km^2 of the Pomorka River close to the source was selected. This area is a post-glacial bottom moraine comprising mainly sandy loam and is gently undulated with a slope of less than 2%. All of the catchment is used as farmland, mostly for wheat and potato production. In the bottom of surrounding valleys ditches were created to drain the surplus water which occurs from time to time. At the village of Paryz the bottom of the main ditch is 1.5 m below ground level and about 1 m wide. The discharge was measured every day and the water was sampled for nitrate and phosphate analysis. The discharge ranged from 0.006 to 0.012 m^3/s per km^2. An example of the variation in nitrate concentration over one year is shown in Figure 27.1.

A strongly developed seasonal cycle in stream water nitrate concentrations with summer minima (July and August) and winter maxima (January and March) was observed. As there is a relatively low intensity of agriculture in the area the pattern may be explained in terms of the seasonal availability of nitrate for leaching within the soil. Generally nitrate concentrations increased with increasing water flows during the late autumn and winter periods and decreased during drier periods of the year. The leaching of nitrate is neither related to the farming activity nor the timing and duration of mineral fertilization of soils unless the fertilizer is applied directly on snow in winter.

Under summer conditions plant uptake of soil nitrate exceeds both the rate of nitrate supply from the atmosphere and the rate of nitrate production through mineralization and nitrification. Summer nitrate concentrations in streams in August are below 0.5 mg/l and are often very similar to or

Experimental catchment data 223

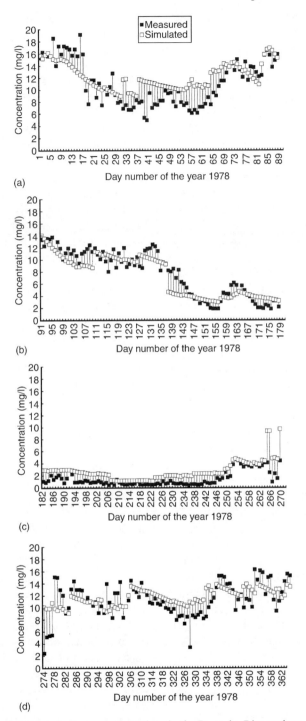

Figure 27.1 Measured and simulated nitrate concentrations in the Pomorka River subcatchment at Paryz village in 1978 (a) January–March; (b) April–June; (c) July–September; (d) October–December. Critical level for drinking water in Poland is 10 mg/l. (Source: Taylor, IMGW, 1980 and Kulbik, 1987.)

lower than those found in rainwater, for example in the range 0.15 to 0.25 mg/l in bulk precipitation noted in upland Wales [6]. These data can be compared with those obtained for Poland: from 0.12 to 0.35 mg/l in northern and western parts of the country, and from 1 to 8 mg/l in the south, close to the big industrial centres [7]. The very low content of nitrate in stream water in summer can be attributed to macrophyte uptake in the river bed, algal activity and denitrification processes. In the autumn run-off, just after the rain in the immediate post-drought summer period at the beginning of October, there is a large input of nitrate leaching into the stream.

During the winter, atmospheric inputs increase and plant uptake declines, which can lead to an increase of nitrate in soil water and leaching. Freeze–thaw cycles within catchment soils can affect the pattern of leaching losses. A similar effect of flushes of nitrate following snow-melt has been observed in March. Soil freezing modifies the hydrological pathways which, in turn, influence stream water chemistry. The result is a much higher nitrate concentration in streams during winter than during summer.

According to the standards for nitrate in drinking water, the quality of river is low. During certain periods of the year the nitrate concentration is unacceptably high. This creates a health risk for humans and animals because the river is used as a drinking water supply. The water may seem clean, but it contains excessive amounts of nitrate.

The origin of the nitrate and a way of predicting its concentration in the stream were investigated. Leaching losses of nitrate are dependent upon the availability of water to transport it. A mathematical model of the catchment containing a soil chemical component was developed. It seems that the climatic conditions are a major variable driving the whole system, more important than fertilization, land use and atmospheric deposition.

27.4 GENERAL FEATURES OF HYDROLOGICAL MODELS

A catchment is a complex system and a large number of processes influence the chemical composition of water during its movement through the soil layers and into surface waters [8]. The effects of precipitation on soil, ground-, and surface waters has attracted the interest of many research teams around the world [9]. In attempts to study the complex interaction between rainfall, soil-water balance, groundwater and run-off, computer-based models have been developed. These models vary in configuration and complexity depending on the purpose of the study. Models are useful in improving understanding of various processes and they are necessary when predictions are to be made.

Bergstroem and Lindstroem [9] reviewed several conceptual models applicable on a basin or regional scale, which integrate the water balance and water quality; most of these refer to short-term simulation on a catchment scale. The most interesting aspects for planners and politicians are the long-term effects of various scenarios of emission and deposition.

The development of the integrated hydrological and hydrochemical model has its roots in earlier rainfall–run-off modelling [9]. The models were modified by the introduction of water storage in the root zone, water in the intermediate (subsoil) zone and non-drainable water in the saturated zone. The volume of this extra storage of water is of the order of several hundred millimetres.

It is well known that a good run-off simulation can be obtained using water balance models with quite different structures. It seems reasonable that these conceptual models should be able to simulate not only run-off but also the water balance in more detail, including such variables as soil moisture content, groundwater fluctuations and residence time. Residence time (the amount of time the water spends in various environments) is a concept that has attracted little interest.

Most models have empirical parameters that are set by calibration. This means that the model is adjusted to an observed set of data, and optimum parameter values are sought, often by trial and error [9]. The calibration procedure is a curve-fitting procedure. It requires the number of events in the observed data set to be much greater than the number of degrees of freedom of model adjustment. In order to validate the model, a final independent simulation must be performed, using data not included in the calibration, to show that the model yields correct results under different conditions. The opportunity for calibration is limited because of the relatively short periods of time for which observed data are normally available [9]. This means that a model's ability to describe long-term changes is normally not verified. In this case, sensitivity analyses play an important role in identifying the range of uncertainty in the predictions.

27.5 SIMULATION OF THE WATER BALANCE OF FARMLANDS AND GRASSLANDS

The pioneering work of Makkink and van Heemst [10] produced one of the first simulation models of the water balance of arable land and pastures. Simulation gives the depth of the water-table, quantity of water in the soil above the water-table and the quantity of run-off to surface water. It is based on standard synoptic meteorological data, physical properties of the soil, field hydrology and pertinent data on the crop cover throughout the year (the crop rotation and measurements of crop height). Validation of the model was possible because the water balance of a small Rottegarts Polder in the Netherlands had been recorded and analysed for many years. The principle of Makkink and van Heemst model was followed; however, the task in this study was to construct entirely new computer programs to reflect the different structure of the modelled river basin.

In the simulation the system is subdivided into compartments (imaginary reservoirs), which contain water. The rates of transfer between the compartments is dependent on a quantitative formulation of the relevant processes. At any moment these rates of transfer depend on variables such as: the meteorological data, the crop cover and height, and the water content of the compartments [10]. The rates of transfer are computed at an instant in time and hold for a small time interval only. After this, the new state of the system is calculated. This is the principle of the simulation of state-determined systems.

Makkink and van Heemst paid particular attention to the choice of time interval. The time constant, defined as the ratio of amount of water to the rate of change, may be of the order of minutes. Hence far too many steps would be needed to simulate the water balance throughout a year. A time interval for integration of one day would necessitate the introduction of some artificial procedures to treat those processes that require a smaller time constant. In order to avoid time steps that result in more water being removed from a compartment than is contained or more water being added than can be stored, limiting capacity functions of reservoirs can be introduced.

The same authors indicate the necessary division of the soil and the river basin into compartments [10]. In each small compartment it may be assumed that the water is equally distributed. This is not so in a large compartment and, therefore, some assumptions are necessary about the water status in the soil. To describe the condition of water in the soil, the physical forces moving water through the soil are considered. Only drainable and plant-available water is considered. In the model of Makkink and van Heemst, if there is more water in the soil than at hydrostatic equilibrium this water can be anywhere in the unsaturated soil [10]. Because it is not possible to ascertain its position it is assumed to be evenly distributed over the unsaturated soil. If there is less water than at hydrostatic equilibrium, it is assumed that this missing water has left the soil by evapotranspiration and comes from the layer of unsaturated soil in which plants are assumed to root. Plants cannot extract all the water from the soil. When the suction of the soil is 15 atm plants are at their permanent wilting point and are unable to extract water from the soil.

In such a model there seems to be an inconsistency between the vertical and horizontal resolutions [9]. The soil profile is described in detail, whereas the natural areal mosaic of different soil characteristics, land use, recharge and discharge has to be approximated by integrating thin slices of soil horizons over the whole basin.

The problem is that some processes have to be simplified. In general, rate calculations are based as far as possible on the knowledge of the processes involved. For this study some parameters were taken from field observations and others were calibrated by comparing observed and simulated results. Makkink and van Heemst were able to use observed data corresponding to prolonged wet and dry periods for calibration purposes [10]. The parameters of the water balance equation for each compartment were calibrated during periods when their influence was decisive. The system was subdivided into compartments for water storage (and in set of equations) defined as:

- AT – atmosphere with evaporation demand (drying power of the atmosphere);
- S – precipitation, including frozen water such as snow and hail. This remains on the field before it ends up in drainage water. The capacity of S is unlimited;
- A – water adhering to the vegetation. This comes from the dew formed overnight and that part of the precipitation not reaching the soil (intercepted water). The capacity of A depends on the quantity and type of crop (up to 2 mm);
- P – puddles (soil ponding) formed on heavy clay or on frozen ground (up to 1.5 mm). These form frequently and drain slowly by surface run-off to the ditches;
- U – unsaturated soil (that part of the soil above the water-table). The upper part is called the transpiration zone T, the lower is the unsaturated subsoil. The crop removes water for transpiration from zone T. The capacity of U depends on the kind of soil and the depth of the water-table. The latter is continuously changing and therefore the capacity of U also changes every time interval.
- G – saturated soil (that part of the soil beneath the water-table down to an assumed basic surface below the lowest water depth). The capacity of G is required for the calculation of the water-table; it changes for the same reason as U.
- DITCH – water storage in the ditches or small streams, related to the run-off from the system. The surface water is supplied by surface runoff and groundwater flow.

The model simulates the relationship between rainfall (S) and run-off (by DITCH), modified by the state of the system (actual capacities of A, P, U and G). A full listing of the simulation program (in FORTRAN) is given in [10].

Calibration of a large number of parameters is impractical. In the models of Makkink and van Heemst [10], Kulbik [11] and Bergstroem et al. [12] calibration of a large number of parameters was unnecessary because many parameters were obtained from plant physiology and physical soil measurements and hydrological considerations.

27.6 MODELLING OF THE POMORKA RIVER SUBCATCHMENT

Because of the many variables and relationships the calculations were performed on the computer following the principles given by Makkink and van Heemst [10]. The starting equations were taken from the water balance equations, taking into account precipitation, evapotranspiration, soil water retention and water run-off.

The retention was further subdivided into: plants, soil surface, topsoil, subsoil and groundwater. Run-off or outflow was the sum of: evaporation of intercepted water, transpiration, surface run-off, subsurface run-off and basal flow from groundwater to river.

The catchment was modelled after Grip [12] and Rustad et al. [8] using homogeneous compartments (imaginary reservoirs) for the zones of water retention and water flow (Figure 27.2). Three compartments were considered: the first compartment was the topsoil, i.e. the root zone in the soil profile, the second compartment was the subsoil, the unsaturated zone between the topsoil and the groundwater level, the third compartment was the groundwater zone. Many of the complex processes had to be simplified, and the empirical expressions and their coefficients found by optimization methods. The model was run using daily time steps with inputs of precipitation and evapotranspiration demand.

Different flow paths are the most important reasons for variations in water quality. Precipitation (rain or snow) supplies the topsoil reservoir, although part of the water is removed by interception. There is a simple interception storage with a capacity of 1 mm, which is emptied by evaporation.

Snow can accumulate until the temperature rises and it starts to melt. Melt-water can be routed to the

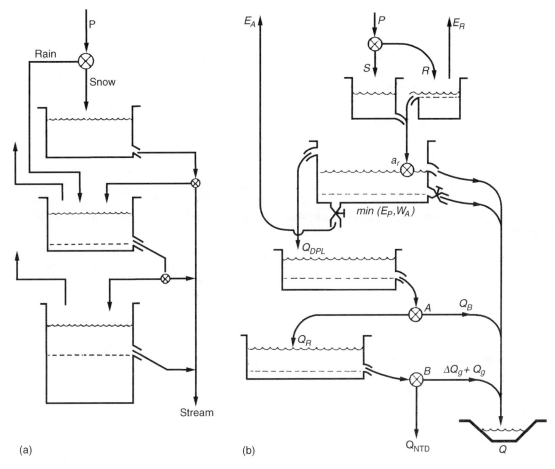

Figure 27.2 Multi-reservoir hydrological model, showing the concept of the model of (a) Rustad *et al.* and Kulbik; (b) Kowalik and Kulbik. (Sources: (a) Model of stream water chemistry of a tributary to Harp Lake, Ontario, *Canadian Journal of Fisheries and Aquatic Sciences*, 1986; Kulbik, 1987; (b) Kowalik and Kulbik, 1994.)

stream directly or to the topsoil, depending on the degree of frozen soil.

Water can evaporate, infiltrate or create overland flowing water. The water can be stored in the topsoil, be taken up by roots for transpiration, percolate to the subsoil or create surface supply to the river. Direct and rapid surface run-off is also allowed for (Figure 27.2). Soil water retention influences transpiration, because drier soil decreases the root water uptake from the topsoil (feedback between transpiration and soil water retention). The actual evapotranspiration decreases as the soil dries out [13]. Water can also be stored in the subsoil, percolate to the groundwater or create subsurface run-off to the rivers (Figure 27.2). All excess water from the soil moisture zone is collected in the saturated zone and drained at two levels using two recession coefficients (switches), which account for rapid superficial run-off and slower, deeper groundwater drainage. This rapid subsurface run-off seems to be the most critical for river water quality and needs further investigation. Soil water content strongly influences the subsurface run-off (second feedback between subsurface run-off and soil water retention). In the groundwater reservoir the water can flow out of the system to the deeper geological layers or can supply the river by direct contact between groundwater and river bed (Figure 27.2). The model was constructed with very little quantitative information on the soil reservoirs; the concentration of nutrients in the outflows of the reservoirs were not verified by observation.

The simulation of run-off is made in four steps:

- snow accumulation and melting
- evapotranspiration demand
- soil moisture accounting
- generation of run-off.

The dynamics of the process are described by the equation for the water retention in the catchment:

$$R(t) = R_0(t_0) + \Delta R(t) \quad (1)$$

where

$R(t)$ is the actual water retention of catchment;
$R_0(t_0)$ is the initial water retention (water content in time t_0);
$\Delta R(t)$ is the change in the water retention of the catchment as a result of rainfall, evapotranspiration and run-off in the period between initial time t_0 and actual time t.

As the catchment is divided into compartments, the equation can be written for each compartment in the form:

$$Rj(t) = R_0j(t_0) + \Delta Pj(t) - \Delta Ej(t) \pm \Delta Qj(t) \quad (2)$$

where,

j is an index of the reservoir;
$\Delta Pj(t)$ is an input from precipitation;
$\Delta Ej(t)$ is an output for evapotranspiration;
$\pm \Delta Qj(t)$ are inflows and outflows of water between the different compartments (depicted as arrows in Figure 27.2).

The model is driven by the time-dependent input variables of precipitation and evapotranspiration demand $E(t)$. The output variable, or reaction of the catchment, is the outflow of water $Q(t)$. As there are few reservoirs, the total outflow $Q(t)$ should contain the sum of outflows from all compartments into the stream. A set of balance equations creates the mathematical model of the catchment, with inputs $P(t)$ and $E(t)$ and output $Q(t)$, the latter being a result of summation:

$$Q = \Delta Q_q + Q_{pp} + Q_B + Q_g + \Delta Q_g \quad (3)$$

where,

Q is the total discharge from catchment to stream;

ΔQ_q is the rapid surface run-off occurring in the same day as rainfall P;

Q_{pp} is the subsequent subsurface run-off generated by the temporary water storage at the soil surface as a result of rainfall P.

Q_B is the temporary subsurface run-off, probably from the water flowing on the top of the dense subsoil. This is below ploughed topsoil, but still within the unsaturated zone of the subsoil (very often this component is related to land drainage and the existence of clay-pipe drainage water in areas of farmland or grassland);

Q_g is the ground water run-off (basal flow) from groundwater to the river, which is more-or-less permanent;

ΔQ_g is the temporary increase of the groundwater run-off as a result of the percolation of water from above and the momentary increase in the level of the groundwater in the catchment (this can be a second component of drainage water in areas where land drainage has been installed);

The balance equations in the model have five unknown parameters: a time-constant for each of the three reservoirs, representing the delay of the water outflow from the reservoir – expressed in days, and two switch parameters, indicating the exchange between the high and low levels of the model. Switch A is to drive the water flow to the river (rapid wash-out of nitrate) or to the deeper layers in the subsoil (by percolation). Switch B is to divide the flow to the groundwater reservoir storage or to the river as a basal flow. These parameters were estimated using the optimization methods of Hooke-Jeeves [14,15].

The importance of hydrological flow patterns is stressed, as is the fact that an understanding of the model requires a knowledge of the hydrogeology of the basin. The water balance was validated only against run-off. The use of additional sources of information, such as groundwater and soil moisture observations [9], would help to confirm the model structure in the future.

Normally the water quality in the compartments is unknown. After some measurement of the groundwater chemistry it was assumed that the groundwater is of the highest quality and the drainage water the lowest. In small catchments the groundwater that supplies the stream as base flow probably has a low nitrate content. The division into three separate compartments clearly affects the hydrochemical simulations. The flow pattern (mainly water pathways in the soil) strongly affects the final water quality of the stream. The leaching of nitrate occurs mainly by transport by percolating water through the unsaturated zone by so-called subsurface run-off (Figure 27.2). Part of the dataset was used for identification of the parameters, and part was used for validation of the model. Data from 1981–2 were used for calibration, because of the considerable variations in meteorological and hydrological conditions experienced during that year. Independent data from 1977–8, 1978–9, 1979–80 and 1980–1 were used for validation. There was good agreement between measured and simulated hydrological data for all years (Figures 27.3 and 27.4).

Validation of the model has been restricted so far to hydrochemical variables and run-off in small catchments. Because of the relative importance of internal fluxes of the model, the future observation of groundwater levels and quality may offer an important independent basis for validation.

27.7 SIMULATION OF NITRATE AND PHOSPHATE LEACHING

For the modelling of the leaching of nitrates from the catchment it was assumed there is a direct relationship between run-off and nutrient concentration. This is a very crude approximation, which makes the method applicable for a first assessment and primary analysis only. The model has no potential for long-term scenario simulation.

The large amount of measured data on river flow and nitrate and phosphate concentrations can be related in many different ways. The simplest approach is to take the exponential function:

$$C = aQ^{b-1} \qquad (4)$$

where,

Q is the river flow;
C is the nitrate concentration in the river;
a and b are empirical parameters.

The results give quite good agreement between measured and calculated values of C depicted in Figure 27.1. The model was quite accurate for calculation of transport peaks of nitrate during flood

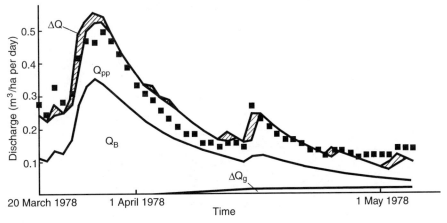

Figure 27.3 Results of the water modelling of floods during spring for small catchment. The squares show the measured discharges and the lines the simulated discharges from topsoil (ΔQ and Q_{pp}), Subsoil (Q_B) and groundwater (ΔQ_g).

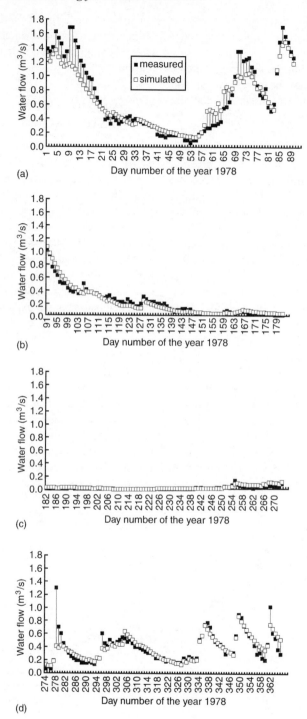

Figure 27.4 Comparison of measured and simulated water flows in the Pomorka River subcatchment at Paryz village in 1978 (a) January–March; (b) April–June; (c) July–September (d) October–December. (Sources: Taylor, IMGW, 1980 and Kulbik, 1987.)

situations. It has already been shown that many of the hydrochemical events can be simulated by a relatively advanced hydrological model with an oversimplified hydrochemistry [9]. It was found that run-off data (Q) are crucial for estimates of the losses of nitrate from arable land into rivers, lakes and coastal waters. According to Bergstroem and Lindstroem [9] a great deal of uncertainty still remains concerning the soil water pathways for solute transport. It is probable that a further examination of the role of recharge and discharge areas would improve the performance of models.

The flow rates of water are much more variable than the hydrochemical concentrations. For this reason it was accepted that there was a direct relationship between river flow and nitrate concentration [11, 15 and 16]. Apart from the nutrient content in the soil, the main factor determining nutrient losses at a given time is the amount of run-off from the catchment considered [17].

The authors' model, analogous to that of Bergstroem et al. [18] does not use deposition data as input. The idea is that, in a short time perspective, the major causes of the variations in hydrochemistry are the variations in water pathways in the soil of the catchment. The hydrochemical characteristics of these pathways are assumed to be stable in time. This is the reason why the model cannot be used for scenario simulations of changes in deposition, but the sensitivity analyses can help in the evaluation of the alternative predictions.

27.8 RESULTS AND CONCLUSIONS

The concentration of nitrate is a function of the river flow. The agreement between measured and simulated nitrate concentrations in the Pomorka River sub-catchment was satisfactory for the daily values for the period 1977–9.

Nitrogen (nitrates) leaching was 14.6 kg/ha per year for the small (4 km^2) catchment with loamy soil; 4.3 kg/ha per year for a medium (72 km^2) catchment of the Pomorka River; but only 1.2 kg/ha per year for the large (236 km^2) catchment of the Wietcisa River with sandy soil and an extensive agricultural land use (Table 27.1). In Scotland, for example, the fluxes of nitrate in streams are equal to 1.5 kg/ha per year [19] or 3.9 and 6.8 kg/ha per year [20]. The authors' small catchment carried 14.6 kg/ha per year, but larger hydrological units give 4.3 or 1.2 kg/ha per year, which is very similar to the Scottish data.

If these numbers are compared with the amount of nitrogen in the soil, estimated at about 5000 kg/ha, and the atmospheric fallout of nitrogen of 15–20 kg/ha per year, it is clear that the effects of climate and weather seem to be the most important processes for the leaching of nitrate. This underlines an important point about the frequency of sampling and the measurements of water quality. During low flows the periods between sampling can be quite long, but during high flows sampling should be done more often to detect extreme concentrations.

The low leaching losses of phosphate of between 0.1 and 0.5 kg/ha per year can be compared with the situation in so-called phosphate-saturated soils in the Netherlands, where the rate of leaching of phosphorus is as high as 2.5 kg/ha per year from poorly and moderately drained soils [21].

From the presented research it seems advisable to protect the water supply intake during the highest flows because of the high concentrations of nitrate. The extreme concentrations of nitrate occur in early spring after the snow has melted and after rains in late autumn, whereas low concentrations occur during dry summer weather. It is a mistake to confine sampling to dry weather and low flows. The results can be extremely inaccurate and potentially dangerous for future decision-making.

Table 27.1 Leaching losses from small, medium and large catchments

Catchment	Fertilizer application rate (kg/ha)	Nitrate losses (kg/ha per year)	Phosphate losses (kg/ha per year)
Small (loam, 4 km^2)	280	14.6	0.516
Medium (loam, 72 km^2)	280	4.3	0.088
Large (sand, 236 km^2)	100	1.2	0.179

In periods of high nitrate concentrations water intakes should be closed; in many cases water treatment processes are not designed for such high loads and effluent nitrate levels would exceed the required 10 mg/l if intakes were left open. Other measures would be to recirculate the drainage water somehow, or close drainage water outlets during critical periods.

REFERENCES

1. Dunnette, D.A. (1992) Assessing global river water quality, overview and data collection, in *The Science of Global Change, the Impact of Human Activities on the Environment*, (eds D.A. Dunnette and R.J. O'Brien), ACS Symposium Series 483, American Chemical Society, Washington DC, pp. 240–59.
2. Taylor, R. (1984) The runoff of nitrogen and phosphorus compounds from selected agricultural regions in the Vistula and Odra drainage basins, *Oceanologia*, 18, 135–48.
3. Taylor, R., Balcerska, M., Dyduch, R. and Kowalewska, K. (1980) Research of influence of non-point pollution upon surface water quality related to the mathematical modelling. Internal report (in Polish) PR 7.04.02.08, IMGW.
4. Greenfelt, P. and Hultberg, H. (1985) Effects of nitrogen deposition on the acidification of terrestrial and aquatic ecosystems. IVL Publ. B 801.
5. Falkowska, L. (1986) Nitrogen and phosphorus supply from the atmosphere to the Baltic Sea. PhD thesis (in Polish), University of Gdansk.
6. Roberts, G., Hudson, J.A. and Blackie, J.R. (1984) Nutrient inputs and outputs in a forested and grassland catchment at Plynlimon, mid-Wales, *Agricultural Water Management*, 9, 177–91.
7. Pawlik-Dobrowolski, J. (1983) Rainfall and water pollution. *Aura*, 8, 13–15.
8. Rustad, S., Christophersen, N., Seip, H.M. and Dillon, P.J. (1986) Model of stream water chemistry of a tributary to Harp Lake, Ontario, *Can. J. Fish. Aquat. Sci.*, 43, 625–33.
9. Bergstroem, S. and Lindstroem, G. (1989) Models for analysis of groundwater and surface water acidification, a review. National Swedish Environmental Protection Board Report 3601.
10. Makkink, G.F. and van Heemst, H.D.J. (1985) Simulation of the water balance of arable land and pastures. Simulation Monograph 7, published by PUDOC, Wageningen.
11. Kulbik, M. (1987) Influence of hydrometeorological conditions on the outflow of non-point pollution from small agricultural catchments in postglacial landscape. PhD thesis (in Polish). Technical University of Gdansk.
12. Grip, H. (1982) Water chemistry and runoff in forest streams at Kloten. Rep. No. 58. Naturgeografiska Institutionen, Uppsala University.
13. Fleming, P.M. (1964) A water budgeting method to predict plant response and irrigation requirements for widely varying evaporation conditions, in *Proc. 6th Intern. Congress of Agricultural Engineering*, Lausanne, Switzerland, **2**, pp. 21–7, published by ISAE.
14. Findeisen, W., Szymanowski, J. and Wierzbicki, A. (1977) *Theory and Computational Methods of Optimalisation*. Polish Scientific Publishers, (PWN), Warsaw.
15. Karlsson, G., Grimvall, A. and Löwgren, M. (1988) River basin perspective on long-term changes in the transport of nitrogen and phosphorus, *Wat. Res.* **22**(2), 139–49.
16. Kowalik, P.J. and Kulbik, M. (1994) Mathematical modelling of water resources of small agricultural catchments in postglacial landscape. Monografie Komitetu Gospodarki Wodnej PAN, z. 5, pp. 3–59.
17. MONITOR (1989) Climate and the natural environment. National Swedish Environmental Protection Board, Solna.
18. Bergstroem, S., Brand, M., and Gustafson, A. (1987) Simulation of runoff and nitrogen leaching from two fields in southern Sweden, *Hydrological Science Journal*, **32** 191–205.
19. Farley, D.A. and Werrity, A. (1989) Hydrochemical budgets for the Loch Dee experimental catchments, southwest Scotland 1981–1985, *J. Hydrol.*, 109, 361–8.
20. Edwards, A.C., Creasey, J. and Cresser, M.S. (1985) Factors influencing nitrogen inputs and outputs in two Scottish upland catchments, *Soil Use and Management*, **1**, 83–7.
21. Breeuwsma, A. and Reijerink, J.G.A. (1992) Phosphate-saturated soils: a 'new' environmental issue, in *Chemical Time Bombs*. Proceedings of the European state-of-the-art conference on delayed effects of chemicals in soils and sediments (eds G.R.B. ter Meulen, W.M. Stigliani, W. Salomons, E.M. Bridges and A.C. Imeson), 2–5 September 1992, Veldhoven, the Netherlands, pp. 79–85.

WATER QUALITY MODELLING OF THE REGA RIVER AND SURROUNDING COASTAL WATERS

by A. Lewandowski and B. Ołdakowski

28.1 INTRODUCTION

The Rega River (200 km long) is located in the northern part of Poland and flows directly into the Baltic Sea. The total catchment area is 2723 km^2. The water quality of the Rega River is poor with pollution originating from municipalities and agricultural activity. In addition, the high concentration of *E. coli* bacteria in the mouth of the Rega River results in unhealthy conditions in the adjacent coastal waters.

To identify the major pollution sources and to describe present water quality in the Rega catchment, water quality environmental modelling tools were applied. The MIKE 11 and MIKE 21 systems were used for river and coastal studies respectively. Based on a calibrated water quality model, the environmental impacts of alternative implementation schemes were evaluated. The final objective was to establish a priority list for the implementation of wastewater treatment facilities with respect to different aspects of river water quality and pollutants discharged to the Baltic.

28.2 BACKGROUND

In April 1991 Danish legislation concerning subsidies for environmental activities in East European countries came into force by decision of the Danish Parliament. One of the projects identified for funding by the Danish Environmental Protection Agency (DEPA) and the Polish Ministry of Environmental Protection, Natural Resources and Forestry was the restoration of the Rega River. A pre-feasibility study concerning the possible rehabilitation and improvement of wastewater treatment facilities in the municipalities of Gryfice, Łobez Nowogard, Płoty, Resko, Świdwin and Trzebiatów and major industries in the Rega catchment area was initiated in July 1991 by A/S Samfundsteknik. This was financed by DEPA on behalf of the Rega Union, a co-operative that handles all matters concerning environmental protection on behalf of the seven municipalities.

The pre-feasibility study was finalized in November 1991. In December a consultancy agreement was entered into between DEPA and A/S Samfundsteknik concerning basic design of the possible improvements of the wastewater treatment facilities in the seven municipalities and the major agricultural industries. The County of South Jutland entered this agreement in June 1992 with the responsibility for assessing the biological baseline conditions in the Rega River as subconsultant to A/S Samfundsteknik.

In September 1992, a consultancy agreement was made between DEPA and the Water Quality

Institute (VKI) in collaboration with Danish Hydraulic Institute (DHI), and with Geoscience and Marine Research & Consulting Co. (GEOMOR) as the Polish consultant. This agreement covered the assessment of the environmental impacts of improvement of the sewage treatment facilities both in the Rega River and the coastal region around its mouth in the Baltic Sea.

28.3 OBJECTIVES

The overall objective of the Rega River Restoration Project was to restore and protect the water quality and ecological conditions in the Rega River and in the coastal regions at its mouth, and to reduce the loading of pollutants discharged into the Baltic Sea.

The specific objectives of the assessment studies were:

- to identify the major pollution sources in the Rega catchment area and evaluate the present sewage treatment facilities;
- to describe the present water quality and biological conditions resulting from the pollutant loadings;
- to identify feasible technologies for improving the sewage treatment facilities;
- to evaluate the environmental impact of alternative improvement schemes; and
- to draw up a plan and list of priorities for the successful implementation of improved sewage treatment facilities based on the feasible technology identified, the environmental improvements expected and the economic implications for the seven municipalities and industries involved.

28.4 HYDROLOGY OF THE RIVER

The Rega River discharges to the Baltic Sea at Mrzeżyno (Figure 28.1). The mean flow at Trzebiatów, which is situated 6 km from the coast, is 20.5 m^3/s calculated over a 20-year period between 1971 and 1991. The lowest and highest mean daily flows at Trzebiatów are 9.1 m^3/s and 53.2 m^3/s respectively. During the period 1956–70 average precipitation was 705 mm, from measurements taken at seven stations within the catchment. The discharge in the Rega River is controlled by surface and groundwater run-off from urban and rural areas. At times of minimal flows, the discharge is primarily controlled by the groundwater contribution and sewage discharges from urban areas.

In the present study, the hydrological year 1991 (November 1990 – October 1991) was chosen for the calibration of the model. This year was selected because water quality parameters were available, and 1991 was a reasonably dry year. Dry periods are considered critical for evaluating water quality problems in rivers. Water levels and discharges were available at 26 hydrological gauging stations from the Institute of Meteorology and Water Management (Figure 28.1).

The tidal range of the Baltic Sea off the Polish coast is small, but considerable variations in the sea level are often seen caused by strong winds and variations in atmospheric pressure. Atmospheric conditions often induce powerful currents in the direction of the wind and cause considerable build-up of water levels along the coast. The prevailing winds are from the west and the south-west, which means that the winds, and therefore the currents, are very often parallel to the Polish coastline.

The salinity of the Baltic Sea remains fairly constant throughout the year. The salinity at the surface progressively decreases from about 7.5 ppt in the extreme south-west, to about 3 ppt in the north-east.

28.5 POLLUTANT LOADINGS

The total loadings to the Rega River arise from the inflows of tributaries (mainly non-point sources in open land), the inflow at the source of the river, and point sources from urban areas and industry. The estimated total loadings to the Rega River from the three different sources are shown in Table 28.1. In general the BOD loading from the diffuse sources constitutes 50% of the total BOD loading, and the urban areas and industry the remaining 50%. The loading of inorganic nutrients (ammonium nitrogen, nitrate nitrogen and phosphate phosphorus) originates mainly from the non-point sources. The *E. coli* loading is solely associated with the urban areas. The concentrations of BOD and nutrients from diffuse sources are estimated to be equal to the total loading less the point sources and inflow from over the border.

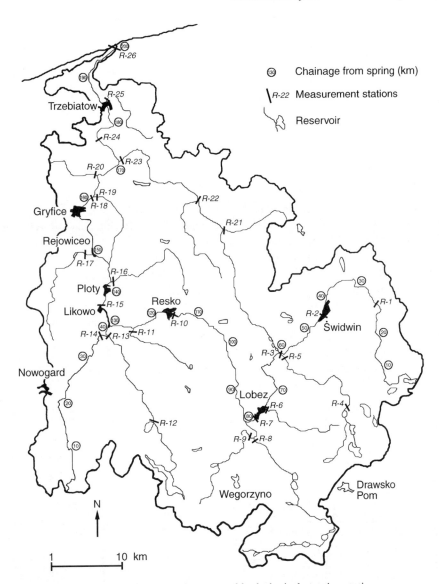

Figure 28.1 Rega River catchment with sub-catchments and hydrological gauging stations.

28.6 ASSESSMENT OF ENVIRONMENTAL IMPROVEMENTS

28.6.1 MODELLING APPROACH

The main factors that determine the quality of the Rega River and the surrounding coastal waters are described in an integrated model system. The present water quality situation is described and compared with the effects of alternative measures and abatement schemes for the reduction of pollutant loadings. The main components in the integrated model system are the one-dimensional MIKE 11 and two-dimensional MIKE 21 models used for forecasting the river and the coastal areas respectively. In brief, the MIKE systems provide a fully integrated description of hydrodynamics, advec-

Table 28.1 Loading to the Rega River used in the water quality model

Total loading	Diffuse inputs (estimated)	Input from over the border	Point sources
Ammonium nitrogen (t/yr)	218	22	78
Nitrate nitrogen (t/yr)	693	110	0
Nitrogen BOD_5 (t/yr)	490	26	495
Phosphate phosphorus (t/yr)	170	18	65
Phosphorus BOD_5 (t/yr)	17	1	17
Total BOD_5 (t/yr)	1689	132	1707
E. coli (pcs/y)	1E+15	9E+15	5E+17

tion–dispersion processes and water quality (Figure 28.2).

The output from the MIKE 11 part of the model system covering the Rega River (the concentrations of different pollutants) is used as the input to MIKE 21 when simulating the impact of pollution reduction in the Rega River system on the coastal environment.

The MIKE 11 system simulates the flow and water quality conditions in rivers and channels. It takes into consideration the non-point sources, the overflow at dams and weirs (and the resultant flooding and drying of low-lying areas), the sources of pollutants and the processes which affect the ecological state of the river.

The MIKE 21 system simulates hydrodynamic and water quality conditions in the coastal area in a two-dimensional rectangular grid taking into account wind direction and boundary conditions. The advection–dispersion part simulates spreading of dissolved or suspended substances. The water quality part, which is integrated in the transport

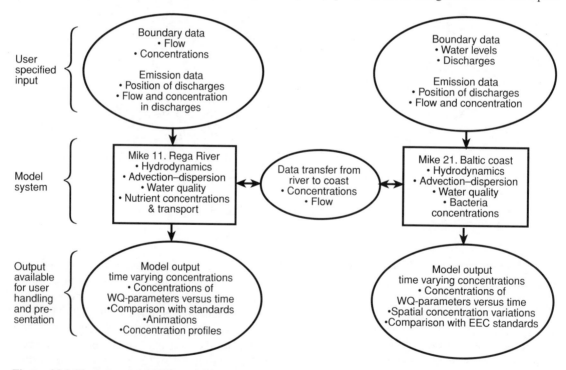

Figure 28.2 The integrated MIKE modelling system.

model, describes the resulting concentration of bacteria, i.e. bathing water quality, oxygen conditions, BOD degradation and nutrient recycling.

The calibration of the model involved adjustments of the model coefficients (within a realistic range) in order to obtain an acceptable agreement between simulated and observed values. The model was calibrated for the hydrodynamic aspects by comparing simulated and observed data for water levels and discharges. For the water quality aspects, dissolved oxygen, temperature, ammonia, nitrate, BOD, dissolved phosphorus and *E. coli* data were used. The results of the calibrations are shown in Figure 28.3. and Figure 28.4.

The overall mass balance for the Rega River is shown in Table 28.2. The discharges to the Baltic Sea were calculated by multiplying the simulated water flow by the concentration in the river mouth

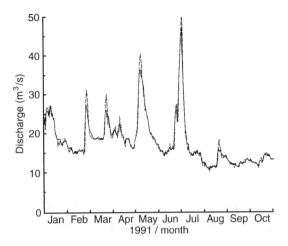

Figure 28.3 Comparison between the simulated and measured flow rates for the Rega River at Trzebaitów. Solid line shows simulated, dashed line measured data.

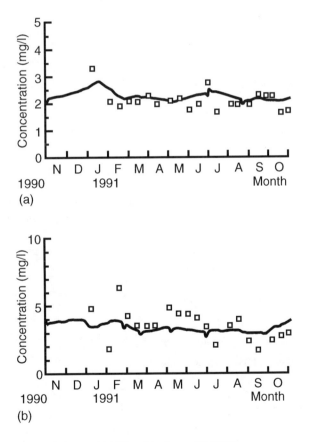

Figure 28.4 Measured and simulated data at 126.8 km (a) nitrate; (b) BOD.

Table 28.2 Mass balance for the Rega River from 1 November 1990 to 31 October 1991

	Total loading	Discharge to Baltic Sea	Retention or removal
Total nitrogen (t)	2131	1790	342 (16%)
Total phosphorus (t)	288	188	100 (35%)
BOD (t)	3528	1559	1969 (56%)
E. coli (pcs)	5E+17	3E+17	2E+17 (40%)

using half-hourly values. The residual values represent the retention or removal of each of the components.

Retention includes adsorption, plant uptake and bacteriological uptake of nutrients, whilst the processes that remove pollutants from the river are degradation and denitrification. From Table 28.2 it can be seen that the self-purification was relatively high for BOD (56%), *E. coli* (40%) and phosphorus (35%) whereas the retention or removal of nitrogen (16%) was more moderate.

28.6.2 MODEL SIMULATIONS

The calibrated water quality model is used as an instrument for the evaluation of the environmental impact of alternative improvement schemes. The effects on the water quality in the Rega River and reduced loading from the river to the Baltic Sea, obtained by technological improvement of the wastewater treatment plants, were described according to three different scenarios of emission from sewage treatment plants:

- current emission standard in Poland (extension stage 1);
- expected emission standard in Poland after the year 2000 (extension stage 2);
- current emission standard in Denmark (extension stage 3).

The emission standards are shown in Table 28.3. The only difference between extension stage 2 and 3 was the *E. coli* loading from point sources.

At all three stages it was assumed that, for the design of the sewage treatment plants, future loadings during stable periods would be reduced. However, to ensure improvements in water quality were not overestimated, maximum concentration demands were used.

The calibrated model was run for all three stages in each of seven towns in the catchment, and additionally for improved treatment in three of the towns. It gives a total of 24 simulations, but only a few results are presented. The present and future BOD concentrations along the river after the implementation of improvements to the wastewater treatment plant in one of the towns (Świdwin) are shown in Figure 28.5.

The reduction in point loading and loading to the Baltic Sea was calculated for each of the three stages. The results for stage 2 are shown in Table 28.4.

28.6.3 MODEL RESULTS FOR COASTAL WATERS

The effect of the improved wastewater treatment on the water quality along the Baltic coast was calculated using the MIKE 21 water quality model by simulating present and future *E. coli* concentrations (Figure 28.6). To calibrate the coliform plume, field investigations were carried out in which 70 samples were taken from the surface layer and from mid-depth. Also, certain environmental conditions (tem-

Table 28.3 Emission standards from sewage treatment plant used in the water quality simulations

Concentration in treated wastewater	Extension stage 1	Extension stages 2 and 3
BOD_5 (mg/l)	30	15
Ammonium nitrogen (mg/l)	6	2
Phosphate phosphorus (mg/l)	5	1.5

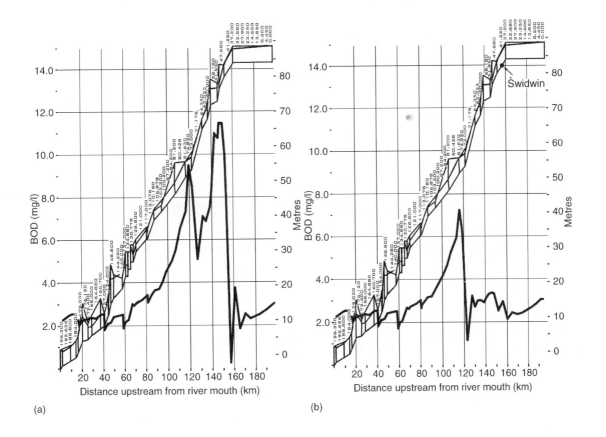

Figure 28.5 Biochemical oxygen demand along the length of the Rega River (a) before and (b) after the implementation of the extension to Świdrin treatment works.

perature, salinity and light intensity) were included to describe E. coli decay parameters.

28.7 ECONOMIC ANALYSIS

The tariffs for future users for the funding of loans and the operation of wastewater treatment plants were calculated. The tariffs were based on the total investment in wastewater treatment plant and sewerage systems, and the assumption that all discharges from treatment plants would need to comply with current Polish emission standards. Finally, the seven municipalities were ranked by size of tariff and by reduction in pollution achieved per invested ECU. These rankings are presented in Table 28.5.

28.8 CONCLUSIONS

Implementation of the improvements to wastewater treatment facilities to meet predictions would secure an acceptable water quality in both the river and coastal region around the Rega River outlet to the Baltic Sea. Biological conditions would be improved in those sections of the Rega River which at present are markedly affected by the

Table 28.4 Comparison of reduction in local point loading with reduction in loading to the Baltic Sea in the simulated period 1 June 1991 to 31 August 1991 following stage 2 extension to wastewater treatment plants

Extension stage 2	BOD (tonnes/3 months)		Total nitrogen (tonnes/3 months)		Total phosphorus (tonnes/3 months)		E. coli (1E+16 pcs/3 months)	
	Local	Baltic	Local	Baltic	Local	Baltic	Local	Baltic
Gryfice	56.4	31.2	17.8	15.7	4.8	3.6	2.3	1.6
Łobez	104.3	12.2	31.5	17.7	1.8	1.1	2.3	0.1
Nowogard	8.7	1.5	3.2	2.2	0.9	0.4	0.4	0.1
Płoty	22.1	8.9	6.8	5.5	1.2	0.7	0.7	0.1
Resko	15.1	3.7	4.4	3.5	0.5	0.3	0.5	0.1
Świdwin	78.3	6	24.2	12.9	4.2	1.1	2.3	0.1
Trzebiatów	99.3	78.5	30.7	29.1	1.2	1.2	2.3	1.9
All	348.1	141.8	118.7	86.3	14.5	8.5	11.5	3.7

wastewater discharges downstream of the seven municipalities.

A priority list for implementation should be drawn up based on the rankings. The prioritization will depend on which aspects are regarded as most important, e.g. financing or reduction of bacteriological pollution in the coastal area.

Table 28.5 Ranking of priorities related to environmental and economic aspects (1 = higher priority; 7 = lower priority)

	Tariffs	BOD-removal per invested ECU	N-removal per invested ECU	P-removal per invested ECU	Biological river quality improvement downstream	Reduction in nutrient loading to the Baltic	Reduction in bacteriological pollution of coastal region
Gryfice	1	4	5	2	3		1–2
Łobez	3	1	2	3	5	3	7
Nowogard	2	7	7	4	2	7	7
Płoty	7	5	6	5	6	5	6
Resko	6	6	3	6	7	6	7
Świdwin	4	3	1	1	1	4	7
Trzebiatów	5	2	4	7	4	1	1–2

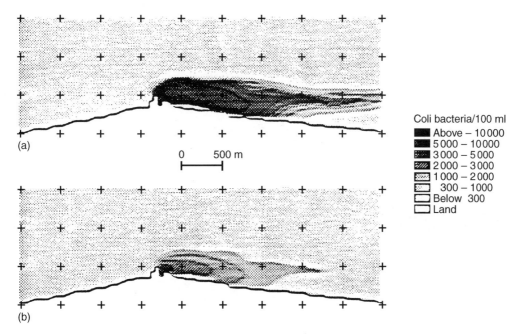

Figure 28.6 Results of a simulation of *E. coli* concentrations with wind blowing from the west (a) existing situation; (b) following 91.5% reduction of *E. coli* at the river mouth.

ENVIRONMENTAL PROTECTION IN THE VISTULA RIVER VALLEY

by Z. Kajak

29.1 INTRODUCTION

The Vistula is probably one of the few rivers in Europe with little regulation and a relatively undeveloped valley environment [1, 2 and 3]. The river is heavily polluted along almost its whole length [2–7]. Approximately half the pollutant load originates from point sources; the other half originates from diffuse sources, mainly from agriculture [8, 9 and 10], which occupies about 65% of the river catchment area. For point sources of pollution, the main remedy is to build sewage purification plants. The obstacles to this are mostly financial. This chapter deals to some extent with non-point pollution and possible measures for reducing it. However, it is concerned mainly with environ-mental protection in the river and its valley.

29.2 REDUCING THE POLLUTANT LOAD FROM NON-POINT SOURCES

There are a number of measures which would help to decrease pollutant loads from dispersed sources in agricultural areas:

- afforestation, especially in higher parts of catchment area;
- creation of patches of permanent plant cover within agricultural areas, to consume substances otherwise transported towards surface waters;
- various measures to prevent soil erosion by water and to slow down rapid drainage of surface water;
- deep tillage (to retain more water in soil and to slow down its outflow);
- increasing the percentage of organic matter in the soil by application of manure, compost etc. (it has been suggested that this could result in retaining as much water as is presently held in existing reservoirs);
- creation of protective belts of permanent plant cover (trees, shrubs, meadows, i.e. buffer zones, to protect rivers from diffuse sources of pollution) along surface waters, to consume substances which otherwise would enter these waters [3, 11, 12 and 13].

The Vistula River itself has such buffer zones in the form of dense plant cover between lateral embankments (levees) although most of its tributaries do not have them.

All these measures would increase the retention of both water and substances (suspended and in solution) 'on the spot' – where they are needed for both soil and plants. The retention of water usually also means the retention of various substances (the soil particles, nutrients, pesticides etc.), because they are transported mainly in water.

29.3 DIVERSITY OF ENVIRONMENTS AND BIOTA IN THE VISTULA RIVER AND ITS VALLEY

The river bed is very variable and comprises deep and shallow stretches with steep and flat shores, many branches with varying depths, currents and types of sediment etc. and many islands [2, 3]. Some of these islands are rather small and transient, often existing only from one high flow to the next which destroys them or moves them to another place; they can be completely bare, sandy or covered partly with pioneer vegetation. Others can be permanent, covered with old trees and shrubs. Intermediate stages are also found. The aquatic environment around islands is also very varied.

The life in the river (plankton, benthos, fish) is rich and quite diverse [2, 3, 6 and 14–18].

The area between levees is also very variable along the river (the slope, type of soil and water conditions), both temporally (mostly due to changes in water level and other conditions) and spatially across the valley (surface and underground water level, type of substrate, distance from open water – the presence of old river beds, small water bodies, swamps, peatlands etc. and various plant associations). The plant cover in many places resembles a jungle. The total width of plant cover ranges from a few hundred metres in the upper reaches of the river to 600–700 m in the middle and low reaches. So the total area along the whole river is several hundred square kilometres.

There are many large forest areas along almost the whole length of the Vistula (Figure 29.1) and numerous smaller forests in-between. All these forests, the rich plant cover between the levees, the river itself and many water bodies and swamps (Figure 29.2) make an ecological corridor along the whole country – from the Baltic Sea in the north to the Carpathian Mountains in the south. This ecological corridor forms an important element in Europe's environmental protection network. As much as 80% of the area along the Vistula is either protected or planned for protection in the near future. All these factors encourage the creation of a landscape park along the least developed middle reach and upper part of the lower reach of the river.

The Vistula valley is inhabited by 67% of the Polish bird population (195 species). Their main nesting places are: islands (including bare, sandy ones which provide the only nesting environment for several endangered species), areas between levees and the river, and steep shores. Almost 95% of the total number of *Larus canus* in Poland (3500

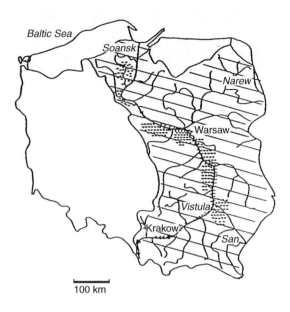

Figure 29.1 Map of Poland showing Vistula drainage basin (hatched); proposed landscape park (dashed area); riverside forest (dotted); reservoirs (triangles); other drainage areas (unmarked). The middle reach of the Vistula is situated between the San and Narew rivers.

pairs), 60% of *Sterna hirundo* (700 pairs) and large percentages of several other species live here [2, 3, 19, 20 and 21].

Many species of birds from Scandinavia and northern Russia migrate southward along the Vistula corridor, then through either the San–Dniester or the Moravian Gap between the Sudetian and Carpathian Mountains and along the Danube to the Black Sea and Mediterranean Sea. As many as 30 species of birds are named in the Polish 'Red Book' of animals (i.e. are threatened with extinction). These include migrating birds and those that winter along the Vistula.

The valley is also inhabited by at least 27 species of mammals, among which the beaver and otter are listed in the Red Book.

29.4 PROTECTION OF THE ENVIRONMENT

The living world depends on environmental conditions, but also influences and creates them. The organisms living in these sites have adapted to their

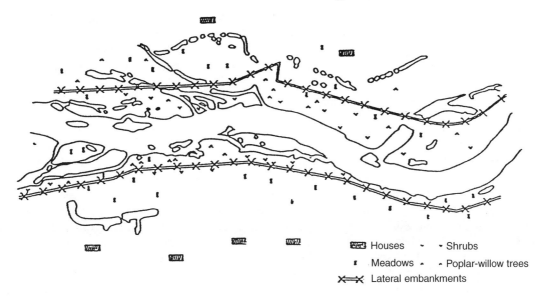

Figure 29.2 Section of the Vistula River, 10–20 km south of Warsaw, with branches, islands and old river beds. Permanent vegetation exists between the lateral embankments and the river; beyond the lateral embankments exist marshes, meadows, shrubs and forests, in addition to agricultural areas and some villages.

variability and changeability, which they need for their normal existence and functioning. These conditions must be preserved if the biological community is to be retained. Some specialists say the richness and the diversity of the community is comparable to those found in tropical rain forests. In Europe 62% of all inland bird species are found in these riparian environments. Any change in water level or timing of its fluctuations by intervention would essentially change these communities or make them disappear.

The human interference in the sites among levees is generally very small.

Nature conservation, to the general public, often implies large spectacular species of plants and animals. It is often forgotten that there are thousands of other species, many microscopic (from bacteria and protozoans to worms and arthropods), which are much more important to the functioning of the environment than large species. Besides, there is no basis or justification for valuing one species over another, as all need protection if the biological diversity is to be maintained.

This new approach to stewardship of the Earth's biota was expressed strongly at the Rio Conference. The resolutions from the Rio Conference should be implemented as well as Poland's own internal regulations on environmental conservation and species protection. The richness of the living world and its biodiversity (including all the endangered species) in the Vistula valley need to be preserved by maintaining the existing environmental conditions.

Protection of the riparian environment ensures protection of many organisms. Often, the importance of these organisms is not appreciated. Many species are becoming extinct, some of which are of great importance to the human race. There are probably many others already extinct; their importance, potentially enormous, will never be known.

29.5 DAMAGE AND THREATS TO THE ENVIRONMENT FROM ENGINEERING WORKS

In the upper Vistula there is one large reservoir (Goczałkowice – 88×10^6 m^3) and six small ones (each with a dam several metres high), four of which are finished and two of which have been abandoned. Goczałkowice Reservoir is very important as a source of water for the industrial, densely inhabited Silesian region although the quality of water in the reservoir is deteriorating. The small ones near Kraków and the large one at Włocławek (400×10^6 m^3) in the lower Vistula (Figure 29.3), are the remnants of the Vistula cascade (about 30 dams along the whole river) planned in the 1970s but abandoned due to political changes and the economic situation. These existing reservoirs have little use except for some hydroelectric power generation. They do however give rise to significant water quality and other environmental problems. The water in the upper Vistula is extremely polluted, and in the reservoirs near Kraków full oxygen depletion occurs periodically [22]. In Włocławek Reservoir on the lower Vistula, ecological catastrophes due to oxygen depletion also occur from time to time: one of these occurred a few years ago, in summer, and resulted in a total fish kill [16]. These problems arise because construction of the reservoirs necessitated the breaking of the general rule not to dam a heavily polluted river.

The Włocławek Reservoir also accounts for other serious problems to the ecology. There have been alterations to the plant associations along the reservoir and also changes to the fauna. These effects have been studied principally in fish and birds. The dam completely eliminated the annual migrations of *Salmo trutta* (and consequently its existence in this river) and some other migratory fish (especially *Vimba vimba*). The other migratory species – *Salmo salar* and *Acipenser sturio* – disappeared as a result of earlier pollution and regulation [15, 23].

The composition of species and the quantitative relationships within the fish community have deteriorated strongly. Some valuable riverine fish species completely disappeared in the Włocławek Reservoir and the total catch in the river below the dam decreased by 85% following dam construction. Also, the number of moderately valuable fish decreased, whilst the number of low value fish significantly increased, both in the reservoir (which became stocked with low value fish only) and below it, as a result of damming of the river (Table 29.1).

Most of those nesting birds species which are either protected or endangered have disappeared in the area of Włocławek Reservoir. As a relative estimate, assuming an unchanged condition above the reservoir as 100%, there are now 25% in the reservoir area, and 25%–50% in various sections below the reservoir (Chylarecki *et al.*, unpublished data). The possibilities for improving the situation for these birds in the reservoir seem unlikely. The creation of artificial islands, for example, is considered unrealistic. The total number of birds in the reservoir at present is higher than in the river, however these are common species, which are numerous everywhere.

Recently a new engineering solution has been considered. This is to build a new cascade of dams in the middle and lower reaches of the Vistula, from Warsaw to the Baltic Sea (Figure 29.3). The main aim of this cascade would be the production of hydroelectricity, up to 6% of the electrical energy needs of the country. There is controversy between supporters of this idea and the ecolo-

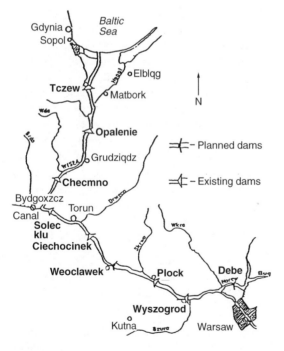

Figure 29.3 Location of dams on the lower and middle Vistula.

Table 29.1 Percentage of total catch of most abundant fish species in the lower Vistula before and after construction of Włocławek Reservoir. (Source: Backiel and Bontemps, unpublished data)[a]

Fish groups	Before damming 1956–66		After damming 1977–89			
	Lower Vistula		Vistula below dam		Reservoir	
Most valuable	29	S.tr. 11 V.v. 11 B.b. 7	4.3	S.tr. 2.3 V.v. 2	0	
Least valuable	48	A.b. 43 R.r. 5	74	A.b. 38 Bl.b. 19 R.r. 17	95	A.b. 83 Bl.b. 1 R.r. 11
Intermediate value	5	E.l. 5	13.2	E.l. 8.6 P.fl. 2.3 Sil.gl. 2.3	2.6	St.l. 1 Sil.gl. 0.6 L.c. 1

[a] S. tr. – *Salmo trutta*; V.v. – *Vimba vimba*; B.b. – *Barbus barbus*; A.b. – *Abramis brama*; R.r. – *Rutilus rutilus*; Bl.b. – *Blicca björcna*; E.l. – *Esox lucius*; P.fl. – *Perca fluviatilis*; Sil. gl. – *Silurus glanis*; St. l. – *Stizostedion lucioperca*; L.c. – *Leuciscus cephalus*.

gists/naturalists, who are against destruction of the environment. It is also possible that this cascade could be left unfinished, as happened to the one planned in the 1970s. The only effect then would be damage to the environment.

29.6 CONCLUSIONS

There is very rich vegetation and many types of water bodies and swamps between the levees, along almost the whole length of the Vistula. In addition there are a large number of forest areas (as well as many smaller ones in-between) outside the levees, which provide an ecological corridor from north to south. This forms an important element in Europe's environmental protection network. Plankton, benthos and fish are abundant and relatively diverse, despite heavy pollution of the river. The Vistula River and its valley are especially important for breeding and overwintering birds, and are a corridor for their migrations from northern Europe towards the Black Sea and Mediterranean Sea.

It is highly desirable and urgent for the protection of the environment of the Vistula River and its valley, to:

1. create buffer zones of permanent plant cover along all tributaries for accumulating much of the particulate and dissolved substances before they reach open waters (the Vistula itself has already got such belts between the levees);
2. retain more water in the drainage area by more afforestation, patches of permanent plant cover (trees, shrubs and meadows) and appropriate agrotechnical measures in rural landscape, to reduce erosion and to stimulate recycling of waste matter on land (this measure requires environmental education and training of relevant workers);
3. halt the construction of more dams and water regulation schemes;
4. increase, where possible, the distance between the levees, create polders (outside levees) and maintain the contact of levees and various water bodies with the river;
5. conserve islands (including temporary, bare, sandy ones) in the river bed and, wherever possible, the plant cover on the islands between levees and outside them;
6. purify sewage from point sources.

Most of the points above would need some legal decisions, for example on the creation of a Vistula landscape park at the most valuable and least degraded middle part of the valley from Sandomierz to Płock (with the possible inclusion of some elements of the Ramsar Convention, which deals with aquatic and wetland environments). Some of the most valuable sections of the valley, from an environmental conservation aspect, should be nominated a National Park. Recreational centres could be created along the middle part of the Vistula, without jeopardizing the aims of the landscape park.

The recreational advantages of the Vistula valley arise from its rich and diverse environment and also from its convenient central location. By contrast, the other main recreational areas are the mountains in the very south, or the Baltic Sea and lakelands in the north.

If this ambitious programme could not be achieved, then a minimal measure should be accepted, that is leaving the river (at least the middle reach) and its valley as it is now, with no dams or water regulation schemes.

Saving the river and its valley as it is, with no more damage, would be an important investment for the future of Poland and Europe. The value of such environments with their biodiversity will undoubtedly be appreciated increasingly in the future.

REFERENCES

1. Augustowski, B. (ed.) (1982). The valley of the lower Vistula (in Polish, English summaries). Ossolineum, Wrocław.
2. Kajak, Z. (1992). The Vistula River and its floodplain valley (Poland) – its ecology and importance for conservation, in *River Conservation and Management* (eds P. Boon, G. Petts and P. Calow), J. Wiley and Sons Ltd, Chichester, p. 35–49.
3. Kajak, Z. (1993). The Vistula river and its riparian zones, *Hydrobiologia* 251, 149–57.
4. Dojlido, J. and Woyciechowska, J. (1985). Changes in the quality of surface waters in Poland during the last half century (in Polish, English summary). Gospodarka Wodna 45, 39–44.
5. Dojlido, J. and Woyciechowska, J. (1989). Water quality classification of the Vistula River basin in 1987. *Ekol. pol.* 37, 405–17.
6. Kajak, Z. (ed.) (1983). *Ecological Basis for the management of the Vistula and its Drainage Basin* (in Polish, English summaries). Polish Scientific Publishers (PWN), Warsaw – Łódz.
7. Dynowska, I. and Maciejewski, M. (1991). *Upper Vistula Drainage Basin* (in Polish). Polish Scientific Publishers (PWN), Warsaw – Kraków, Parts I and II.
8. Kajak, Z. (1979). *Eutrophication of Lakes* (in Polish). PWN, Warsaw.
9. Taylor, R. (1987). Factors forming the level and dynamics of nutrient runoff from an agricultural watershed. *Pol. Arch. Hydrobiol.* 34, 285–95.
10. Pawlik-Dobrowolski, J. (1990). Protection of waters against agricultural non-point pollution in *Proc. of Sem. 27*, (in Polish), Institute of Land Reclamation and Grassland Farming (IMVZ), Falenty.
11. Melčanov, B.A. (1988). Determination of the forest zone width for the water runoff purification at the river valley. *Water Resources* 1, 157–161.
12. Nikitina, A.P. and Spirina, A.T. (1989). Protection of surface waters by forest plantations against fertilizers and toxins (in Russian). *Vodnye resursy* 6, 104–9.
13. Pauliukievicius, G., Baconis, M. and Pakalnis, R. (1987). Water protecting afforestation (in Russian, English summary), in *Ecological optimization of agrolandscape*, Nauka. Moscow pp. 207–10.
14. Backiel, T. (1983). Fisheries and fishes of the Vistula River (in Polish, English summary), in *Ecological Basis for Management of the Vistula and its Drainage Basin* (ed. Z. Kajak), PWN, Warsaw – Łódz, pp. 511–43.
15. Backiel, T. and Penczak, T. (1989). The fish and fisheries in the Vistula River and its tributary, the Pilica River, *Can. J. Fish. Aquat. Sci.*, 106, 488–503.
16. Giziński, A., Błędzki, L. Kentzer, A., *et al.* (1989). Hydrobiological characteristics of the lowland, rheolimnic Włocławek Reservoir on the Vistula River. *Ekol. pol.* 37, 359–404.
17. Klimowicz, H. (1981). The plankton of the River Vistula in the region of Warsaw in the years 1977–1979. *Acta hydrobiol.*, 23, 47–67.
18. Kowalczewski, A., Perłowska, M. and Przyłuska, J. (1985). Seston of the Warsaw reach of the Vistula River in 1982 and 1983: 3. Phyto- and zooplankton, *Ekol. pol.* 33, 439–54.
19. Nowicki W. and Kot, H. (1993). Avifauna of the middle Vistula and its main tributaries – unique features and conditions for their preservation, in *Nature and Environment Conservation in the Lowland River Valleys of Poland* (in Polish, English summary), Institute of Environmental Protection, Polish Academy of Science, Kraków, pp. 81–96.
20. Pinowski, J. and Wesołowski, T. (1983). Influence of Vistula River regulation on avifauna, in *Ecological basis for management of the Vistula and*

its drainage basin (ed. Z. Kajak), PWN, Warsaw – Łódz, pp. 543–68.
21. Tomiałojć, L., Dyrcz, A. (1993). Importance of great rivers and their valleys for nature in Poland in the light of ornithological research (in Polish, English summary), in *Nature and Environment Conservation in the Lowland River Valleys of Poland* (ed. L. Tomiałojć), Institute of Environmental Protection, Polish Academy of Science, Kraków, pp. 13–38.
22. Kownacki, A. (1988). A regulated river ecosystem in a polluted section of the upper Vistula: 10. General considerations. *Acta hydrobiol.*, 30, 113–23.
23. Backiel, T. (1985). Fall of migratory fish populations and changes in commercial fisheries in impounded rivers in Poland, in *Habitat Modification and Freshwater Fisheries* (ed. J.S. Alabaster), Butterworth, London, pp. 28–41.

BIOLOGICAL BARRIERS IN WASTEWATER TREATMENT

by I. Kulik-Kuziemska and K. Czerwionka

30.1 INTRODUCTION

The occurrence of periphytonic algae on the overflow of secondary settling tanks and in ditches draining purified wastes into the receiving water is a common phenomenon in municipal wastewater treatment plants. This fact indicates that these facilities create the specific hypertrophic system with good conditions for algae development. The treated wastes are the source of nutrient substances. The open settling tanks provide free access to light and the wastewater flow ensures water movement.

In a natural environment, the following factors determine the periphyton development: light intensity, temperature, water current, presence of micro- and macro-nutrient components as well as bed characteristics and the occurrence of natural animal predators.

The flow velocity is a particularly important factor in the development of periphytonic algae in watercourses. It determines the rate of growth of the biomass as well as its access to the nutrient substances [1, 2]. A weak current facilitates the colonization of algae on the bottom sediments and on objects submerged in water. A velocity in the range of 0 to 0.5 m/s, causes an increase in biomass development but more intensive growth is observed in the upper range [1, 3]. The growth of periphyton biomass is limited if a velocity of 0.5–0.6 m/s is exceeded. This is associated with increasing friction and detachment of biomass fragments from the algae mat.

The second important factor related to the flow velocity is the availability of nutrients. In an aquatic environment, the water layer flowing round the algae is the source of nitrogen and phosphorus for the periphyton. In low flow conditions, a decrease in the supply of nutrients has been observed. The increase in water flow ensures the continuous inflow of nutrients to the surface layer of the algae mat and also accelerates the diffusion of nitrogen and phosphorus to its deeper layers. The rate of nutrient diffusion deep into the algae mat depends on the concentration of dissolved reactive phosphorus and on the flow velocity [2]. The thickness of the mat is the limiting factor for diffusion. An increased flow velocity limits the thickness of the mat. At the same time, the depth of the laminar boundary layer to which the nutrients are diffusing is reduced. Stevenson [4] and Sand Jenson [5] each reported that the water flow in the boundary layer of algae mat is almost completely reduced, and therefore the access to nutrients is greatly limited. The light penetration into the internal layers of the mat is also limited. As a result, the process of photosynthesis does not proceed in these layers, although respiration does. This leads to maceration (the process of destruction) of the laminar boundary layer of periphyton. The ratio of nitrogen to phosphorus in the water environment is also a limiting factor for the rate of nutrient substances uptake. The depletion of one nutrient substance results in a limitation of uptake of the other.

International River Water Quality. Edited by Gerry Best, Teresa Bogacka and Elżbieta Niemirycz. Published in 1997 by E & FN Spon, London, ISBN 0419215409

The photic and temperature optima for periphyton algae are similar to that for microalgae [6]. The increase in light intensity accelerates the biomass development while the opacity limits the processes of photosynthesis and also can reduce the variety of species. The limiting temperature range for most algae is above 40°C (denaturation of protein) and below 4–6°C (inhibition of enzymatic processes).

In the literature, there appears to be no information on investigations into using periphytonic algae for the removal of nitrogen and phosphorus from wastewater. However, these organisms can develop perfectly well in the oxygen stabilizing pond as well as on some treatment plant structures. One probable reason for the lack of interest in this form of algae is that only the external layer of periphyton, which has a direct contact with water flowing around it, constitutes the active part.

The aim of these studies was to improve the efficiency of periphyton at removing biogenic substances from wastewater through an enlargement of the active sorptive surface.

30.2 ASSUMPTIONS

It was assumed that the active biosorption of nitrogen and phosphorus occurs in the whole biomass of periphyton. The condition mentioned above is achieved if the periphyton algae colonize those surfaces over which the wastewater flows. This phenomenon takes place when the algae are allowed to develop on screens and this results in the formation of biological barriers [7].

30.3 MATERIALS AND METHODS

The channel stream used in the laboratory had the dimensions 600 × 150 × 120 mm and was made from organic glass (Figure 30.1). The active volume in the vessel when filled to 100 mm was 9 l. Three polyethylene screens 8 mm thick were installed along the longitudinal axis of the reactor and inclined to the bottom surface at an angle of 60°. The supply wastewater was introduced into the reactor by peristaltic pumps. After treatment, the wastewater flowed into a vertical settling tank.

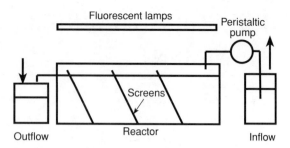

Figure 30.1 Schematic diagram of algae reactor with periphyton screens.

At the beginning of the experiment, the reactor was filled with a flocculated suspension of periphyton algae (concentration of 300 mg/l of dry algae mass) and with wastewater from the stabilizing pond at the wastewater treatment plant in Swarzewo, Poland. To ensure rapid colonization of the algae on the screens, the contents of the reactor was recirculated in closed circuit until the algae flocs disappeared from the effluent.

30.3.1 PERIPHYTONIC ALGAE

Periphyton was collected from the oxygen stabilizing pond incorporated in the municipal wastewater treatment plant in Swarzewo, Poland. The dominant periphyton were the filamentous forms of *Tribonema, Stigeoclonium, Phormidium* and *Oscillatoria* as well as the massive forms of *Pleurocapsa*. Moreover, the microalgae such as *Scenedesmus, Chlorella, Ankistrodesmus* and some species of diatoms also occurred. Before the colonization on the screens began in the reactor, the algae were kept in the wastes from the pond. The wastes were aerated with coarse bubble diffusers until the flocculated suspension was obtained.

30.3.2 WASTEWATER

Experiments were carried out on the biologically treated wastewater (activated sludge process) which came from the wastewater treatment plant in Swarzewo, Poland. As shown below, the chemical composition of the wastewater varied greatly during the course of the experiments:

- BOD_5 from 1.9 to 13.2 mg/l;
- COD from 20.8 to 35.8 mg/l;
- suspended solids from 2.4 to 34.4 mg/l;
- ammonium–nitrogen from 0.1 to 30.4 mg/l;
- phosphates from 14.1 to 32.7 mg/l.

In the laboratory the wastewaters were kept at 4°C. The analyses were performed according to Polish standards.

30.3.3 EXPERIMENT DETAILS

Fluorescent lamps, 'Flora' type, were used as the source of light for the periphyton. During the day the reactor was illuminated with a light intensity of 300 lx. The flow velocity of wastewater through the reactor was 5.2×10^{-4} m/s, which corresponded to a 48-hour retention time. The parameters investigated were: concentrations of phosphates, ammonium–nitrogen, dissolved oxygen and the pH of the inflow and outflow of the reactor. The measurements were made every 48 hours. The analyses were undertaken according to Polish standards.

30.4 RESULTS

30.5.1 PERIPHYTON

The period of complete periphyton settlement on the screens was 10–12 days. During the total experimental period changes in the periphyton community were noted. In the first period (colonization) no qualitative changes were observed. After two weeks of experiments *Tribonema* disappeared from the periphyton community. *Stigeoclonium tennue*, *Oscillatoria limosa* and *Pleurocapsa minor* remained dominant. However, later, there was a rapid development of the diatoms with the species *Gomphonema olivaceum* dominating.

30.4.2 PHOSPHATE UPTAKE

The results of phosphate content are shown in Figure 30.2. Two phases of phosphate removal in the reactor were observed. The first one occurred during the period of colonization of the periphyton. The average phosphate concentration in the reactor influent was 24 mg/l. This decreased to about 7.25 mg/l in the effluent, a 63%–73% reduction. In the second phase (periphyton stabilization) the average phosphate concentration in the influent was similar to that of the first phase, but the average residual phosphate concentration in the effluent was only 1.1 mg/l. The phosphate reduction efficiency was as high as 94.7%.

30.4.3 AMMONIUM-NITROGEN UPTAKE

During the whole experimental period, the quantity of ammonium-nitrogen absorption was maintained at a constant level (Figure 30.3). The average ammonium-nitrogen concentrations in the reactor influent and effluent were 19.17 mg/l and 0.38 mg/l respectively. This represents a 98% reduction of this nutrient.

30.4.4 OXYGEN CONCENTRATION IN THE REACTOR

The periphyton development in the reactor had a positive effect on the wastewater oxidation. The average concentration of dissolved oxygen in the wastewater feeding the reactor was 5.8 mg/l, but the concentration in the effluent was almost double at about 10.8 mg/l. As a result of temperature fluctuations from 22 to 32 °C during the whole experimental period, the effluent was greatly supersaturated with oxygen (Figure 30.4).

30.4.5 pH

The pH values of the reactor influent and effluent fluctuated insignificantly. The average difference between these values was 0.7.

30.5 DISCUSSION

Macrophytes submerged in rivers constitute a huge surface area that is colonized by microphytes and invertebrate organisms, creating biological membranes. The nutrients are taken from the flowing water and self-purification of the watercourse is promoted. In highly degraded rivers, where submerged vegetation has been damaged, artificial barriers (e.g. screens) with large surface areas,

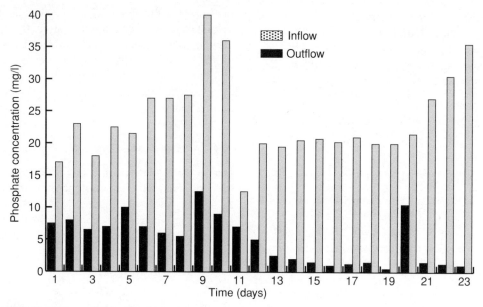

Figure 30.2 Phosphate reduction on the periphyton screens.

have been installed. These barriers enable the colonization of biological membranes, which actively participate in the self-purification process in the water [7]. Periphytonic algae develop in eutrophic waters which are rich in phosphates and mineralized nitrogen. The algae compete with phytoplankton for access to light and nutrients [8]. The periphytonic algae also occur in stabilizing ponds and on structures in wastewater treatment plants [9–13].

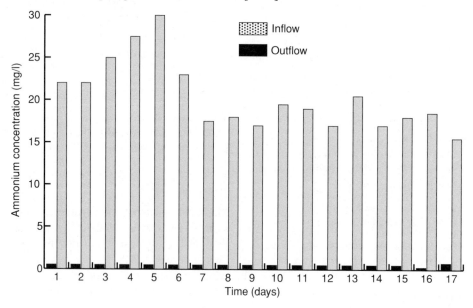

Figure 30.3 Ammonia reduction on the periphyton screens.

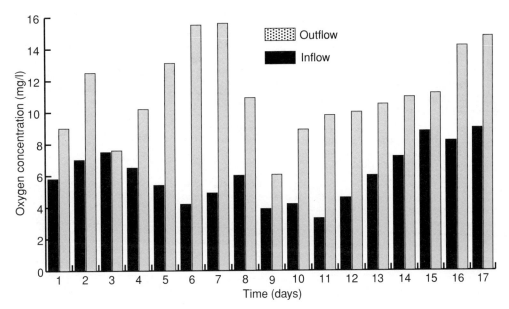

Figure 30.4 Dissolved oxygen concentrations in the reactor.

In the studies of the sorption capacity of periphyton for phosphates and ammonium–nitrogen from the biologically treated wastes, artificial barriers coated with algae were used in a flow model reactor. The ammonium–nitrogen and phosphates were reduced from 19.17 to 0.38 mg/l and from 24 to 1.1 mg/l respectively; i.e. the reduction was very high.

These results were obtained for a retention time of 48 hours in the reactor. The flow of a thin wastewater layer through the whole surface of the barrier is essential for the development of the periphyton barrier on the screens. It enables the active sorption of nutrients by algae.

In practice, use of the periphyton barrier method should be limited to small wastewater treatment plants with low flows. The necessary retention time can be controlled by adjusting the flow velocity and the number of periphyton barriers, as well as by the amount of available light and temperature. In every algae wastewater treatment process, the ratio of nitrogen to phosphorus is very important and should be about 20/1 and not less than 7/1 because it ensures a high reduction of the nutrients. In this, as with other processes, supervision of trained personnel is necessary. At high temperatures the intensive periphyton development can block the barrier and make the flow-through difficult, eventually resulting in blockage of the effluent channel. When this occurs, the barrier which is overgrown with algae can easily be exchanged.

REFERENCES

1. Horner, R.R. and Welch, E.B. (1981) Stream periphyton development in relation to current velocity and nutrients, *Can. J. Fish. Aqua. Sci.* **38**, 449–57.
2. Horner, R.R., Welch, E.B., Seely, M.R. and Jacoby, J.M. (1990) Responses of periphyton to changes in current velocity, suspended sediment and phosphorus concentrations, *Freshwater Biology*, **24**, 215–32.
3. McIntire, C.D. (1966) Some effects of current velocity on periphyton communities in laboratory streams, *Hydrobiologia*, **45**, 559–70.
4. Stevenson, R.J. (1983) Effects of current and conditions stimulating autogenetically changing microhabitats on benthic diatom migration, *Ecology*, **64**, 1514–24.
5. Sand-Jensen, K.A. (1983) Physical and chemical parameters regulating growth of periphytonic communities, in *Periphyton of Freshwater Ecosystems* (ed. R.G. Wetzel), pp. 63–71, Dr W. Junk Publisher, The Hague.

6. Kulik-Kuziemska, J. and Czerwionka, K. (1995) Algal system in tertiary wastewater treatment. Symposium of Biological wastewater treatment in Poland, May 1995, Kielce.
7. Szlauer, L. and Świerczyńska, I. (1988) The role of artificial substrate in a polluted stream. Scientific papers of Agricultural Academy in Szczecin, Department of Fisheries, Oceanography and Ocean Protection.
8. Hansson, L.A. (1988) The effects of competitive interaction of the biomass development of plankonic and periphytonic algae in lakes, *Limnology and Oceanography*, **33**, 121–8.
9. Davis, L.S., Hoffman, J.P. and Cook, P.W. (1990): Seasonal succession of algal periphyton from a wastewater treatment facility, *J. Phycol.*, **26**, 611–17.
10. Davies, L.S., Hoffman, J.P. and Cook, P.W. (1990) Production and nutrient accumulation by periphyton in a wastewater treatment facility, *J. Phycol.* **26**, 617–23.
11. Otake, H., Aiba, S. and Sudo, R. (1978) Growth and detachment of periphyton in an effluent from secondary treatment plants of wastewaters, *Jap. J. Limnol.*, **39** 163–9.
12. Traaen, T.S. (1975) Biological effects of primary, secondary and tertiary sewage treatment in lotic analog recipients, *Verh. Int. Ver. Limnal.*, **19**, 2064–9.
13. Wood, A. (1987) A simple wastewater treatment system incorporating the selective cultivation of a filamentous algae, *Wat. Sci. Tech.*, **19**, 1251–4.

HYDROBOTANICAL SYSTEMS: CHARACTERISTICS AND EXAMPLES IN POLAND

by H. Obarska-Pempkowiak

31.1 INTRODUCTION

About 20 years ago, a rapid deterioration in the quality of sea water occurred in the Bay of Gdansk and the Vistula Lagoon. This was attributed to insufficiently purified wastewater, polluted rivers which received inadequately purified sewage (usually only primary treatment) and pollutants in surface water run-off. The results of this deterioration were especially evident in the case of surface, near-shore waters. A review of the temporal changes in the percentage of rivers of various classes of water quality – using an *E. coli* index – showed that the most dramatic situation occurred in the 1970s; there are now only 21% of rivers classified as class I or II compared with 75% in 1970. When chemical parameters are applied to the classification the situation is as bad. At present there are no rivers in the Gdansk region with water fulfilling the criteria for class I (Chapter 14 by Kowalik and Obarska-Pempkowiak).

Small rural purification plants have difficulty in meeting the requirements of the Ministry of Environmental Protection, Natural Resources and Forestry (Act No. 503, 5 November 1991). According to these regulations, the limits for nutrients are: 30 mg/l nitrogen and 5 mg/l phosphorus in sewage discharged to rivers. The limit for phosphorus in sewage discharged to sea is 1.5 mg/l.

Hydrobotanical methods are thought to be a solution to these problems because they are inexpensive (no skilled personnel are needed), safe for the environment and suitable for very small communities. For these reasons and because of tough regulations against nutrients in sewage discharged to surface waters, there is much interest in ecological methods of purification in Poland. Of course the technique also has disadvantages:

- two to three years' construction time needed;
- large land area required per population equivalent (PE);
- low efficiency in winter (in periods when temperature is less than 5°C).

Various hydrobotanical systems are presently being tested in Poland.

31.2 CHARACTERISTICS OF HYDROBOTANICAL SYSTEMS

The elimination of organic matter and nutrients in hydrobotanical systems occurs because of the specific redox conditions and sorption on particles of submerged soil. The processes depend on the quantity and quality of micro-organisms and plants inhabiting the system.

Hydrobotanical systems are divided into two groups: natural wetlands and artificial wetlands.

258 *Hydrobotanical systems in Poland*

Wetlands are areas in which the water-table is at or above the ground level long enough each year to maintain saturated conditions and the growth of suitable vegetation. Most often wetland vegetation consists of reeds, sedges, cat-tails, bulrush and willow. The vegetation contributes to:

- the transport of oxygen to the rhizosphere;
- the stabilization of the hydraulic conductivity of the soil;
- the support for the growth of micro-organisms which enables the degradation of organic matter and the growth of nitrifying bacteria.

The two main types of artificial wetlands are:

- Flow Water System (FWS), where water flows on the surface. This system is similar to a wastewater pond.
- Vegetated Submerged Bed (VSB). In contrast to FWS systems, water flows laterally through the medium (native soil, sand or gravel) and the root zone of vegetation. This system is similar to a flooded filter.

There are several examples of hydrobotanical systems in operation in Poland. The performance and efficiency of pollutant removal of some of them are presented in this chapter.

31.3 SECONDARY WASTEWATER TREATMENT IN FROMBORK USING NATURAL WETLAND

Municipal wastewater from the town of Frombork is collected in a sewer and brought to an Imhoff tank situated on the outskirts of the town some 450 m from the shore of the Vistula Lagoon (Figure 31.1). The flow of sewage averages 850 m^3/day (about 2600 PE), and the retention time in the tank is some two hours. The sewage is largely of domestic origin, characterized by concentrations of organic substances and nutrients in the ranges: BOD_5 from 147 to 264 mg/l; COD from 388 to 459 mg/l; suspended solids from 201 to 513 mg/l; and pH from 6.8 to 8. The removal of organic substances from sewage in the tank is in the range 20% to 30%.

In 1985 a secondary biological treatment stage was introduced in the form of a natural wetland. Part of the coast of the Vistula Lagoon is naturally overgrown with dense common reed. An area of these reeds was separated with artificial embankments and used as the natural wetland. Sewage is drained to the field through a 200 mm diameter pipe (Figure 31.1). After passing through the reed bed, the sewage is directed to the Vistula Lagoon via a ditch running along the north-east boundary of the field. The total area separated with embankments is about 10 ha, but sewage spreads over just 2.2 ha.

In the first years of operation the system was found to reduce concentration of organic substances by 70%. Unfortunately at that time there were no measurements taken of nitrogen and phosphorus. During 1990, however, nitrogen and phosphorus were measured in order to establish the performance of the system regarding nutrient removal. The average monthly concentrations of the nutrients in inflowing and outflowing sewage are presented in Figure 31.2. The permissible concentrations of contaminants in class I, II and III waters, according to Polish standards, are also shown. Total nitrogen is the sum of ammonium nitrogen, organic nitrogen, nitrates and nitrites.

The system is characterized by a very poor annual removal of nitrogen and phosphorus. The removal of nitrogen is about 26% whilst for phosphorus it is only 3%. The poor performance is caused by overloading of the system, i.e. applying too much sewage to too small an area (about 8.5 m^2/PE), and uneven spreading of sewage [1].

31.4 TERTIARY WASTEWATER TREATMENT IN WIEŻYCA USING ARTIFICIAL WETLAND

The wastewater treatment system in Wieżyca was constructed to service several recreational centres located in the Kaszubian Lake District, in the upper part of the Radunia River catchment area. The Radunia River is the source of drinking water for the city of Gdansk (Chapter 13 by Walkowiak and Korzec) and, because of this, more stringent requirements apply to effluent from wastewater treatment plants in this region. The artificial wetland is a pond inhabited by emergent hydrophytes (mainly reeds). Sewage is directed to the pond after

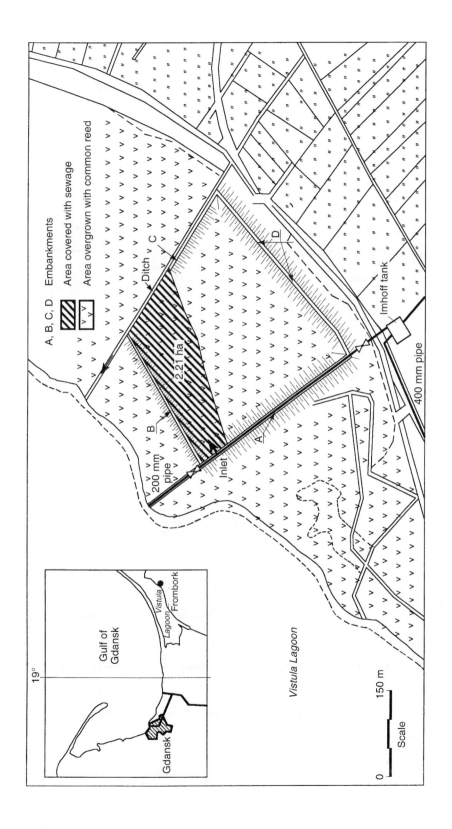

Figure 31.1 Location of natural wetland in Frombork.

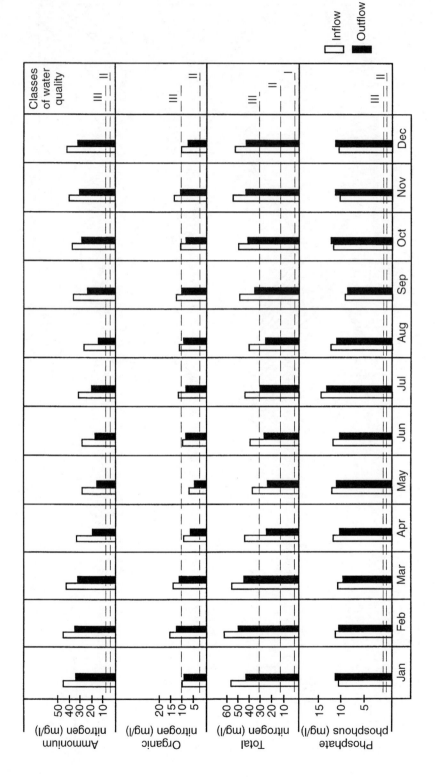

Figure 31.2 Average monthly concentrations of nutrients (N and P) in inflowing and outflowing sewage.

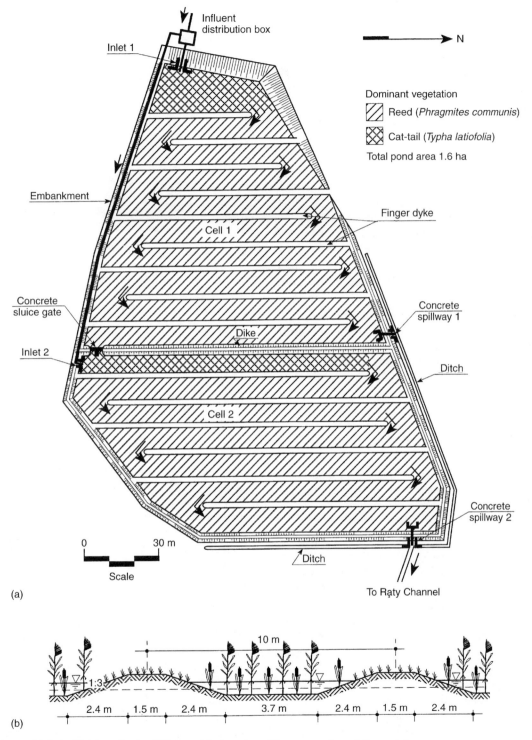

Figure 31.3 Artificial wetland at Wieżyca (a) plan view; (b) cross-section of finger dykes.

mechanical and biological treatment. Biological treatment is carried out in a package treatment plant based on rotary filters with primary aeration and sewage recirculation.

The pond consists of several inner dykes, which cause the wastewater to flow along ditches and ensures a long retention time – about 49 days (Figure 31.3). The surface area of the pond is about 2 ha, which is about 20 m²/PE; the water depth is 0.3 m and the hydraulic loading rate is about 0.015 m/day. The average flow in the holiday season is 120 m³/day, however it fluctuates over a wide range. Wastewater flowing into the pond is periodically much more concentrated due to troubles with starting up and operating the secondary treatment [2]. The pond inflow after secondary treatment is characterized by: suspended solids – 25 to 40 mg/l; BOD – 30 mg/l; ammonium nitrogen – 8 to 12 mg/l; nitrate nitrogen – 10 to 23 mg/l; total nitrogen – 30 mg/l; and phosphate – 10 mg/l [2].

The results of tertiary treatment using the artificial wetland are given in Table 31.1. The data show that this hydrobotanical system can ensure efficient removal of organic substances (as measured by BOD_5 and SS).

The removal of nutrients is 85% and 82% respectively for total nitrogen and phosphates. Despite this, the concentrations of phosphates in the effluent exceed the permissible levels. This is caused largely by inadequate purification of sewage during secondary (biological) treatment.

31.5 HYDROBOTANICAL SYSTEMS AT OLIWA ZOO

The main point sources of pollution to the Bay of Gdansk are streams. Between Gdynia and Gdansk there are about 20 streams, and one of the major ones is the Jelitkowski Stream. The main tributary of the Jelitkowski Stream is the Rynaszewski Stream whose middle reach passes through Oliwa Zoo. Figure 31.4 shows the study area with the main sources of pollution marked.

Measurements of concentrations and loads of pollutants between 1989 and 1991 show that there was a substantial increase in pollution in the Oliwa Zoo part of the stream, mostly from organic nitrogen and *E. coli* bacteria (Table 31.2).

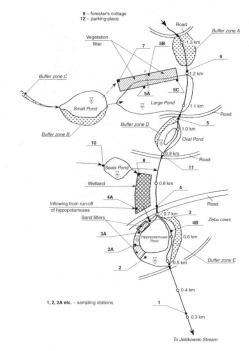

Figure 31.4 Location and general layout of study area

An additional investigation at the zoo (Figure 31.5) established the location of the sources of pollution and estimates of their respective loads of organic nitrogen (Table 31.3).

The following areas were considered of particular interest: Large Pond, Small Pond, Oval Pond, Seals Pond, Hippopotamus Pond and run-off from animals located along the stream [3]. Table 31.4 gives the annual average concentrations of organic substances and nutrients in the inflow and outflow of Oliwa Zoo in 1991. Based on the findings, the following approach was adopted for protecting the stream from pollution originating in the zoo.

- Pollutants – wastes with organic nitrogen should be retained as close to their source as possible; this would prevent dilution which makes their removal much more difficult;
- The retention times of wastewater should be extended; this is because, although the wastewater has a low BOD_5, it still contains high concentrations of nitrogenous organic matter which is biochemically relatively stable;

Table 31.1 Results of tertiary treatment using artificial wetland at Wieżyca

Date	Suspended solids mg/l (% removed)	BOD$_5$ mg/l (% removed)	Ammonium nitrogen mg/l (% removed)	Nitrite nitrogen mg/l (% removed)	Nitrate nitrogen mg/l (% removed)	Total nitrogen mg/l (% removed)	Phosphate mg/l (% removed)
Influent							
26 Aug 1993	48	35.2	18.2	2.4	13	32.9	15.2
29 Sep 1993	–	325.7	1.4	0.07	0.85	21.2	10.3
29 Oct 1993	520	315.7	8.12	0.184	1.28	50.38	17.5
08 Feb 1994	14		5	0.49	5.6	22.16	8.5
Influent Mean value	25–40	30	8–12		10–23	30	10
Effluent							
26 Aug 1993	4.0 (91)	9.2 (74)	0.31 (98)	0.05 (98)	1.9 (–46)	3.79 (84)	2 (87)
29 Sep 1993	–	19.1 (94)	0.19 (86)	0.003 (96)	0.65 (23)	2.5 (88)	0.28 (97)
29 Oct 1993	22 (95)	1.38 (99)	4.8 (41)	0.0014 (92)	0.55 (57)	7.41 (85)	trace (99)
08 Feb 1994	4 (71)	–	0.5 (90)	0.03 (94)	1.7 (69)	3.8 (83)	4.5 (47)
Effluent Mean value	10 (85)	9.89 (89)	1.45 (78)	0.097 (95)	1.2 (26)	4.37 (85)	1.69 (82)
Class II water quality requirements (Act No. 503, 5 Nov 1991)	≤30	≤8	≤3	≤0.03	≤7	≤10	≤0.6

- the treatment of wastewater coming from various sources should be carried out in such a way that the quality of the landscape is retained.

The implementation of these principles was carried out in the following way. In order to reduce the loads of pollution from point sources, artificial wetlands (natural vegetation: alder tress, willow, reeds), sand filters and vegetation filters were constructed in the areas marked in Figure 31.5. In order to reduce loads originating from surface sources, a series of buffer zones containing willow were constructed at areas

Table 31.2 Average concentrations of organic substances, nutrients and *E. coli* in outflow from zoo in 1991 and comparison with regulations

Parameter	Outflow from the zoo	Classes of water quality		
		I (clean)	II (moderately clean)	III (polluted)
Total nitrogen (mg/l)	10.1	≤5	≤10	≤15
Ammonium nitrogen (mg/l)	1.6	≤1	≤3	≤6
Organic nitrogen (mg/l)	8			
Nitrate nitrogen (mg/l)	0.4	≤5	≤7	≤15
Nitrite nitrogen (mg/l)	trace	≤0.02	≤0.03	≤0.06
Phosphate (mg/l)	0.6	≤0.2	≤0.6	≤1
COD–Mn (mg/l)	10	≤10	≤20	≤30
BOD$_5$ (mg/l)	3.5	≤4	≤8	≤12
E. coli index	0.06	≥1	≥0.1	≥0.01

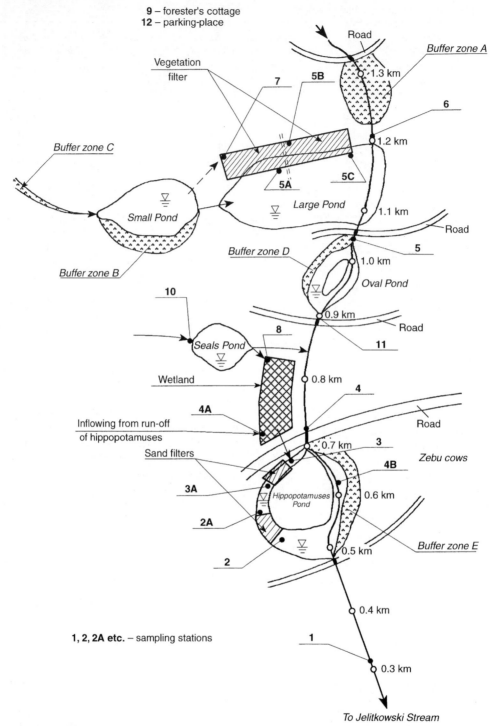

Figure 31.5 Configuration of hydrobotanical systems and location of sampling stations along the Rynaszewski Stream at Oliwa Zoo.

Table 31.3 Sources of pollution of Rynaszewski Stream at Oliwa Zoo.

Source of pollution	Location/distance from Jelitkowski Stream (km)	Estimated load of total nitrogen (kg/day)	Type of source
Run-off from deers	1.2	2.5	surface
Small Pond, waterfowl and run-off from goats	Inflow to Large Pond	6	point
Oval Pond	0.9	42	point
Pond and Seals Pond	Inflow to Seals Pond	5	point
Hippopotamus Pond	0.7	13.5	point
Run-off from Zebu cows	0.6	47	surface
Rynaszewski Stream outflowing from zoo	0.4	60.5	point

indicated in Figure 31.5. There are five buffer zones: A, B, C, D and E with a total area of 6650 m^2.

The measurements of water quality were carried out again in 1993–4 after construction of the hydrobotanical systems (Table 31.4). Assuming a comparable flow of water of 0.07 m^3/s in 1991 and in 1993–4, the load of organic nitrogen retained by the system is 46.6 kg/day (17 t/yr).

31.6 CONCLUSIONS

- Vegetation filters and buffer zones ensure the effective removal of organic substances and nutrients from sewage and river water;
- For ecological and chemical reasons hydrobotanical systems are attractive alternatives to traditional plants;
- A prerequisite for the efficient removal of pollutants is ensuring sufficient surface area per population equivalent.

REFERENCES

1. Obarska-Pempkowiak, H. (1991) Seasonal variations in the efficiency of nutrient removal from domestic effluent in a quasi-natural field of reed (*Phragmites communis*), in *Ecological Engineering for Wastewater Treatment* (eds. C. Etnier and B. Guterstam), Bokskogen, Sweden, Gothenburg, pp. 239–47.
2. Obarska-Pempkowiak, H., Mierzejewski, M. and Toczyowska, I. The application of surface flow

Table 31.4 Average annual concentrations of organic substances and nutrients in inflow and outflow at Olina Zoo

Parameter	1991 inflow (mg/l) / outflow (mg/l)	1993–4 inflow ± δ(mg/l) / outflow ± δ(mg/l)
Total nitrogen	3.1 / 10.1	2.1 ± 0.7 / 2.4 ± 0.6
Ammonium nitrogen	0 / 1.6	0.20 ± 0.08 / 0.09 ± 0.04
Organic nitrogen	3.1 / 8	1.26 ± 0.51 / 1.77 ± 0.72
Nitrate nitrogen	0 / 0.4	0.58 ± 0.21 / 0.53 ± 0.14
Nitrate nitrogen	trace / trace	0.013 ± 0.01 / 0.011 ± 0.01
Phosphate	0.42 / 0.6	0.21 ± 0.04 / 0.13 ± 0.03
COD – Mn	2.9 / 10	7.4 ± 1.8 / 4.8 ± 1.2
BOD$_5$	2.5 / 3.5	2.5 ± 1.2 / 2.8 ± 1.3
E-coli index	not determined / 0.06	not determined / 0.8

wetlands for the treatment of municipal wastewater – two full scale systems. Swedish – Polish Workshop on wastewater treatment by irrigated vegetation. Uppsala, Sweden, 9–10 June 1994 (in press).

3. Obarska-Pempkowiak, H. The application of willow and reed vegetation filters to the protection of a stream passing through a zoo. International conference on municipal wastewater and sludge purification using willow vegetation filters. Uppsala, Sweden, 5–8 June 1994 (in press).

A MULTIFUNCTIONAL WATER MANAGEMENT INFORMATION SYSTEM (MIS)

by W. Szczepański and W. Jarosiński

32.1 INTRODUCTION

Poland has considerable problems concerning threats to its environment, and pollution of water is one of the most important issues. Environmental degradation is particularly marked in the Upper Silesia region, mainly in Katowice Voivodeship. For example, Katowice Voivodeship is home to 22 of the 80 most noxious industries in Poland. The voivodeship is also inhabited by almost 4 million people, about 10% of the Polish population.

The range of water management problems can be described by both quantitative and qualitative parameters. For example, quantitative problems can be expressed in terms of the water resource index in m^3/year per capita as shown in Figure 32.1.

The qualitative problems in Upper Silesia can be summarized by the water quality classifications of the Odra and Vistula river basins in Katowice Voivodeship (Figure 32.2). Approximately 97% of the Odra and 87% of the water in the Vistula river basins are unclassified.

There are many factors which contribute to the present catastrophic condition of the quality of water resources in Upper Silesia. The most relevant are:

1. its geographical location on the upper catchment of the Odra and Vistula rivers – which have low flow rates in that area;

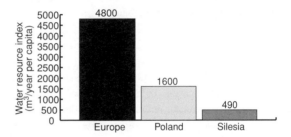

Figure 32.1 Water resource indices for Europe, Poland and Silesia.

2. overconcentration of heavy industry and industrial exploitation – 98% of Polish coal mining, 100% of Polish lead ore and zinc ore mining and processing;
3. overcrowding – 10% of the Polish population live in 2.1% of the total land area (over 600 inhabitants per square kilometre);
4. neglect of the fundamental requirements of water quality protection – 40% of wastewaters discharged to rivers untreated.

These conditions have resulted in various activities aimed at improving the water quality and the introduction of effective rules for water resources management in Katowice Voivodeship and the rest of Poland.

International River Water Quality. Edited by Gerry Best, Teresa Bogacka and Elżbieta Niemirycz. Published in 1997 by E & FN Spon, London, ISBN 0419215409

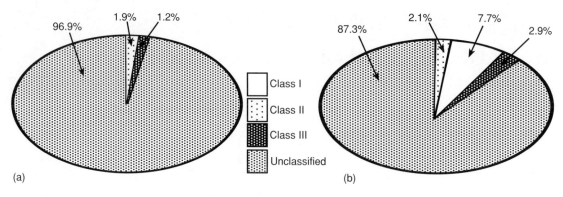

Figure 32.2 Water quality classifications of the (a) Odra and (b) Vistula river basins.

Efficient management is based on adequate information, coordination, decision-making and finance, so work on a computer-based multifunctional water management information system (MIS) has begun.

In the mid-1980s monitoring of water quality and the identification of the effects of pollution were relatively well developed, but there was a complete lack of monitoring of sources of water pollution. This lack of monitoring of sources contributed to the deterioration of water quality and resulted in a lack of investment in pollution control.

One of the main goals of the MIS was to create a fully pragmatic programme which could meet expectations of people and institutions responsible for policy-making and water resources management. The programme was intended to be complex, flexible and open to new events, tasks and requirements.

In order to prepare such a pragmatic programme, theoretical assessments were based on actual data such as the hydrographical network, engineering structures and all aspects of water resources consumption and water management. That was done initially in an area of 11.5×10^3 km^2 covering the Katowice Voivodeship – including water consumers and the neighbouring voivodeships because of the sources of water supply. Later, the programme was developed nationally via the seven hydrographical regions and their Regional Boards of Water Management.

It was possible for the work on MIS, particularly the part concerned with monitoring of point sources of pollution, to be enhanced and developed because of:

1. political and socio-economic transformation processes in Poland;
2. The activity of the National Inspectorate of Environmental Protection (PIOS) and national environmental monitoring, including water monitoring;
3. The involvement of water management in the process of transformation of the national economy;
4. Professional and financial support from the PIOS;
5. Successful co-operation with the Institute of Meteorology and Water Management – co-promoters of the programme.

The usefulness of the work on MIS was confirmed by a policy document of the Ministry of Environmental Protection, Natural Resources and Forestry [1], particularly paragraphs 11 and 12 concerning the economics of the ecological policy – in essence, this is summarized as 'the polluter pays'.

The timeliness of these activities has also been confirmed by the conclusions of the second United Nations Conference on the Environment and Development in Rio de Janeiro in 1992: first of all in the part concerning the guidelines for quality protection and distribution of water resources, and secondly in the use of the integrated approach to the problems of development, management and exploitation of water resources.

32.2 MANAGEMENT INFORMATION AVAILABLE

The computerized information system on the quality and quantity of water intakes and wastewater discharges has been implemented and tested for five years in Katowice and Bielsko voivodeships. Subsequently the MIS has been used and tested nationally with contributions from the Ministry of Environmental Protection, Natural Resources and Forestry and the 49 Voivodeship Inspectorates of Environmental Protection. The Regional Boards of Water Management and other institutions involved in water management have also helped.

Using the MIS, data on the quality and quantity of water intakes and wastewater discharges, previously measured to provide a large data set, can be called up and processed in various configurations. The range of information can be limited to basic data such as:

- identification of industrial plants;
- number and location of wastewater discharges and water intakes;
- lists of rivers. It can also be widened to include selected information such as:
- complete identification of engineering structures;
- amount and characteristics of discharged wastewater;
- licences;
- hydrographical network, and many others.

As far as catchment areas are concerned, data can be selected from:

- the whole catchment;
- a selected river;
- a given section of a river.

Taking into consideration the manner of exploitation of watercourses by industry, it is possible to specify:

- intakes and discharges together;
- intakes or discharges separately.

These specifications can additionally include data concerning:

- actual and permitted values together;
- actual or permitted values separately.

Information on discharged wastewaters can include:

- the total amount of the discharge;
- concentrations of selected pollutants – singly or in groups.

Using the MIS it is possible to:

- identify the industry responsible for a given type of pollution;
- identify the main sources of water pollution with respect to the catchment area and administrative units;
- analyse the impact of discharged wastewaters on river water quality;
- balance all the pollution components in a given area and define the location of treatment plants;
- obtain information on the types of water use by industry and identify large consumers;
- plot the whole or selected fragments of the hydrographical network;
- define the legal status of consumers' discharges;
- identify the legal regulations relating to water consumption, including issues resulting from the contravention of regulations and the collection of fines;
- acquire complete information on every industrial plant in the form of statistical information;
- assess the impact of new treatment facilities on river water quality using simulations;
- prepare statistical data on the main sources of pollution in the Vistula and the Odra catchment areas. Print-outs of basic and processed data can be obtained in textual and graphical forms.

32.3 INPUT DATA REQUIRED

At present the MIS comprises eight linked databases. These are:

- intakes and discharges (monitoring of sources)
- engineering structures (monitoring engineering structures)
- monitoring (a system of control–measurement stations)
- water-level indicators
- industries
- river network

- hydrographical networks
- water-level indicators and flows.

The first two databases contain the basic data. The remaining six databases are important because they provide the data for the first two.

32.3.1 INTAKES AND DISCHARGES

All quantitative and qualitative information required by the MIS has been gathered in the form of charts. Each chart corresponds to a particular water intake or wastewater discharge. Every industrial plant has a chart for each of its discharge and water intake points. The data entered into the MIS are almost identical to the data collected via a questionnaire filled in at the industrial plant during inspection; they include:

- the location of the intake or discharge in relation to measurement points and water-level indicators;
- administrative data;
- data on the industrial plant (analogous to the industries database);
- data on the official and legal status of the intake or discharge;
- data on the quantity and quality of abstracted water or discharged wastes according either to the licence or to the actual situation discovered during inspection;
- a description of the intake or discharge (including which side of the river it is located, its distance from the source, category of river, etc.);
- additional technical information on the water or wastewater treatment process(es).

The introduction of charts to the MIS is made easier by the presence of the additional databases mentioned above, from which certain data about water-level indicators, control-measurement points, industries and the river network are assigned to the correct positions.

Databases are linked via a number (REGON). As soon as REGON is called up, data are automatically taken from the industries database. By typing the name of the river and number of kilometres from the source, information about the class of the river, the measurement cross-sections and the water-level indicators are automatically displayed.

32.3.2 ENGINEERING STRUCTURES

Information about monitored engineering structures has been gathered in the form of charts engineering structures database. In these charts information on the objects is grouped into the following blocks:

- data on the company;
- location in the river basin (the side of the river and category of the river basin);
- the type and the description of the structure;
- anti-filtration devices;
- control–measurement devices;
- data on the design documentation;
- data on the maintenance of the structure;
- evaluation of the technical condition of the structure.

As in the intakes and discharges database, data on the authorities responsible for the engineering structures and the location of the structures in the river basin are automatically taken from the rivers, industries, monitoring and water-level indicators databases.

32.4 APPLICATIONS

A fully tested MIS will be valuable for the activities and guidelines of the national environmental monitoring programme. As soon as MIS has been implemented nationally it will enable rational management of all water resources and facilitate:

- systemic management of water resources with respect to river basin systems;
- rational decision-making on investments and water exploitation;
- optimization of decision-making in extreme conditions such as drought, floods and ecological calamities;
- determination of fees and fines for violation of standards;
- analysis of quality changes in the river when wastewater discharges exceed the prescribed limit or when a new wastewater treatment plant is opened.

Thus it will be possible for the system to be used widely by:

- the Ministry of Environmental Protection, Natural Resources and Forestry and other cen-

tralized units involved in water management for statistical purposes;
- specialist sections of the national administration dealing with water management;
- Regional Boards of Water Management;
- Voivodeship Inspectorates of Environmental Protection;
- Central Statistical Office;
- national environmental monitoring;
- design offices;
- universities and research institutes.

32.5 EVALUATION

In 1992, the MIS was evaluated by specialists who analysed existing information systems and those being developed for water management. The outcomes of this analysis were as follows.

- The range of information was critically evaluated to eliminate inputs of dubious value;
- It was relatively simple for a new user to work with the system and extract data;
- There was clear textual presentation of system output, which met the requirements of the user;
- The computer graphics allowed the presentation of spatial relationships as simplified schemes;
- Validation of the system is required for practical use.

REFERENCES

1. National Ecological Policy (1990) Ministry of Environmental Protection, Natural Resources and Forestry.

THE STATE, MANAGEMENT, USE AND PROTECTION OF NATURAL WATER RESOURCES IN THE PROVINCE OF GDANSK

by J. Błażejowski, D. Grodzicka-Kozak and M.S. Ostojski

33.1 DESCRIPTION OF GDANSK PROVINCE

The Province of Gdansk is centrally situated in the northern part of Poland (Figure 33.1). It occupies an area of 7394 km^2, which constitutes 2.4% of the territory of Poland. The province can be divided into four regions, namely: the Baltic Coast, Cashubian Lake District, Starogard Lake District and Vistula Fens (Figure 33.2). The province is home to 1.44 million people, 76% of whom live in towns and cities. Gdansk and Gdynia have over 100 000 inhabitants each. People are mostly employed in industry and agriculture, but tourism also provides many jobs.

Because of the relatively long coast line (199 km, of which 72 km comprise the Hel Peninsula) various elements connected with the maritime economy such as shipping, fishing, shipbuilding and port services have been developed. Other industries in Gdansk Province are power production, the manufacture of food, chemicals, timber and building materials, and electrochemical industries. Most of these industries are located in and around the Gdansk–Sopot–Gdynia (Three-City) agglomeration.

The Cashubian and Starogard Lake Districts are mostly rural areas. However, their picturesque landscapes and beautiful lakes and rivers create ideal conditions for tourism and recreation. Agriculture is the main occupation in the Gdansk lowlands. Fifty two percent of the territory of the Province of Gdansk is used for agriculture and 33% for forest production.

Figure 33.1 Location of Gdansk Province.

International River Water Quality. Edited by Gerry Best, Teresa Bogacka and Elżbieta Niemirycz. Published in 1997 by E & FN Spon, London, ISBN 0419215409

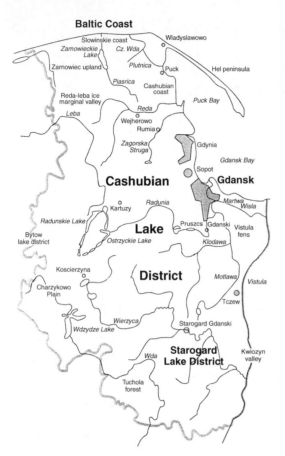

Figure 33.2 Geographical map of Gdansk Province.

The region is relatively rich in water. Inland surface water occupies 3.7% of the total area. The total length of rivers is around 1500 km, and the total area of lakes is 178.7 km^2. There are 591 lakes with an area greater than 1 ha, of which 29 are over 50 ha and are important reservoirs for the country. There are also smaller lakes that are of unique environmental value or are important for ecological and economic reasons.

33.2 STRATEGIC POLICY FOR WATER MANAGEMENT

The strategic policy for the management of water resources in Poland is conducted by Regional Water Management Boards. There are seven such Boards in the country, and they are central government institutions (Figure 33.6). The Regional Water Management Board in Gdansk controls the water resources of three provinces (Gdansk, Elbląg and Toruń) and partially controls a further six provinces (Słupsk, Bydgoszcz, Olsztyn, Włocławek, Ciechanów and Płock) (Figure 33.7). The Board's area also includes the Vistula (Wisła) River, downstream from the dam at Włocławek. The Regional Water Management Board in Gdansk controls an area of 36 300 km^2, which amounts to 11.6% of the territory of Poland. The number of inhabitants, approximately 3 946 000, is 10.3% of the population of the country. Further details are given in Table 33.1.

In general, Polish law provides for three ways of using natural water resources: general, common and special. According to general rights, people can take advantage of surface waters which are under government control, e.g. for recreational purposes. Citizens have the common right to use water from sources on their land to provide for their own needs. Within a given province, the Voivode has the authority to give so-called special rights, which enable natural water resources to be used as water supplies and the natural surface water to be used for wastewater disposal.

33.3 POLLUTION OF NATURAL WATER RESOURCES

33.3.1 WATER QUALITY OF RIVERS

The state of some of the rivers in the Province of Gdansk (Figure 33.3) has changed over the past ten years. The quality of water has deteriorated in the Motława, Wda, Łeba, Piásnica, Czarna Woda, Reda and Zagórska Struga rivers, whilst it has improved in the upper and middle reaches of the Radunia (water for the Gdansk agglomeration is supplied partly from this river). The quality of water in the remaining rivers of the province has not changed.

Five years of systematic monitoring (1988–93) of about 70% of the total length of rivers in the province revealed that there was none in class I, only 1.4% in class II, 44.1% in class III and 54.1%

Pollution of natural water resources 275

lower. This deterioration in the quality of water is caused mainly by an increase in bacteriological contamination and phosphorus content. The fact that the quality of water has decreased in the upper reaches of many rivers which were class I in the 1970s is of particular concern.

It should also be noted that the water of the Vistula (Wisła), along the 94 km of its lower reaches before it enters the estuary with the Baltic Sea, was so polluted that it is unclassified. In 1991, it was class III, but increased bacteriological and chemical (phosphorus and nitrate) pollution caused it to deteriorate.

33.3.2 WATER QUALITY OF LAKES

The monitoring of water quality in the period 1988–93 (Figure 33.5) revealed that 90.2% of the total area of lakes in the province were class II, 4.9% class III and 4.9% unclassified. The main sources of pollution were phosphorus and nitrogen compounds, while bacteriological contamination was rare. Most of the unclassified lakes were those into which wastewater was discharged in the past. Examples of these are the lakes around Kartuzy, which have received untreated municipal waste for many years. The state of water in these lakes has not improved despite elimination of the main sources of pollution three years ago and intense recultivation.

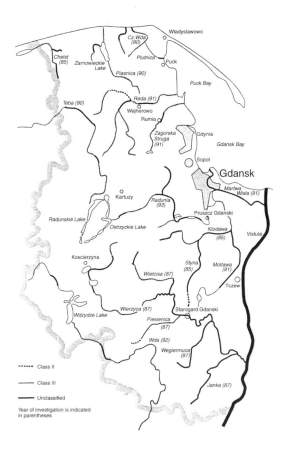

Figure 33.3 Rivers of Gdansk Province.

unclassified (Figure 33.4) [1]. The classification system used to assess water quality in Poland is described by Korol *et al.* in Chapter 22.

When the river classes are compared with those for earlier years, the data show that, in many cases, class I and II waters deteriorated to class III or even

33.3.3 SOURCES OF POLLUTION

The main sources of pollution in the Province of Gdansk are municipal and industrial wastewater. In 1992 the quantity of wastewater was 121.8×10^6 m^3. These discharges, treated and untreated, con-

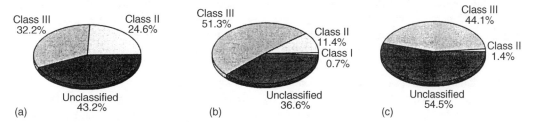

Figure 33.4 Quality of water in the rivers of Gdansk Province according to (a) physico-chemical criteria; (b) bacteriological criteria; (c) physico-chemical and bacteriological criteria.

Figure 33.5 Lakes of Gdansk Province.

the province was drawn up (eight of these are located in Gdansk, five in Gdynia and one each in Czarna Woda, Grabowo Kościerskie, Pelplin, Starogard Gdański, Strzebielino, Tczew and Władysławowo). Of this total of 22 industrial users 14 constitute a threat to inland and sea waters.

33.4 SEWAGE COLLECTION AND TREATMENT

Municipal wastewater from the major urban areas in the Province of Gdansk are collected in 16 sewerage systems. The catchment areas of these systems vary greatly. Some of them collect wastewater from small towns, others from several cities and towns, e.g. the system connected to Dębogórze sewage treatment plant in Gdynia collects wastewater from Gdynia, Reda, Rumia and Wejherowo, whilst the system connected to the sewage treatment plants in Gdansk (Zaspa and Wschód) collects wastewater from Gdansk, Sopot and Pruszcz Gdański.

In most of the treatment works, sewage is given only mechanical treatment before it is discharged into the surface water. Biological purification is provided at only a small number of treatment plants: these include Zaspa, Dębogórze, Chmielno, Jastarnia, Jastrzębia Góra, Kartuzy, Kościerzyna, Krokowa, Skarszewy, Somonino, Swarzewo, Tczew and Wdzydze. Although 15.3% of municipal wastewater is biologically treated, none of the sewage plants reduce biogenic substances to the level required by law.

The amount of industrial wastewater discharged directly to the environment was reduced from 19.4 $\times 10^6$ m^3 in the period 1988–92. This is equivalent to a reduction in the load of pollutants, measured as COD, from 5300 t to 3100 t. This improvement was achieved by connecting three large polluters to the central sewerage systems and sewage treatment plants, by the partial reuse of some wastewater (e.g. from sugarbeet refineries) and by reduction of biogenic substances in highly-polluted wastewater from four factories.

In the last few years new investments have been made. At Dębogórze sewage treatment plant a biological process has been introduced, at Kartuzy and Skarszewy sewage treatment plants the biological

tained pollutants with a total BOD equal to 32 000 tonnes O$_2$, discharged to either surface waters or Gdansk Bay. Up to 90% of the total load of organic pollutants is of municipal origin and only 10% comes from industry. The total load of pollutants in the municipal wastewater as COD was 29 700 tonnes. In the period 1988–92 the amount of wastewater produced increased by 3.9% [2].

In the late 1980s, 80 dischargers were identified and listed as posing particular threats to the environment of Poland. These included two industrial users in Gdansk, Siarkopol (Enterprise for the Treatment and Shipment of Chemical Minerals) and a phosphorus fertilizer plant, although Siarkopol has since been removed from the list. In addition, a list of the 20 most serious polluters in

Figure 33.6 Areas served by Poland's Water Management Boards.

stage has been extended, whilst in Chmielno, Jastrzębia Góra, Kościerzyna, Krokowa, Somonino, Tczew and Wdzydze completely new sewage plants have been constructed. Moreover, chemical treatment of sewage was installed at the

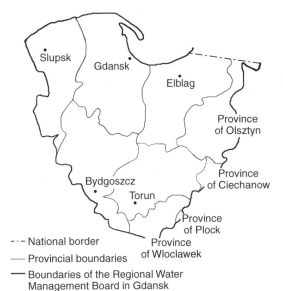

--- National border
— Provincial boundaries
— Boundaries of the Regional Water Management Board in Gdansk

Figure 33.7 Areas served by the Regional Water Management Board in Gdansk.

Wschód sewage plant, whilst the biological process at the Zaspa sewage treatment plant was modernized. The effects of all of these efforts should be seen in the coming years.

33.5 WATER SUPPLY

In cities and towns, the water for human needs is supplied through public water-pipe systems. Larger enterprises, heat and power plants and hospitals have their own water supply systems based on underground water intakes. About 20% of the water for the Gdansk agglomeration originates from the surface water intake on the Radunia River (The quality of this river is described by Walkowiak and Korzec in Chapter 13.) The remaining 80% of water is drawn from underground sources located in water-bearing strata at three levels: water bearing: Cretaceous (24%), Tertiary (3%) and Quaternary (53%) (5% is drawn from shallow drainage systems). In recent years, the quality of the underground water has deteriorated steadily, particularly in the Quaternary level, but also in the Tertiary and even deeper Cretaceous levels – the latter usually being considered the cleanest. Characteristic differences in water from the Quaternary level are increased salinity and concentrations of iron and manganese, and sometimes the presence of hydrogen sulphide. Polycyclic aromatic hydrocarbons (PAHs) have occasionally been detected in this water. More than half of the water originating from the Cretaceous level has excess colour, exhibits increased concentrations of fluoride and ammonia, and sometimes contains hydrogen sulphide. Water from surface intakes usually contains excessive organochlorine compounds, pesticides, heavy metals and PAHs.

Water from surface and shallow drainage intakes must be subjected to continuous disinfection to bacteriological contamination. On occasion during the last four years it has also been necessary to disinfect water from underground sources.

Water for human use in rural areas is supplied mainly from underground sources. Public water-pipe systems exist in only two communities and they are in a poor state. As water in rural areas is not purified before use, it is sometimes polluted with iron

Table 33.1 Statistical data on the area controlled by the Regional Water management Board in Gdansk

Province	Territory of the province included (%)	Number of cities and towns	Number of local communities	Number of inhabitants
Gdansk	100	19	43	1 431 600
Elblag	100	15	37	478 900
Toruń	99	13	40	652 300
Słupsk	66	8	35	277 800
Bydgoszcz	58	11	41	637 700
Olszłyn	37	4	15	275 700
Włocławek	37	5	16	160 700
Ciechanów	7		2	30 800
Płock	0.1		1	800

and manganese compounds and contains excessive fluorides, nitrates and occasionally bacteria.

33.6 RECENT ENVIRONMENTAL PROTECTION PROJECTS

Despite Poland's difficult economic situation, numerous investments directed towards protection of the environment have been undertaken in Gdansk Province in the past three years. These investments have been financed from the Voivode's budget, the Voivodeship Fund for Protection of the Environment, by private and state enterprises, and from the National Fund for Protection of the Environment (Figure 33.8). The latter is a government fund and gives preference to those investments which are of benefit to the whole country. This fund is contributing to the building and modernization of sewage plants in Gdynia (Dębogórze) and Gdansk (Wschód), and it is hoped that both plants will be operating fully by 1998.

Details of the sources of finance for environmental investments in the Province of Gdansk are given in Table 33.2. It is difficult to evaluate how far these funds will help to improve the state of the natural water resources. Approximately half of this money is directed towards modernization and construction of sewerage and sewage treatment systems.

33.7 PROSPECTS FOR FUTURE IMPROVEMENTS

The main activities in the future will focus on slowing down the degradation of the environment, which will create conditions for the regeneration of surface water reservoirs and rivers, as well as the waters of Gdansk Bay. This will require the modernization and construction of sewerage systems and sewage treatment plants, improvements in the handling of municipal and industrial solid wastes, implementation of 'clean technologies' and improvements to the operational safety of ports and associated industries. In Chapter 26 Gromiec describes how water quality management models can be used to prioritize investment in treatment plants.

The present ecological problems are mainly the consequence of rapid development of ports and industry, and associated urbanization, particularly the cities of Gdansk and Gdynia. The communal infrastructure did not parallel the rapid industrialization of these agglomerations and is now the cause of numerous environmental problems.

Several activities will be undertaken in the years leading up to 2000 to protect and improve the quality of water in the Province of Gdansk [3]:

1. Restoration of the Radunia River, which as mentioned earlier, provides drinking water for the city of Gdansk;
2. Development of hydrobiological models for this river;
3. Protection of underground water reservoirs in the Radunia River valley;
4. Improvement of the quality of water in Gdansk Bay and restoration of its recreational value. Within the programme called the 'Purity of Gdansk Bay – 2000' it is planned to complete the construction of Wschód and Dębogórze

Prospects for future improvements 279

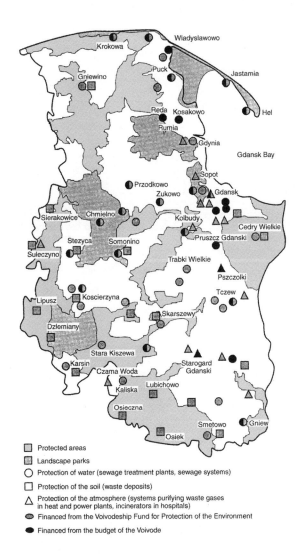

Figure 33.8 Principal investments for environmental protection in Gdansk Province 1991–93 (new construction and modernization).

Table 33.2 Sources of finance for environmental investments in Gdansk Province 1991–3

Source	Amount (billion zlotys)		
	1991	*1992*	*1993*
Voivode's budget	109.5	148	161.2
Voivodeship Fund for Protection of the Environment	30.9	58.4	97
National Fund for Protection of the Environment	9.1	21.5	83

[a] 1 billion zlotys equals approximately US$44 500.

sewage treatment works together with the modernization of the sewerage systems;
5. Improvement of the quality of water in the Vistula (Wisła) and Wierzyca rivers by building or modernizing sewage treatment plants in Tczew, Starogard Gdanski and Pelplin;
6. Evaluation of underground water resources in all the hydrobiological watersheds of the province;
7. Protection against flood and drought;
8. Introduction of clean technologies with implementation of water recycling
9. Protection of lakes;
10. The construction and modernization of municipal sewerage systems in provincial towns;
11. Improvement of monitoring systems.

REFERENCES

1. P105, W105 (1993) Report on chosen elements of the state of the environment of Gdansk Province for the period 1988–1992. National Inspectorate of Environmental Protection and Gdansk Voivodeship Inspectorate of Environmental Protection, Gdansk.
2. P105, W105 (1994) Report on the state of the environment in Gdansk Province in 1993. National Inspectorate of Environmental Protection and Environmental Monitoring Library of Gdansk Voivodeship of Environmental Protection, Gdansk.
3. W105 (1994) Aims in the field of environmental protection in Gdansk Province in the years 1994–2000. Department of Environmental Protection of Gdansk Voivodeship, Gdansk.
4. Ministry of Environmental Protection, Natural Resources and Forestry (1994) Brochures on water management in Poland, Warsaw.

NATIONAL ENVIRONMENTAL MONITORING PROGRAMME

by A. Mierzwiński and E. Niemirycz

34.1 DEVELOPMENT OF MONITORING OF SURFACE WATERS IN POLAND

River water quality has been the subject of interest as far back as the last century. In 1876 Mendelejew, to support his proposal for a new water supply and sewage system for the city of Warsaw, analysed the quality of the Vistula water [1]. He was concerned mainly with determining the elements responsible for water hardness. The results of his measurements, therefore, cannot be considered as a basis for the assessment of the quality of Vistula water in the last century.

The first comprehensive investigation of the water quality of the Vistula was made by Kirkor in 1923–4 [2]. He determined concentrations of 14 parameters in samples of river water taken every two weeks. He found that the Vistula above Warsaw was satisfactorily clean despite its opacity and yellowish colour. The mean values of pollutants were rather low e.g. BOD_5 was 2.2 mg/l; iron, 0.24 mg/l; ammonium nitrogen, 0.1 mg/l; etc. Only the amount of suspended matter was very high and it was noticed that an increase in the suspended matter was dependent on an increase in the water flow.

The establishment of the National Inspectorate for Water Protection in the mid-1950s and the setting up of water testing laboratories in the provinces were the first steps towards systematic monitoring of surface water. The frequency of inspection depended on the economic importance of the river and varied from a few inspections over a five-year period to twelve inspections per year. At first the scope of inspection was restricted to dissolved oxygen, BOD_5, COD, suspended matter and salinity. Based on the archival data of the Voivodeship Inspectorate of Environmental Protection in Gdansk (formerly the Centre of Environmental Investigation and Control), a significant increase in the pollution of the Vistula occurred in the 1960s [3].

In addition to these inspections, special monitoring was undertaken to determine specific substances and the movement of selected elements, etc, which was not possible from routine monitoring. Examples of special monitoring were the investigations of polychlorinated pesticides [4] and nutrients [5] carried out in 1972–3: measurements were made twice a week, yielding a large data set.

Regular monitoring of the cross-section of the mouth of the Vistula has been carried out twice a week since 1975 for a wide range of measured parameters as part of a long-term programme [6, 7]. This monitoring covers: temperature, colour, pH, dissolved oxygen, BOD_5, COD, organic carbon, Kjeldahl nitrogen, ammonia, nitrate, nitrite, phosphorus – total and suspended, phosphate phosphorus – total and dissolved, chlorophyll-*a*, ether extract, non-polar substances of the extract, volatile phenols, anion detergents, PCB, DDT and metabolites, DMDT, γ-HCH, iron, manganese, zinc, copper, lead, cadmium, mercury, sodium, potassium, magnesium, calcium, total hardness, chlorides, sulphates, dry residue, dissolved substances, and faecal coli.

34.2 THE NATIONAL RIVER MONITORING NETWORK

Abundant information has now been accumulated on the outflow of pollutants from the drainage area as a result of the Vistula investigation programme in the lower reaches [8] and also from the routine river monitoring. Since 1987 this has resulted in a new concept of monitoring.

The national river monitoring network [9] consists of three measurement and control programmes:

- benchmark (the baseline river monitoring network);
- basic (the basic river monitoring network);
- border (the border river monitoring network).

34.2.1 THE BENCHMARK PROGRAMME

The main tasks of the benchmark programme are:

- to provide data on the pollutants discharged into the major rivers and the Baltic Sea;
- to provide data on the quality of the major rivers;
- to forecast changes in water quality in relation to the hydrological conditions;
- to verify the water quality forecasting models.

Figure 34.1 The sampling points of the baseline river monitoring network (20 cross-sections).

The baseline network consists of 20 measurement and control cross-sections (five in the Vistula basin, five in the Odra basin and ten in the Pomeranian river basins). The sampling sites are located either upstream of the estuaries of these rivers or downstream of regions of special industrial importance (Figure 34.1).

The results of the analysis obtained from the baseline network are used in the preparation of reports, every ten days or monthly, on the quality of Polish rivers. The results are compared with the mean values for the different classes of water quality, which are defined in the regulations issued by the Ministry of Environmental Protection, Natural Resources and Forestry on 5 November 1991. (These values form the basis of the classification of surface waters and the conditions that have to be met by the wastewaters discharged into surface waters and the ground [10].)

34.2.2 THE BASIC PROGRAMME

The main tasks of the basic programme are:

- to provide data on the quality of water in Polish rivers for the annual assessment of water quality;
- to provide data for (a) the analysis of the physical and chemical processes taking place in river basins and (b) making the rational decisions concerning the use of water in the basin;
- to provide data for the comparison of the changes in water quality over the long term.

The basic network consists of 640 measurement and control cross-sections located on the rivers that are considered economically important.

In the annual reports on the water quality of the country's rivers, the assessment of water quality is carried out by the 'competent concentrations' method and the 'statistical' method.

34.2.3 THE BORDER PROGRAMME

In the border network, the locations of the cross-sections, the measuring programmes, the harmonization of methods and the means of providing the results are discussed bilaterally with other countries and described in the detailed agreements.

The measurements of water quality in the national monitoring network are carried out by the Voivodeship Inspectorates of Environmental Protection and the Institute of Meteorology and Water Management.

The design of the surface water monitoring programme takes into account liaison with:

- the Regional Water Management Boards, which are structured on river water catchments;
- the Voivodeship Departments of Environmental Protection and the Voivodeship Inspectorates of Environmental Protection which are structured on administrative areas.

34.3 WATER QUALITY OF POLAND'S RIVERS

The assessment of river water quality for 1991 was based on the results obtained from the basic network [11]. The required conditions of water purity in the rivers as well as the classification of the rivers is dependent on the type of pollution. The classifications are shown in Figures 34.2–34.8. According to the total assessment (physico-chemical and bacteriological data) there were no river waters which satisfied the conditions for class I water quality. The proportion of unclassified river waters (i.e. the most polluted) was 78% of the total (Figure 34.8). Comparing the assessment of the riverine water quality made in 1991 with the ones for 1985 and earlier, it can be stated that the river water quality is unsatisfactory (Table 34.1).

The location of most industrial plants (the main sources of pollutants) in the upper reaches of the river catchments is a significant factor in the quality of Polish rivers. These sources of pollutants have a strong influence on the rivers for a long distance downstream.

34.4 MAIN SOURCES OF SURFACE WATER POLLUTION

The main factors causing degradation of the surface waters are:

Table 34.1 Variation of river water quality 1974–91. (Source: Walewski; National Inspectorate of Environmental Protection, 1992)

Year	Type of parameter	The length of study river (10^3 km)	Water quality class			
			I	II	III	Unclassified
1974–7	physico-chemical	17.8	9.6	30.7	26.5	33
1978–83	physico-chemical	16.2	6.8	27.8	29	36.4
1984–8	physico-chemical	17.6	4.8	30.3	27.8	37.1
	bacteriological	16.5	0	3.9	20.3	75.8
1990	physico-chemical	10.1	6	27.9	30.3	35.8
	bacteriological	10.1	0	3	16.8	80.2
1991	physico-chemical	10.5	2.3	32.7	30	35
	bacteriological	10.5	0	3.7	18.4	77.9
	physico-chemical and bacteriological	10.5	0	3.3	14.5	82.2

Figure 34.2 Required riverine water quality classification.

Figure 34.3 Classification of riverine water quality based on organic pollutants.

- wastewaters discharged from industrial plants and towns (point sources);
- organic and inorganic fertilizers, pesticides, etc. (non-point sources);
- wastewaters discharged from rural areas;
- saline mine-waters.

The wastewaters discharged from rural areas are a serious threat to surface waters. Only 2% of villages have a wastewater treatment plant, 29% of villages a water supply system and 5% a sewerage system. Wastewaters discharged without treatment from livestock farms and households into receiving waters are a sanitary and epidemiological hazard.

The wastewaters from mines and quarries contain large quantities of chlorides and sulphates and constitute a hazard to surface waters. They adversely affect the biological life of a river, cause corrosion of structures and installations, and prevent the use of these waters for industrial purposes.

Data on the sources of surface water pollutants are obtained from reports by industrial organizations and other economic and administrative units. A system for recording point sources of pollution exists for all wastewater outflow of more than 20 m^3/day.

The licences have conditions for the control of substances which are discharged into the water; the substances controlled will depend on the type of wastewater discharged. The control is carried out by the Voivodeship Inspectorates of Environmental Protection.

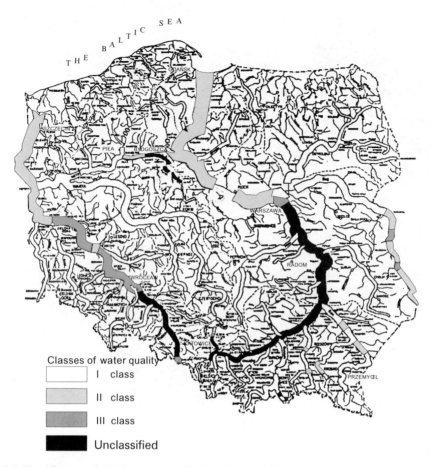

Figure 34.4 Classification of riverine water quality based on salinity.

34.5 CONCLUSIONS

1. Since 1987, the water quality control of Polish rivers has been carried out according to the national environment monitoring programme.
2. The baseline monitoring network comprises 20 measurement and control cross-sections. About 50 parameters of water quality are regularly measured.
3. The basic river monitoring network includes about 650 measurement and control cross-sections. This monitoring provides basic river water quality data only. The frequency of measurement is once per month.
4. The condition of the river water in recent years has not been satisfactory. Physico-chemical parameters cause 35% of Polish rivers to fail to meet the class III standard, and bacteriological parameters cause 78% to fail.
5. Improvement of the rivers will require a high input of Polish and foreign capital on modernization and construction of new treatment plants.

REFERENCES

1. Dojlido J. and Wojciechowska, J. (1974) Changes in physico-chemical parameters of the Vistula water quality at the measurement and control cross-section in Warsaw during the period 1945–1970.
2. Kirkor, T. (1928) Results of systematic water analyses in the Vistula River. Chemical Industry, 12th Annual, c.6, Lwów.

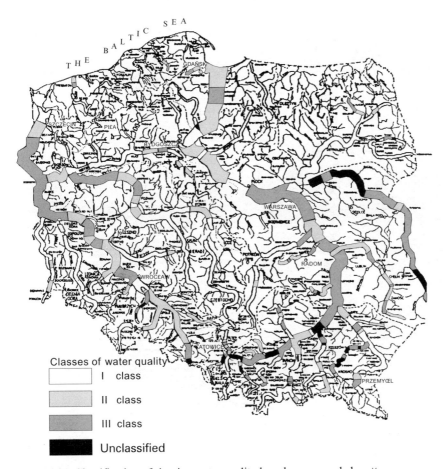

Figure 34.5 Classification of riverine water quality based on suspended matter.

3. Archival data of the Voivodeship Inspectorate of Environmental Protection in Gdansk.
4. Taylor, R. and Bogacka, T. (1979) Transport of pesticides to the sea by the Vistula River, *Oceanology* **11**: 129–38.
5. Szarejko, N. and Bogacka, T. (1980) Study on the water quality of the Vistula River at the cross-section localized in Tczew/Kiezmark. Mat. IMWM, Gdansk.
6. Niemirycz, E., Rybiński, J. and Korzec, E. (1980–4) The Vistula input of specific pollutants with particular regard to substances of agricultural origin. Mat. IMWM, Gdansk, Annuals since 1975.
7. Januszkiewicz, T., Kowalewska, K., Szarejko, N. and Żygowski, B. (1974) Effect of fertilization on the Vistula water quality, Mat. IMWM.
8. Niemirycz, E. and Borkowski, T. (1993) Riverine Input of Pollutants in 1992. Environmental conditions in the Polish zone of the southern Baltic Sea during 1992. Maritime Branch Materials, Institute of Meterology and Water Management, 205–47. Annual Reports since 1987.
9. Walewski, A. (1992) National environmental monitoring programme, Warsaw, National Inspectorate of Environmental Protection.
10. Ministry of Environmental Protection, Natural Resources and Forestry (1991), Legislation Gazette 116. (dated 5 November).
11. Library of Environmental Monitoring (1993) State of water quality in rivers, lakes and the Baltic Sea on the basis of results of the state environmental monitoring in 1991–1992. State Inspectorate of Environmental Protection.

288 *National environmental monitoring programme*

Figure 34.6 Classification of riverine water quality based on nutrients.

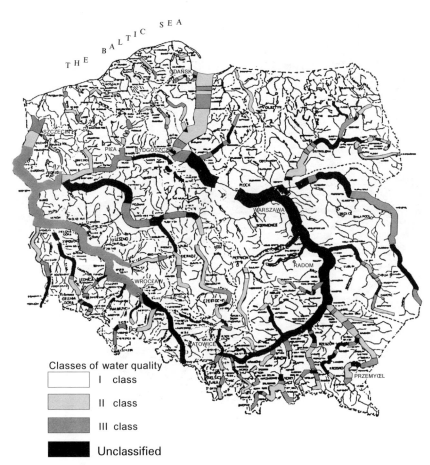
Figure 34.7 Classification of riverine water quality based on physico-chemical parameters.

290 *National environmental monitoring programme*

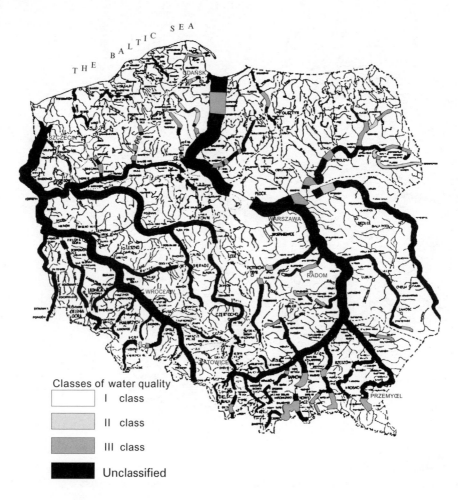

Figure 34.8 Classification of riverine water quality based on physico-chemical and bacteriological criteria.

THE LEGAL FRAMEWORK OF WATER QUALITY MANAGEMENT IN ENGLAND AND WALES

by B. Barrett

Since this chapter was written, the control of pollution of the environment in the UK underwent a fundamental reorganization with effect from 1 April 1996 (see Warburton, Chapter 6 and the note at the end of the reference section).

35.1 INTRODUCTION

The UK has a long history of regulatory legislation. Legislation relating to water quality can be traced back to the early 19th century. The Municipal Corporations Act 1835 gave borough councils and urban district councils control of public utility undertakings, including sanitation and water supply. When cholera was endemic, the priority was the provision of wholesome drinking water, free from contamination from sewage [1]. Local government authorities undertook responsibility for the provision of sewers and the treatment of sewage; Under current legislation local authorities may still perform sewerage functions on behalf of the water company [2]. Clean water was supplied either by local authorities or by commercial water companies.These were incorporated by an Act of Parliament which usually contained standard clauses [3–5]. Meanwhile heavy industry created its share of environmental pollution.

The Water Act 1973 did much to rationalize the water industry. Regional water authorities were created to manage the hydrological cycle: in England and Wales ten authorities assumed the responsibilities of 157 water undertakings and 1393 sewerage authorities; they also assumed the responsibilities (largely for land drainage) of 29 river authorities. Local authorities lost their responsibilities for supply of water, but 28 of the statutory water companies remained, and the water authorities were required to accept them as agents [6]. The water authorities had a duty to take all reasonable steps to make water available to the companies to meet the foreseeable demand of consumers. The water authorities became responsible for sewerage services; they were also empowered to prosecute for pollution offences. However, over time, expectations changed and the water authorities proved to be ill-equipped to meet EU (or EEC as it was first known) requirements.

This chapter will:

1. outline the relationship between the traditional UK system and European initiatives;
2. consider the present system for water management in England and Wales after 1989;
3. outline how well the legislative system is operating and;
4. suggest the challenges that may have to be addressed.

International River Water Quality. Edited by Gerry Best, Teresa Bogacka and Elżbieta Niemirycz. Published in 1997 by E & FN Spon, London, ISBN 0419215409

35.2 COMPARISONS BETWEEN THE UK AND EUROPE

35.2.1 UK TRADITIONS

The traditional UK approach prior to 1 April 1996 was to deal separately with atmospheric and water pollution. For atmospheric pollution there were different regulatory systems for different types of emission but in many instances a duty was placed on the owner of industrial works to 'use the best practicable means' for preventing the escape of noxious or offensive gases, and for rendering such gases harmless when discharged [7]. In the event of pollution, the industrialist had to demonstrate that the best practicable means was being used: this system allowed standards to be improved as technology advanced.

Water pollution was controlled through a system of discharge consents, which authorized discharge of pollutants into inland waters, of the type, in the quantities and by the persons authorized. Discharges that were not authorized were criminal offences. The system of consents enabled the situation to be reviewed as circumstances, including technological advances, were appropriate, while at the same time allowing for adjustment where the environment was either sensitive or robust. After the Water Act 1973 the water authorities in England and Wales granted both discharge consents and prosecuted for pollution offences.

35.2.2 EUROPEAN INITIATIVES

The EU is concerned with improving the environment. The original Treaty of Rome, which established the EEC, was based on the premise that no member state secured an unfair economic advantage by neglecting the environment. Recently the potential for conflict between economic growth and environmental protection has been realized and the EU is developing a policy of promoting sustainable growth which respects the environment. A new Title VII was added to the Treaty of Rome by the Single European Act, Article 25: the Treaty of European Union; Article 2 introduces sustainable growth as an objective of the EU. A programme 'Towards Sustainability' was presented by the Commission in 1993 [8].

As a member of the EU the UK has been involved in formulating policies and in determining the standards for member states. In the end, however, the UK is but one of many member states and has to honour the EU's requirements. In any case the UK empathizes with international environmental programmes although it has, so far as inland water is concerned, no experience of transnational pollution.

In environmental matters the EU normally legislates through directives. Directives set out what member states have to achieve: each state has to implement particular directives within the framework of its regulatory system.

European Union provisions relating to improvement of water quality can be categorized as measures dealing with:

- dangerous substances
- water for specific uses
- specific industries
- sources of pollution.

(a) Dangerous substances

In 1976 the EEC adopted a framework directive [9] on the discharge of dangerous substances into the aquatic environment. The substances concerned are listed in the annex of the directive: List I contains substances which are particularly dangerous because of their toxicity, persistence and bio-accumulation, and List II contains somewhat less harmful substances. For List I substances emission standards must be set by member states and these must not exceed limit values set by the EEC; alternatively member states can opt for emission standards based on quality objectives for the receiving waters. Emission limits are favoured in practice, partly because the quality objectives have to be matched to water use. Limit values have been set for substances such as mercury [10], and cadmium [11]. Hexachlorocyclohexane [13], carbon tetrachloride, DDT, pentachlorophenol [14], aldrin, dieldrin, endrin, isodrin, hexachlorobutadiene and chloroform [15], dichloroethane, trichloroethylene, perchloroethylene and trichlorobenzene [16] have also been covered. Member states have to establish programmes to reduce pollution by substances in List II. The substances in List I for which no EEC limit values and quality objectives have been established have to be considered as List II substances.

(b) Water for specific uses

Quality standards for water intended for specific purposes are sought by particular directives aimed at:

- water intended for the abstraction of drinking water [17];
- the methods of measuring and frequencies of sampling and analysis of such water [18];
- bathing water [19];
- water intended for human consumption [20];
- the quality of fresh waters to support fish [21]; and
- the quality of shellfish waters [22].

Member states have to ensure that quality requirements are met within the specified time limits.

(c) Specific industries

Directives have so far focused on the titanium dioxide industry [23, 24], including one on procedures for harmonizing the programmes for the reduction and eventual elimination of pollution caused by the industry [25].

(d) Sources of pollution

Directives aimed at particular sources of pollution include:

- provisions concerning batteries and accumulators containing certain dangerous substances [26];
- urban wastewater treatment [27];
- the protection of waters against pollution caused by nitrates from agricultural sources [28]; and
- hazardous waste [29].

Underlying these EU initiatives are certain themes: firstly there should be clear regulatory standards; secondly standards should be set as high as they can taking into account available technology; thirdly a balance should be achieved between environmental improvement and the cost involved, by selecting 'the best available technique not entailing excessive cost'.

35.2.3 UK RESPONSES

In the meantime the UK had not been complacent. Since the 1970s there had been a standing Royal Commission on Environmental Pollution which had investigated various aspects of environmental pollution. By the end of the 1980s the UK accepted that it was time for change. The Water Act 1989 therefore changed the structure of the water industry and the Environmental Protection Act 1990 made provision for an integrated approach to environmental protection. The new regulatory system set up applies only to England and Wales.

35.3 WATER MANAGEMENT IN ENGLAND AND WALES AFTER 1989

The Water Act 1989 was a complex piece of legislation: its principal purposes were to appoint ten companies as regional water and sewerage undertakers and replace the regional water authorities by a National Rivers Authority (NRA).

Improving the quality of both inland waters and drinking water required capital investment in plant and equipment at a time when it was deemed desirable to reduce public expenditure. The Act, by creating commercial water companies, transferred to the private sector the task of financing the investment needed.

The NRA was created to be a watchdog to monitor the quality of waters as required by EU directives [18], and to develop a national policy on water quality. It is responsible for some 20 directives which have direct environmental monitoring requirements; currently in the forefront of these is the Urban Wastewater Treatment Directive. It is noteworthy that the chairman of one of the new water undertakings publicly stated, 'the extent of the regulatory regime now far exceeds anything that existed in pre-privatisation days' [30].

The Water Act 1989 repealed some earlier legislation but much still remained so, in 1991, the 1989 Act was brought together with the older laws giving five new statutes: the most important of these, for present purposes, are the Water Industry Act 1991 and the Water Resources Act 1991. The others are the Land Drainage Act 1991, relevant to the NRA's responsibilities in flood defence; the Statutory Water Companies Act 1991 (concerning the 29 private water companies who still provide clean water for regions within the jurisdiction of, and subordinate to, the ten major companies); and the Water Consolidation (Consequential and Amendments) Act 1991.

The Water Industry Act 1991 and the Water Resources Act 1991 have to be read with another substantial piece of legislation: the Environmental Protection Act 1990. Part I of the 1990 Act introduced the concept of integrated pollution control, the objective of which is to ensure that the disposal of pollutants is done according to the best available techniques not entailing excessive cost [31]. The system of integrated pollution control is concerned with the dangerous substances listed in the framework directive.

The NRA works closely with the Government through the Department of the Environment (England) and Welsh Office (Wales). The NRA is not a servant or agent of the Crown, however, the Government can give it directions with respect to its functions [32]. Its functions include water resource management; flood defence; maintenance and improvement of fisheries; and control of pollution [33]. It is also charged with the duty to promote:

(a) the conservation and enhancement of the natural beauty and amenity of inland and coastal waters and of land associated with such waters
 . . . and
(c) the use of such waters and land for recreational purposes. [33]

35.3.1 WATER RESOURCE MANAGEMENT

It might be argued that the whole purpose of the NRA is water resource management, but within the Water Resources Act 1991 the phrase relates to its duty to manage the demand for water in the context of the available supply. The legislation spells out that it is the duty of the NRA to:

 . . . take all such action, as it may from time to time consider . . . to be necessary or expedient for the purpose —
(a) of conserving, redistributing or otherwise augmenting water resources in England and Wales; and
(b) of securing the proper use of water resources in England and Wales. [34]

Within this general obligation the NRA has to work with water undertakers [35]. However, the Water Resources Act 1991 makes it clear that the NRA's general duty does not relieve any water undertaker of its obligation to develop water resources [36].

The obligations entered into by a water undertaker are subject to Draconian enforcement provisions: namely, an enforcement order non-compliance, which in the last instance could mean the termination of the water undertaker's appointment [37].

The NRA has responsibility for granting licences for the abstraction of water [38]. Crop irrigation involves very heavy use of water and there are special provisions for abstractions for this purpose [39]; leisure facilities, e.g. golf courses, public parks and swimming pools are also heavy users; but one third of all water abstracted is used (and recycled) by electricity generating companies for cooling purposes. Currently the NRA manages about 50 000 licences, reviewing and, where necessary, revising or revoking them and enforcing compliance with their conditions through a programme of inspections. In 1992–3 a total of over 19 000 enforcement checks were carried out; there were 31 prosecutions. Solutions were identified to low-flow problems at 14 sites; five low-flow alleviation schemes were completed. European wide recognition was obtained for a Groundwater Protection Policy [40].

The granting of abstraction licences is related to the power of the NRA to determine minimum acceptable flows in relation to any inland waters [41], a power which must be seen in the context of a general environmental duty:

(a) . . . to further the conservation and enhancement of natural beauty and the conservation of flora, fauna and geological or physiographical features of special interest ... [42]

In determining minimum acceptable flows the NRA must also have regard to any statutory water quality objectives [43, 44].

Water companies and the NRA are empowered to apply to the Secretary of State (i.e. the Department of the Environment) for drought orders when exceptional shortage of rain has caused a serious deficiency in water supplies.

35.3.2 FLOOD DEFENCE

The Water Act 1989 [45] transferred the flood defence functions of the water authorities to the NRA. These functions were set out in the Land Drainage Act 1976, now the Land Drainage Act 1991. The NRA operates through regional and local flood committees, in which relevant local government authorities are heavily involved. Over 50% of the NRA's annual income and expenditure is in relation to flood defence. In 1992–3 165 km of new defences were constructed and 1350 flood warnings were issued. The NRA also contributed to a national study on coastal defence [46]. Arguably, however, flood defence is more a function of land-use management than water resource management; it is also concerned with the protection of life and property.

35.3.3 FISHERIES

The NRA's obligations for the general control of fisheries gives it the duty to maintain, improve and develop salmon, trout, freshwater and eel fisheries [47]. For the purpose of this duty the NRA is empowered to acquire land, dams, weirs etc. [48].

35.3.4 WATER QUALITY

The NRA is playing a major role (working with central government) in setting quality objectives for 'controlled waters'. Controlled waters are defined as:

(a) territorial waters;
(b) coastal waters;
(c) inland fresh waters; and
(d) waters contained in underground strata [49].

The NRA is also responsible for granting (for an appropriate fee) discharge consents and permitting the discharge of specified pollutants, in specified quantities, into controlled waters. Finally it is charged with policing controlled waters to ensure the quality of water is maintained.

Although its jurisdiction is not limited to rivers, the NRA monitors some 40 000 km of rivers and canals through 6000 sampling locations, taking approximately 10 000 samples annually [50].

Under the Water Resources Act 1991, regulations may be made to classify controlled waters. The matters which may be included in such regulations are:

(a) general requirements as to the purposes for which the waters to which the classification is applied are to be suitable;
(b) specific requirements as to the substances that are to be present in or absent from the water and as to the concentrations of substances which are or are required to be present in the water;
(c) specific requirements as to other characteristics of those waters [51].

Within the general classification system particular water quality objectives may be set for any stretch of water [52]. Regulations may require the NRA to determine by sampling whether prescribed requirements are satisfied.

35.3.5 HER MAJETY'S INSPECTORS OF POLLUTION

The role of the NRA has to be seen in the context of the system of integrated pollution control provided for in Part I of the Environmental Protection Act 1990. The concept of integrated pollution control is that the release of harmful substances into the environment should be so controlled as to ensure that the release takes place in the media (i.e. air, water, land) and in circumstances which cause least harm to the environment.

The Act empowers the Secretary of State (in practice the Department of the Environment) to make regulations prescribing any process [53] or substance [54]. Once a process has been prescribed it may not be carried on without an authorization. Regulations prescribing substances may specify the medium to which the release may be made and the amounts that may be released. Regulations may also establish, for any environmental medium, quality objectives or quality standards in relation to any substances.

Authorizations to carry on prescribed processes are granted subject to conditions, which are subject to periodic review. Conditions are imposed to achieve the statutory objectives that are set out [55]. The primary objectives are:

(a) ensuring that, in carrying on a prescribed process, the best available techniques not entailing excessive cost will be used . . .
(b) compliance with . . . any obligations of the UK under Community Treaties or international law relating to environmental protection . . .

If a person contravenes the conditions of an authorization an enforcement notice may be served specifying the contravention and the steps that must be taken, and the time within which they must be taken, to remedy the contravention. In circumstances where continuing a prescribed process involves 'an imminent risk of serious pollution' a prohibition notice may be served: it may be served whether or not there is a contravention of an authorization. Thus a prohibition notice might be served on an organization whose conduct was lawful if, for example, the break-down of a nearby plant was causing pollution to such an extent that further releases into the environment could not be tolerated. Until the prohibition notice is withdrawn it is a criminal offence to carry on the process, or that aspect of it which is subject to the prohibition. A failure to comply with a notice could result in a heavy fine and/or a term of up to two years' imprisonment.

The enforcing agency for these provisions is normally Her Majesty's Inspectorate of Pollution. Where less harmful processes and substances are involved, local authorities have been designated as enforcement agents (through their environmental health departments). The identification of prescribed processes and substances, and their regulation, could therefore produce an uneasy interface between the policing activities of the NRA and the Pollution Inspectorate. However the Environmental Protection Act 1990, provides that authorization shall not be given for release of prescribed substances into controlled waters if the NRA certifies that the proposed release will result in a failure to reach water quality objectives set under the Water Resources Act 1991 [56].

35.3.6 THE ROLE OF THE WATER INDUSTRY

Under the Water Industry Act 1991 each water undertaker has a general duty to develop and maintain an efficient and economical system of water supply within its area [57]. In English law, water undertakers have traditionally had a general duty to supply to premises for domestic purposes '. . . only water which is wholesome . . .' and in the Water Act of 1989 this remained the principal obligation. Within this general requirement it was provided that water would not be considered wholesome if it failed to meet the standards laid down in the regulations. Interestingly, in the consolidation exercise the ordering of the statutory provisions was altered, so that in the Water Industry Act 1991 the first statement is that regulations will prescribe what is meant by wholesome water [58], and the second statement is that the water undertaker has a general obligation to supply wholesome water [59]: this re-ordering symbolizes the move from general responsibilities to precise standards. The water companies have a statutory duty to monitor and record the quality of the water they supply [60]. The statutory provisions provide that regulations may prescribe requirements according to the purposes for which the water is to be deemed suitable. In practice, at the present time, in England, all tap water piped to domestic premises is intended to be suitable for drinking. The water undertakings also have responsibility for sewerage [61].

35.3.7 WATER CLASSIFICATION

The Water Act 1989 made statutory provision for water quality classification and enabled the making of regulations. The Environmental Protection Act 1990 enabled the making of similar regulations for any substance subject to integrated pollution control [62–64]. The process was started by the Surface Waters (Classification) Regulations 1989 [65] which classify inland waters into three categories according to their suitability as drinking waters.

At the time of its most recent annual report the NRA could note that it had completed public consultations [66] about the principles which will underlie statutory water quality objectives and was preparing proposals for consultation and approval at 20 sites.

The NRA has proposed six classifications for controlled waters according to their intended use:

- Fisheries ecosystem: includes standards necessary to support aquatic life, intended as the basis for statutory standards for all rivers;
- Abstraction for potable supply: to allow standards to be applied at strategic locations within catchments to protect downstream resources to achieve standards to meet abstraction needs;
- Industrial/agricultural abstraction: as different industries have different needs site-specific criteria will be drawn up;
- Water sports: to take account of the health risks associated with different kinds of water sport;
- Special ecosystem: the principal risk is nutrient enrichment;
- Commercial harvesting of fish/shellfish for human consumption: standards for the protection both of the fish and the health of consumers.

It is envisaged that these classifications will be incorporated into regulations: then water quality objectives will be set for individual stretches of water, taking into account the current water uses, the desired uses, the benefits, what is necessary to achieve them, what are the consequences including the costs falling on dischargers and others, and what is a reasonable time frame for achievement. The classifications in EU directives are: freshwater fisheries; surface water abstracted for potable supply; bathing waters; dangerous substances; shellfish waters; titanium dioxide.

An important aspect of the new approach is that it moves away from the traditional practice of expressing standards as maximum allowable concentrations, annual averages or a percentile value such as a 95%-ile. The previous approach permitted intermittent, but potentially damaging, fluctuations in water quality, such as may occur in urban storm drainage. In future it is intended that standards will be expressed in terms of the concentration which must not be exceeded at any time. Water quality objectives will in future formally determine the discharge consents which the NRA will grant.

Additionally the NRA has suggested the need for a general quality assessment scheme to provide information on the quality of rivers with respect to their chemical, biological, nutrient and aesthetic status. These tasks are urgent because, due to overabstraction, drought and agricultural activity the quality of inland waters in some regions has actually declined in recent years, in spite of the virtual disappearance of those heavy industries which were traditionally considered as the worst polluters.

35.4 ENFORCEMENT OF THE POLLUTION LEGISLATION IN THE COURTS

35.4.1 CRIMINAL LIABILITY

Pollution of controlled waters is a criminal offence and liability is strict. The current legislation retains the traditional UK approach. A person commits an offence if:

he causes or knowingly permits –

(a) any poisonous, noxious or polluting matter or any solid waste matter to enter any controlled waters [67].

In order to prevent seepage, provision is also made for the control of the discharge of trade or sewage effluent on to land or into waters other than controlled waters (i.e. certain ponds and lakes). A person also commits an offence by causing or knowingly permitting: any matter whatever to enter any inland waters so as to tend . . . to impede the proper flow of the waters in a manner leading or likely to lead to a substantial aggravation of (i) pollution due to other causes; or (ii) the consequences of such pollution [68].

This provision gives statutory form, with criminal sanctions, to the view, long accepted in civil law, that no individual wrongdoer should escape liability for contributing to pollution even though that individual's conduct is relatively innocuous [69, 70].

The strictness of the law has been emphasized by the interpretation which the courts have given to the expression 'causes or knowingly permits' [71]. In Alphacell v. Woodward [72] decided under the Rivers (Prevention and Pollution) Act 1951, pollution was caused when settling tanks at a paper factory overflowed into the River Irwell, without any negligence on the part of the defendants. This was a strange finding since the pumps that should have stopped the overflow were blocked. The defendants were nevertheless found liable: they had carried on the activity which caused the pollution. As long as the activity (in this case running the factory) was intentional all that was needed was that there was a causal link between the activity and the discharge. Some recent cases have taken the same approach. In NRA v. Alfred McAlpine Homes East Ltd [73] liability was attached to a building company for causing a stream to be cloudy downstream of its construction site. Both the site manager and the site agent accepted responsibility explaining that the pollution had been caused by cement being washed into the stream. The court rejected the employer's defence that the wrongdoing of their employees was not the wrongdoing of the company. The case was indistinguishable from Alphacell: the company had to accept responsibility for the actions of its employees. This is technically known as vicarious liability. In justifying this Mr Justice Morland said that he could see no reason why Parliament as a matter of policy should not have placed on companies responsibility for environmental protection as they were best placed to ensure, through discipline, training, supervision and the highest standards of maintenance, that streams were not polluted during their activities by their servants or agents.

The water companies have considerable responsibilities as sewerage undertakers for ensuring that water resources are not polluted. This was demonstrated in NRA v. Yorkshire Water Services Ltd [74]. The defendant water company discharged treated sewage into the River Spen. It had a discharge consent to do this, but the conditions included a prohibition on the discharge of iso-octonal. Iso-octonal was placed in a sewer by an unknown person, travelled through the sewage works and entered the river. The appeal court found that, as the polluted water had been fed into the river from the defendant's plant, the defendant had 'caused' the pollution: it operated a system which failed to detect its presence and remove it. Had the actual wrongdoer been identified he might also have been prosecuted.

In a third case, NRA v. Wright Engineering Co. Ltd [75] heating oil entered controlled waters. The oil had escaped, as a result of vandalism, from a tank owned by the defendants. There had previously been minor vandalism on the defendants' site but nothing involving the heating oil storage tank. The appeal court accepted that the vandalism was not foreseeable and so the defendants should not be liable for the pollution. The defendants appear to have been treated leniently, but the oil was neither discharged by them nor discharged by their employees. The case resembles an earlier one, Impress (Worcester) Ltd v. Rees [76], where a trespasser caused pollution by opening the valve to an oil storage tank owned by the defendants, resulting in pollution to the River Severn.

The recent cases demonstrate that the NRA is rigorously monitoring controlled waters: some 435 successful prosecutions were taken in 1992–3 for both breaches of consent conditions and pollution incidents: of the successful prosecutions 28 were against water companies.

35.4.2 CIVIL LIABILITY

A water company which fails to supply wholesome water may face a civil claim for damages. Recently litigation followed an incident in which a delivery truck accidentally unloaded a chemical into the wrong tank at the treatment works at Camelford: the water company continued to supply water although they knew it to be contaminated. A person who had allegedly become ill as a result of drinking the contaminated water brought a successful claim for damages [77]. It was held that the law did not allow the plaintiff to recover exemplary damages to punish the company for their wilful wrongdoing: the company could only be made liable for the actual damage caused.

In Cambridge Water Co. Ltd v. Eastern Counties Leather plc [79] it was the water company which sued. The defendants were old-established leather manufacturers who used a chemical solvent in their tanning process. In the years before 1971 there had been regular spillage of the solvent on to the concrete floor of the tannery: the total spillage over a period of years being at least 1000 gallons (4.55 m^3). The solvent seeped into the soil until it reached an impermeable stratum 50 m below the surface: from this point it percolated until it reached the stratum from which the plaintiffs extracted water from a borehole for domestic use. The distance between the borehole and the tannery was about 2.1 km and the time taken for the solvent to seep into the borehole was about 9 months. The plaintiffs' claim for damages went on appeal but was unsuccessful: while there is strict liability on a person who allows an escape from land of things likely to do 'mischief' this liability only arises if the defendant knows or ought reasonably to have foreseen that such an escape might cause damage [80]. Whilst this liability exists only where there is a non-natural user of land, their Lordships would apparently have been prepared to regard the tannery as a non-natural user. Since the defendants could not have reasonably foreseen that the seepage from their tannery could have caused pollution to the plaintiffs' borehole they were not liable. There were apparently two reasons why the damage caused was not foreseeable. Firstly, . . . any spillage would have been expected to evaporate rapidly in the air. . .the only harm that could have been foreseen. . . . was that somebody might have been overcome by fumes . . . [81]. Secondly it was not until after the spillage ceased that any concern began to be expressed in scientific circles about the presence of organic chemicals in drinking water. Lack of foreseeability was equally fatal to a claim in nuisance.

To have placed liability on the tannery would have had disastrous implications for industry and/or its insurers: certainly insurance premiums would be high if enterprises could be required to pay for pollution which occurred at a time when the potential for pollution was not appreciated. Exonerating the enterprise left the water company with the cost of finding an alternative water supply and the general public with polluted water. The NRA is given the powers 'of intercepting, treating or disposing of any foul water arising or flowing upon' land and 'of otherwise preventing the pollution' of surface and underground waters [82]. It is not clear that these powers would extend to dealing with polluted water contained underground.

35.5 THE CHALLENGES TO BE ADDRESSED

It is too early to state whether the new system is likely to be effective, but there are challenges which may have to be addressed:

35.5.1 BALANCING SUPPLY AND DEMAND

The task of ensuring a sustainable use of water in England is difficult because in some areas the average rainfall is low in relation to the demand for water. In the heavily populated but prosperous South-east and East Anglia rainfall is only about 610 mm per year. In addition regional rainfall can vary by as much as a third from year to year. In 1992 there was a drought in many areas; the winter of 1993 brought flooding. Flooding has been encouraged by factors such as urban development in flood plains and channelling of rivers [83]. Forecasting water demand is problematic. The rule of thumb has been that demand on the water supply industry increases at about 1% per year [84]. Water metering is being considered, but its use is controversial: firstly meters are expensive to install in existing properties; and, secondly, charging for actual water used can impact heavily on poorer households. Some areas have started to experiment with meters, e.g. the Isle of Wight [85]. Since the drought of 1976 much has been done to link sources and reinforce supply systems, for example, by providing long-distance trunk mains. Major works to increase supply can, however, have environmental costs: building reservoirs requires flooding of land. Another approach is to raise public awareness of the need to conserve water, possibly by pricing structures [84, 86].

35.5.2 MAINTAINING WATER FLOW

Urbanization (leading to flash flooding) and heavy use of water have increased the potential for rivers to flood and/or dry up. The challenge for the NRA is to maintain the flood defence and water abstraction licensing systems that control these extremes.

35.5.3 REGENERATING IN A POST-INDUSTRIAL ERA

In the past, rivers have been used as highways for the carriage of goods; with the closure of industrial enterprises the river banks may be derelict. There is often a high level of unemployment in these areas resulting in a generally run-down urban environment. Investment is needed for regeneration: protecting the waterways from the impact of redevelopment, and using the water for recreation will be challenging. The River Lea, near Middlesex University, has urban regeneration plans which involve these issues.

35.5.4 CLEANING UP AFTER MINE CLOSURES

The closure of mines may result in a build-up of contaminated underground water which, in due course, may break out and pollute watercourses. The current large-scale closures of coal mines may escalate this problem [87]. The challenge will be to identify likely problem areas and then determine whether industry or the state shall pay for pollution that occurs.

35.5.5 PREVENTING AGRICULTURAL POLLUTION

Large-scale, intensive farming has made agriculture one of the major polluters. The challenges are to control the use of pesticides and other chemicals to prevent their seepage into watercourses and also avoid the spread of animal diseases to the human population, through water.

35.5.6 MEETING THE NEEDS OF THE LEISURE INDUSTRY

As society becomes more affluent increasing calls are made upon watercourses for recreation. The challenge will be to achieve and maintain a suitably healthy and attractive environment.

35.5.7 OPERATING A SYSTEM OF QUALITY OBJECTIVES

Carrying out the necessary surveys; agreeing the objectives for particular stretches of water; putting these objectives into practice and monitoring performance will be an expensive, lengthy and challenging task. The UK may yet regret that it has chosen this route to fulfilling EU requirements.

35.5.8 ENFORCEMENT

The problems of enforcement could easily be exacerbated if the roles of the Pollution Inspectorate and the NRA are not clearly demarcated and maintained.

35.5.9 INVESTMENT

Setting up and maintaining an appropriate water environment is costly. Privatization placed much of this cost on the water companies. The challenge for them is to invest for environmental protection while maintaining a service of the quality and at the price that consumers demand.

35.5.10 MEETING PUBLIC EXPECTATIONS

In England today, citizens' expectations are high and expressed both through the media and in the law courts. Neither the water companies nor government departments can hope to escape challenge. Already Friends of the Earth have challenged the Department of the Environment (and indirectly the water companies) about perceived delays in meeting EU drinking water standards [88].

REFERENCES

1. Chadwick, E. (1842) Report on the sanitary condition of the labouring population of Great Britain.

2. Water Industry Act 1991, s.97.
3. Waterworks Clauses Act 1847
4. Waterworks Clauses Act 1863
5. Water Act 1945, Schedule 3
6. Water Act 1973, s.12.
7. The Alkali etc. Works Regulation Act, 1906, ss.2, 7 and 8.
8. OJC 138 of 17 May 1993.
9. Directive 76/464/EEC. Pollution caused by certain dangerous substances discharged into the aquatic environment of the Community.
10. Directive 82/176/EEC.
11. Directive 84/156/EEC. Limit values and quality objectives for mercury discharges by sectors other than the chlor-alkali electrolysis industry.
12. Directive 83/513/EEC. Limit values and quality objectives for cadmium discharges.
13. Directive 84/491/EEC. Hexachlorocyclohexane.
14. Directive 86/280/EEC. Pentachlorophenol.
15. Directive 88/347/EEC. Chloroform.
16. Directive 90/415/EEC. Trichlorobenzene.
17. Directive 75/440/EEC. Quality required of surface water intended for the abstraction of drinking water in the member states.
18. Directive 79/869/EEC. Methods of measurement and frequency of sampling and analysis of surface water intended for abstraction for drinking in the member states.
19. Directive 76/160/EEC. Quality of bathing waters.
20. Directive 80/778/EEC. Quality of water for human consumption.
21. Directive 78/659/EEC. Quality of fresh waters needing protection or improvement in order to support fish life.
22. Directive 79/923/EEC. Quality of shellfish waters.
23. Directive 78/176/EEC. Titanium oxide.
24. Directive 82/883/EEC. Surveillance and monitoring of environments concerned by waste from the titanium oxide industry.
25. Directive 92/43/EEC.
26. Directive 91/157/EEC. Procedures for batteries and accumulators containing certain dangerous substances.
27. Directive 91/271/EEC. Treatment of sewage of municipal origin.
28. Directive 91/676/EEC. Protection of waters against pollution caused by nitrates from agricultural sources.
29. Directive 91/689/EEC. Hazardous waste
30. Southern Water plc (1991) Annual Report.
31. Environmental Protection Act 1990, Part I, s.7(2).
32. Water Resources Act 1991, s.5.
33. Water Resources Act 1991, s.2.
34. Water Resources Act 1991, s.19(1).
35. Ibid., s.20(1).
36. Ibid., s.19(2).
37. Water Industry Act 1991, s.18.
38. Water Resources Act 1991, s.24.
39. Ibid. s.57.
40. NRA (1992–3) Environmental progress made in a business-like way. NRA Annual Report and Accounts.
41. Water Resources Act 1991, s.21.
42. Water Act 1989 s.8, now Water Resources Act 1991, s.16.
43. Water Act 1989 s.22(4)(c).
44. Water Resources Act 1991, ss.83 and 104.
45. Water Act 1989, ss.136–40.
46. NRA Annual Report and Accounts 1992–3.
47. s.114.
48. s.156.
49. Water Resources Act 1991, s.104.
50. NRA Annual Report 1992–3.
51. Water Resources Act 1991, s.82.
52. Ibid., s.83.
53. Environmental Protection Act 1990 s 2(1)
54. Ibid., s 2(5)
55. Ibid., s 7(2)
56. Ibid., s 28
57. Water Industry Act 1991, s.37.
58. Ibid., s.67.
59. Ibid., s.68.
60. The Water Act 1989 s.53; now the Water Industry Act 1991, s.69.
61. Ibid., s.94.
62. Environmental Protection Act 1990, s.3(4).
63. Surface Waters (Dangerous Substances) (Classification) Regulations 1989 (SI 1989 No.2286).
64. Directive 76/464/EEC. Dangerous substances.
65. Surface Waters (Classification) Regulations 1989 (SI 1989 No.1148).
66. Department of Environment/Welsh Office (1992). *River Quality. The Government's proposals: a consultation paper*.
67. Water Resources Act, s.85(1).
68. Ibid., s.85 (1)(e)
69. Crossley v. Lightowler (1966–7) 1 Ch App Cas 478.
70. Pride of Derby Angling Association v. British Celanese Ltd [1952] Ch 149.
71. Wales Resources Act, s.85(1).
72. Alphacell v. Woodward [1972] AC 824.
73. NRA v. Alfred McAlpine Homes East Ltd, Queen's Bench Division, *The Times*, 3 February 1994.
74. NRA v. Yorkshire Water Services Ltd, *The Independent*, 19 November 1993.
75. NRA v. Wright Engineering Co. Ltd, *The Independent*, 19 November 1993.
76. Impress (Worcester) Ltd v. Rees [1971] 2 All ER 357.

77. AB v. South West Water Services Ltd [78].
78. AB v. South West Water Services Ltd [1992] 4 All ER 574.
79. Cambridge Water Co. Ltd v. Eastern Counties Leather plc [1994] 1 All ER 53.
80. Rylands v. Fletcher (1868) LR 3 HL 330.
81. Cambridge Water Co. Ltd v. Eastern Counties Leather plc [1994] 1 All ER 53 *per* Lord Goff at p. 64.
82. Water Resources Act 1991, ss. 158 and 162.
83. *The Times*, 15 January 1994.
84. DoE (1992) Using water wisely. A consultation paper (July). Department of the Environment and the Welsh Office.
85. Dovey, W.J. and Rogers, D.V. (1993) The Isle of Wight water metering trial, *IWEM Journal*, **7** (2), 156–61.
86. NRA (1992) Water Resources Development Strategy. Discussion document ME/43929, Bristol.
87. The Times 4 December, 1993.
88. Regina v. Secretary of State for the Environment, *ex parte* Friends of the Earth Ltd and Another *The Times* 4 April, 1994.

The Government has established two new Agencies – the Environment Agency in England and Wales, and the Scottish Environment Protection Agency in Scotland. The Agencies still have the same powers as have been described in this chapter, but they also control pollution of the air (Her Majesty's Inspectors of Pollution will be incorporated into the Environment Agency), licence solid waste disposal sites so as to prevent contamination of land and groundwater, and are involved in monitoring radio activity and the licensing of radioactive sources.

INDEX

Acid mine water 73
Action plan 49
Advection–diffusion equation 82
Advective transport 83
Agriculture 51, 113, 142, 150, 273
   water quality for 206
Agricultural fertilizers 39
Aguera streams 56
Air pollution 69, 111, 154.
Air purification 115
Algae 72, 85, 116, 251
Algal blooms 34, 71
Alginate factory 78
Amenity 51
Ammonia nitrogen 16, 25, 34, 64, 85, 150, 207, 253
Analytical methods selection 100
Angling 51
Area of Outstanding Natural Beauty (AONB) 51
Artificial barriers 253
Athens Environmental Research Laboratory (USEPA) 215
Atomic absorption spectrometry 89
Atrazine 142
Attribute theory 188
Ayrshire River Purification Board 75

Bacteria, drug resistance 136
Bacteriological
   load 134
   modelling 7
   pollution 69, 71
Baltic Sea xvi, 89, 92, 116, 234
   reduction in polluting load 238, 240
Barley 139
Basic programme of monitoring 283
Basin Water Quality Management 1, 9
Basque country 55
Bathing 118
Bathing waters, bacteriological requirements 134, 135
Bear Creek, Oregon 168
Benchmark programme of monitoring 283
Benefit-cost analysis 191
   criticisms 200
Benefit and costs of conservation measures 196
Benthic community assessment 7
Benzene 130
Biochemical oxygen demand (BOD_5) 16, 108, 207
Black List 95, 292
Brake fluids 113
Breast milk 97
British Steel Corporation 48
Buffer zones 13, 113, 126, 243, 247, 263
Bug River 24
Bull trout
   distribution, Oregon 167
   temperature requirements 168

Cadmium 17, 25, 89, 292
California Urban Water Agencies 197
Campylobacter 134
Canoeing 51
Cashubian Lakeland 157
Catchment management planning 45
   limitations 49
Cause-effect relationships 222
Cellulose-paper wastewater 72
Central Office of Water Economy 105
Center for Exposure Assessment Modelling 215
Centre for Environmental Protection and Control 105
Centre for Marine Biology, Polish Academy of Science 81
Chemical analysis, classification of sample sites 186
Chemical treatment of mine water 77
Characteristic concentration 106
Chief Pollution Control Inspector, Poland 173
Chloride 14, 41, 69, 90, 113
Chlorobenzene 130
Chlorophyll *a* 25
Cholera 46
Chromium 25, 89
Civil liability 298
Classification of river water 29, 211, 275, 287–90
Clean Water Act, 1972 5, 34, 165
Climate, Tyneside 50

Clyde River Purification Board 75
Coal 111
Coal mining 23, 50, 73, 113, 300
Coastal water quality 136
COD-Cr 16
COD-Mn 16
Coliform index 117, 118
Commission on Energy and the Environment 79
Concentration–discharge relationships 158
Conductivity 41, 56
   and flow 59
Consent conditions 75, 79, 295
Conservation measures 196
Conservation tax 203
Consultation report 48
Consumer surplus 197
Control of Pollution Act, 1974 (UK) 73, 78
Coolants 113
Cooperative Water Quality Study 6
Copper 17, 25, 89
Cost effectiveness analysis 191
Council for Environmental Quality 205
Criminal liability 297

Dalquharran colliery 74
   discharge chemistry 76
Dams 246
Dangerous substances 292
Danish Environmental Protection Agency (DEPA) 233
Danish Hydraulics Institute 234
Danish legislation 233
DDT 99
Decane 130
Decision support system 81
Deforestation 12
Delft Hydraulics 81
Delft water quality model (DELWAQ) 82, 85
Delphi method 207
Demand management 191
Department of Environmental Quality (Oregon) 4
Derwent River 46
Detection limits for analysts 207
Detergent phosphorus ban 36
Detergents 40, 71
Dichlorprop 142
Diffuse sources, see Non-point sources
Dioxins 96, 97
Direct aqueous injection 127
Diurnal variations in water quality 63
Dog faeces 136

Dow Chemical Company 97
Drainage basin control 109
Drought policies 198
Dunajec River 216, 217

EC directives 123, 292, 297
EC legislation 52
EC standards, application to Polish rivers 178
Ecological system investigation 7
Economic analysis 239
Economic evaluation of conservation measures 197
Economic feasability 197
Education programme 36
Electrolytes 113
Emission standards 292
   for sewage treatment works 238
Environment Agency 46, 302
Environmental activism 36
Environmental Impact Assessment (EIA) 52
Environmental improvement assessment 235
Environmental Protection Act, 1990 293, 294
Environmental Quality Commision 172
Environmental quality standards 95
Erie Lake 97
Erosion 209
Escherichia coli 134, 233, 240, 257, 262
European initiatives 292
Evapo-transpiration 226
Experimental catchments 145

Faecal coliforms 208
Farming and Wildlife Advisory Group (FWAG) 51, 52
Fat processing plant 121, 125
Ferric hydroxide 75
Ferruginous waters 73
Fertilization rates 40, 222
Fertilizers 63, 71, 113, 142, 209
Fertilizer production 115
Financial feasability 197
   strategy 35
Fish community health assessment 8
Fisheries 51
Fish kill 40
Flood Control Act, 1936 (USA) 198
Flood defence 50
Flooding 300
Flow and water chemistry 55
Flow augmentation 5
Flow changes 55
Fly ash 97

Forest ecosystems 119
Forested catchments 150
Forestry 33, 51, 56, 247, 273
Frombork wetland 258
Fourier transform - infra red (FT-IR) 102
Fucus vesiculosus 118
Fund for Ecoconversion of Polish Debts 118

Gas liquid chromatography 102, 130
Gdansk Bay 81, 83, 117, 121, 133, 257, 278
   bacteriological load 135
Gdansk Province
   description 273
   environmental protection projects 278
   industrial discharges 276
   main sources of pollution 275
   municipal wastewater 276
   water quality of rivers 274
   water resources 273
   water supply 277
Gdansk refinery 129, 130
Gdansk Region
   characteristics 115
   liquid wastes loading 115
   river classification 117
   sewage treatment 116
Geographic Information System (GIS) 181
Geological section, Dailly coalfield 76
Geoscience and Marine Research and Consultancy Co (GEOMOR) 234
Gizdepka River catchment 149
Goczalkowice Reservoir 21, 245
Great Poland Lake District 149
Green Book 192, 199
Grey List 95, 292
Groundwater abstraction 50
Guaranteed concentration 177

Habitat enhancement 52
γHCH 25, 142
Heat and power generating plant 121
Helsinki Commission (HELCOM) 126
Her Majesty's Inspectors of Pollution 295
Hexane 130
Hooke - Jeeves optimisation methods 228
Hydraulic characteristics 215
Hydrobotanical systems 257
Hydro-electric power 50, 246
Hydrogen sulphide 71, 277
Hydrological flow pattern 228
Hydrological models 186, 224

Impoundments 21
Indicative concentration 213
Industrial production, reduction 71
Industrial Revolution 46
Industrial wastewater 125
Institute of Environmental Protection (IOS) 81, 112
Institute of Hydroengineering, Polish Academy of Science 81
Institute of Meteorology and Water Management 214, 268
   Gdansk 14, 81, 89
   Gdynia 81
Integrated data management and decision support system (IDSS) 81
Integrated resource planning 196
International Standards Organisation (ISO) 174
International Water Quality Index 205
Investment 35, 113, 118, 219, 279, 283, 300
Iron 17, 25, 69, 89, 277
Issaquah aquifer 202
Issue paper 167

Jelitkowski stream 125, 262

Kacza River 130, 134
Kielder Water 50, 51
Kirkor 22, 281
Klodnica River 216

Lake quality, Poland 96, 114
Land Drainage Act, 1976 and 1991 (UK) 295
Landfill 114, 126
Land use planning 53
Landscapes 51
Leaching losses 139, 231
   minimisation 142
Lead 17, 25, 89
Lead mining 50
Least cost planning 195
Leisure industry 300
Lime injection 77
Lindane, see γHCH
Lipno Reservoir 39
List 1, see Black List
List 2, see Grey List
Load factor 192
Loadings, metals 89
Loads, instantaneous 90
Lobster Creek, Oregon 169
Low flush toilets 191
Lubricants 113

Makkink and van Heemst model 225
Manganese 17, 25, 89, 141, 277
Manure 13, 113
Marine Fisheries Institute, Gdynia 81
Marginal cost pricing 193
Mass spectrometer 102
Mathematical modelling, N and P leaching 221
Mazowiecka Lowland catchment 149
Mendeleyew 21, 281
Mercury 71, 292
Metalaxyl fungicide 142
Metallurgical plants wastewater 92
Metals in sediments 91
Metals leaching 139
Metal loadings 89, 92
    and flow 91
    variability coefficients 91
Microbiological indicators 133
Middle Pomeranian Lake District 149
MIKE 11 and 21 233, 236
Mining 300
Minister of Environmental Protection, National Resources and Forestry (Poland) 112, 212, 233, 257, 268
Minister of Health and Social Welfare (Poland) 212
Monitoring guidelines 66
Monitoring systems 212
Monte Cassino 134
Motlawa River 121, 129
Multifunctional Water Management Information System (MIS) 267
    applications 270
    configurations 269
    engineering structures 270
    evaluation 271
    input data 269
    intakes and discharges 270
Multi reservoir hydrological model 227
Multi variate analysis 57
Municipal Corporations Act, 1835 (UK) 291

Napoleonic Wars 14
Narew River 24
National Association of Regulatory Utilities Commissioners (US) 195
National Coal Board 75
National control system for measurement of quality 112
National Fund for the Protection of the Environment 278
National Inspectorate of Environmental Protection (PIOS) 268
National Inspectorate for Water Protection 281

National Nature Reserve 51
National Rivers Authority (NRA) 45, 293, 295
    prosecutions 298
National Water Quality Assessment (US) 181
Nature conservation 245
Neman River 69
New Deal, 1930s 198
New supply costs 192
Nickel 89
Nitrate 16, 25, 64, 69, 85, 150, 157, 207
    seasonal cycle 222
Nitrite 16, 64
Nitrogen 40, 83
    balance of arable land 139
    riverine loading 145
Nitrogen/discharge 41, 150
Non governmental organisations (NGOs) 166, 277
Non point sources 83, 91, 145, 186, 243
Non point source modelling 7
North Sea 95
Northumbrian Water Authority 46
North West Power Act (US) 198
Notec catchment 154
Nutrients 281
    from agriculture 12
    leaching 139, 159, 221, 229, 231
    loading 157
    removal 252, 258
    variations with seasons 150

Oats 139
Odra River 89
    quality classification 267
    run off of N and P 147
Oils 113
Oliwa River 134
Oliwa Zoo wetland 262
Ontario Lake 97
Oregon, Department of Environmental Quality 166, 209
Oregon State 165
Oregon State Sanitary Authority 4
Oregon water quality index 206, 208
Organic pollutants 96
Organo chlorine pesticides 25, 277, 281, 292
Organo halogen compounds 98
Organisation for Economic and Cultural Development (OECD) xv
Orlik Reservoir 39
Oxygen, dissolved 64, 71
    depletion 207
    solubility 169

Pathogens 133
PCBs 25, 96, 101
Periphytonic algae 251
Permissible limits 89, 91, 174
Pesticides 69, 96, 99, 101, 185
    contamination of water 251
    leaching 139, 141
    toxicity 100
Pesticides Manual (UK) 100
Petroleum hydrocarbons 96, 98, 113
pH 15, 64, 208
Phenols 96
Phenoxy acids (MCPA) 142
Phosphate/flow 60, 157, 159
Phosphate phosphorus 13, 17, 25, 34, 56, 85, 108, 157, 251
    leaching 221
    uptake by algae 221
Phosphorus fetilizer plant 121, 125
Phosphorus 40, 41, 212
    loss 140, 154
    reduction 36
    riverine loading 145
Phosphorus, total 17, 25, 83, 86, 207
Photosynthesis/repiration ratios 64, 251
Physico-chemical variables, Aguera streams 63
Phytoplankton 42
Planning for additional water supplies 194
Poland
    atmospheric pollution 11
    bird population 244
    contribution of N and P to Baltic Sea 145
    emission balance 112
    environmental policies 219
    forest area 12, 112
    industrialization 22
    law 274
    length of rivers 173
    major industrial plants 13
    monitoring of surface waters 281
    natural environment 111
    pollution of water 111, 283
    population increase 12
    productive land 119
    rainfall 211
    river quality 96
    sewage treatment 112, 213
    ship building industry 14
    towns 112
    wastewater volumes 29, 112
    water quality classification 213, 285
    water resources 211, 267

Polish Ministry of Environmental Protection, Natural Resources and Forestry, see Ministry of EP, NR and F
Polish parameters for drinking water quality 175
Polish Regulations 174
Polish 'Red Book' of rare species 244
Polish river quality reports 283
Polish Sanitary Inspectorate guidelines 136
Polish standards
    for nitrate 221
    on water quality 173
Policy Advisory Committee (PAC) (US) 166
Pollution definition 95
Polluting load to Gdansk Bay 121
Poly chlorinated -p-dioxins 96
Poly cyclic aromatic hydrocarbons (PAHs) 96, 98, 99, 277
Polynomial regression 159
Pomeranian rivers 89
Pomorka River 222, 226
Potassium 40, 89
Precipitation 226
Pregel River 69
    analysis 70
Prescribed processes 296
Preservatives 113
Preventative action 119
Principal component analysis 57
Priority setting 219
Procurator fiscal 75
Producer's surplus 197
Prohibition notice 296
Protection zone 105, 110
Provincial Inspectorate of Environmental Protection 14
Przemsza River 21
Prymorze catchment, run off of N and P 148
Public
    awareness 114
    consultation 48
    expectations 300
    involvement 35, 166, 200, 205
Puck Bay 87, 117
Pulp mill effluent 98
Purge and trap technique 127
Pyrite 74

Quality
    assessment 113
    classes 95
    objectives 292

Raba River 145

Radunia Canal 125
Radunia River 105, 115, 121, 129, 157, 277, 278
   quality of 109
      sampling points 107
      sewage treatment plants 107
      tributary rivers 108
Raftsmen 14
Railway 14
Rainfall - run off modelling 224
Ramsar convention 247
Ramsey pricing 194
Rate setting 194
Recreation and tourism 109, 133, 273
Recreation, water quality for 134
Reda River 115
Red List 95
Redfield ratio 42
Reed beds 78, 258
Regional Water Management Boards 274
Reliable concentration 158, 177
Reservoirs 51
Rega River 133
   hydrology 234
   mass balance 238
   pollutant loading 234, 236
   ranking of priorities 241
Rhine River 99
Rio Conference 245, 268
River basin compartments 225
River Basin Water Authorities 212
River Purification Boards 73
River quality, Poland 96
Royal Commission on Environmental Pollution 293
Rozwojka River 129
Rubbish tips 73, 109
Rudawa River 145

Salinity 113, 277, 286
Salmon 5, 46, 51, 77, 165, 246
Salmonella 134
Sample collection 102
Sand, bacteriological quality 136
Sandy River basin 182
Scheduled Ancient Monuments 51
Scotland, lowlands of 73
Scottish Development Agency 76
Scottish Environment Protection Agency 302
Screen for algae 252
Sea outfall 77
Seasonal
   distribution of metal outflows 91
   variability 55
Sediment oxygen demand 7
Self purification 65, 238, 253
Sewage treatment plants 109
Sheriff's order 76
Ship building 115
Silesia 23, 28, 154, 246
Silesia - Vistula canal 14
Silica 85
Slapy Reservoir 39
Sites of Special Scientific Interest (SSSI) 51
Skeletal abnormalities 9, 71
Sodium 89
Solid waste disposal 50, 114, 117
Soil water retention 226
Statutory Water Companies Act, 1991 (UK) 294
Steelheads 5
Storm water drains 133
Straszyn Reservoir 105
Stream Ecology Group, University of Basque Country 55
Stream Water Quality Model (QUAL 2E) 214
Sulphates 14, 41, 69, 113
Sulphur dioxide 115
Summer houses 109
Surface Water (Classification) Regulations, 1989 (UK) 297
Supply and demand framework 192
Sustainability 292, 299
Sweden 139
Swedish Environmental Protection Agency 142
Szczecin lowland area 150

Target reduction levels 202
Technical advisory committee (TAC) 166
Temperature standard 167
Thermal load 171
Total solids 208
Toxic substances transport 82
Town and Country Planning process 46, 52
Treaty of Rome 292
Trichloro ethylene 130
TRISULA 82
Tropical rainforests 245
Tualatin River 33
   ammonia concentration 36
   improvements 37
   water quality strategy 34
Tyne River 45
   catchment 47
   description 46

effluent disposal 50
  resources, uses, activities 49
  rod catches 47
Tyneside Joint Sewerage Board 46

UK and Europe comparison 292
Uncertainty analysis 214, 216
Unified Sewerage Agency 33
Union of Baltic Cities 126
University of Gdansk 81
Upper Silesia 13, 113, 267
Urban development 50
Urban Wastewater Treatment Directive 293
US Clean Water Act, see Clean Water Act
US Environment Protection Agency (USEPA) 8, 166
Utilities and Transportation Commission 195

Variability of Aguera tributaries 58
Vegetated submerged bed wetlands 258
Villages, influence on water quality 64
Vistula Delta 11
Vistula Lagoon 71
Vistula River
  agricultural irrigation 21
  ammonia 22, 25
  BOD_5 23, 24
  chloride 22-24
  classification of water quality 30, 267
  dams 246
  description 11
  dissolved oxygen 16, 22, 24
  dissolved salts 24
  diversity of environment and biota 243
  ecological corridor 244
  estuary 115
  flow variations 11
  input of nutrients 16
  islands 244, 247
  landscape park 247
  land use 12
  levees 244, 247
  migratory birds 244
  navigability 14
  pH 15, 24
  pollution of Gdansk Bay 83, 86
  power plants 21
  river life 244
  run off of N and P 146
  salinity 13
  shipping 13
  sulphate 24

volatile organo halogens 129
  wastewater treatment 31
  water quality 1945-1970 22
                1968-1978 23
                1989-1992 24
  water quality index 209
  water quality parameters 26
Vltava River 39
  agricultural fertilizers 39, 40
  eutrophication 43
  total phosphorus 39
Voivodship Inspectorate of Environmental Pollution 105, 281
Volatile organic compounds 127, 292
  in natural waters 129
  in surface waters, Krokowa 129

Warsaw Water Works 22
Washington State, Conservation Planning Requirements 192
  public water systems 191
Wastewater treatment plants 22, 29
Wastewater treatment facilities improvements 233
Water abstraction 21, 56, 294
Water Act 1973 (UK) 291
          1989 (UK) 293, 296
          1991 (Poland) xv, 211
Water
  balance of rural land 225
  chemistry and flow 55
  designated uses 165, 293
  management, strategic policy 274
  metering 299
  movements, Gdansk Bay 82
  pollution control 27
  quality classification 25, 74, 213, 296
    improvements 36, 206
    index 205, 208
    of lakes 275
    management 33, 291
    maps 28
    models 81, 83, 214, 233
    objectives 297
    process 85
    problems 35
    review 166
    sampling strategy 183
    standards 165
Water Industry Act, 1991 (UK) 294, 296
Water Purification and Prevention of Pollution Act, 1938 (US) 4

Water Quality Act, 1965 (USA) 165
Water Resources Act, 1991 (UK) 294
Water Resources Council (USA) 191
Water resources management 294
Water run off 226
Water systems components and demand characteristics 193
Water use forecasts 196
Weighted geometric mean function 208
Wetlands 257
Wheat 139
Wietcisa River 222
Wiezyca wetland 258
Willamette River
   agriculture 2
   annual precipitation 183
   basin boundaries 183
   beneficial use 3
   dams and reservoirs 6
   description 1
   dissolved oxygen 4
   ecoregions 183
   flood control system 5
   history of pollution 3
   hydrogeology 184
   land use and cover 184
   restoration 1
   sample site selection 181
   study area 181
   wastewater treatment 4
   water quality index 208
Wooded mountains 149
World Health Organisation 97
World War II 12
Wloclawek Reservoir 21, 246

Year of Clean Technology (Poland) xv

Zinc 17, 89
Zinc mining 50
Zulawy Wimslane area 150